ATLAS OF VIBRATIONAL SPECTRA OF LIQUID CRYSTALS

ATLAS OF VIBRATIONAL SPECTRA OF LIQUID CRYSTALS

Editors

P SIMOVA, N KIROV
Institute of Solid State Physics
Bulgaria

MP FONTANA
University of Parma
Italy

H RATAJCZAK
University of Wroclaw
Poland

Published by

World Scientific Publishing Co. Pte. Ltd.
P. O. Box 128, Farrer Road, Singapore 9128

U. S. A. office: World Scientific Publishing Co., Inc.
687 Hartwell Street, Teaneck NJ 07666, USA

Library of Congress Cataloging-in-Publication data is available.

ATLAS OF VIBRATIONAL SPECTRA OF LIQUID CRYSTALS

Copyright © 1988 by World Scientific Publishing Co Pte Ltd.
All rights reserved. This book, or parts thereof, may not be reproduced in any form or by any means, electronic or mechanical, including photocopying, recording or any information storage and retrieval system now known or to be invented, without written permission from the Publisher.

ISBN 9971-50-613-0

Printed in Singapore by JBW Printers & Binders Pte. Ltd.

CONTENTS

INTRODUCTION

Although the first liquid crystalline compound has been found about a century ago, interest in this class of substances has increased considerably in the last two decades. Besides, the organic chemists and the physicists traditionally involved in synthesizing new compounds and investigating their properties, we now find engineers and technologists, interested in displays, modulators, bistable devices; physicians, who for instance use them as very sensitive temperature indicators for diagnostic purposes; biologists, who find liquid crystalline mesophases useful models for membranes and their structural changes on the molecular level.

Nowdays there are about ten thousand known mesogens, and this number increases constanly. For effective applied or fundamental research in this field, it is necessary to know well the molecular structure of the mesogens, since it determines many fundamental properties such as the dipole moment, molecular anisotropy in the dielectric, optic, magnetic behavior, inter and intra-molecular interactions, molecular stochastic dynamics etc.

Vibrational spectroscopy (Infrared and Raman) has been used for many years to elucidate the structure of thousands of organic and inorganic molecules, polymers, solutions etc. It is therefore natural that many research groups have applied these techniques to the study of liquid crystals.

The detailed study of vibrational spectra and the calculation of vibrational frequencies, the shape and half-width of the spectral bands, their polarization and intensity allow the investigation of phase transitions, molecular dynamics and order parameters in the mesophases.

The location of the frequency maxima is relatively simple. The greatest complication in mesogens is the overlap between the many bands which make up the vibrational spectrum. Such overlap of course is the more likely the larger the number of atoms that make up the molecule. Since mesogen molecules contain in general many atoms (typically 50), most vibrational modes will overlap to some extent. This creates a serious problem in the precise determination of frequencies, intensities and bandshapes.

The vibrational spectrum extends over the range 4000 cm^{-1} to ≈ 5 cm^{-1}. Following the recent recommendations of the IUPAC, it is convenient to divide it into two regions: from 4000 to 200 cm^{-1}, where intramolecular (internal) vibrations are mainly observed, and below 200 cm^{-1} where the main contributions come from intermolecular (external) vibrational modes. The dividing line of course is not a clear one. For certain light molecules with strong intermolecular forces the lattice vibrations may appear above 200 cm^{-1}. On the other hand, some intramolecular vibrations, such as torsional or librational modes, especially present in complex molecules, may feature frequencies lower than 200 cm^{-1}.

The present Atlas collects IR absorption and Raman scattering spectra of over one hundred mesomorphic substances in their several phases - namely isotropic liquid, crystal and liquid crystal phases. They cover the results obtained by the Authors in their investigations since 1972. This Atlas is limited to spectra of thermotropic liquid crystals, since data on lyotropic systems are still too scarce in the literature.

CHAPTER I

Structure and Properties of Liquid Crystals

I.1 Classification of liquid crystals

The term "liquid crystal" indicates a condensed state intermediate between the perfectly ordered crystalline structure and the disordered structure of isotropic liquids. The terms "mesomorphic phase" or "mesophase" are also widely used in the literature. In the liquid crystalline phase all effects which are typical of normal fluids are observed: fluidity, Kerr effect, electrodynamic and hydrodynamic instabilities etc. Mesogens in their liquid crystalline phase have significant anisotropy in their mechanical, optical, electrical and magnetic properties, and in this sense they resemble crystalline solids; they differ from these however since they lack effects such as charge injection and transport, uncoupled electrons etc.

The structure of the liquid crystalline phase is very labile, and even relatively small external electric and magnetic fields may alter macroscopic properties of liquid crystals - and this is of course the basis for their wide range of applications.

Mesophases fall naturally into two main classes: amphiphilic and non-amphiphilic. Amphiphilic organic substances, i.e. compounds containing localized lipophilic and hydrophilic groups within the same molecule, have a tendency to confer solubility in organic solvents and water. They form a mesophase in some characteristic compositional range either at room temperature or at higher temperatures, and usually lead to structures containing a considerable amount of water and/or organic solvents. Because of this such mesophases are labeled "lyotropic". Since lyotropic liquid crystals are not included in this Atlas, they shall not be considered further.

Non-amphiphilic mesogens have been studied mainly as pure compounds and the transitions among the several possible mesophases and phases are induced thermally. Non-amphiphilic mesophases are thus labeled "thermotropic".

I.2 Classification of thermotropic liquid crystals

Thermotropic liquid crystals are subdivided into three main groups: nematics, cholesterics, smectics.

I.2.1 Nematics

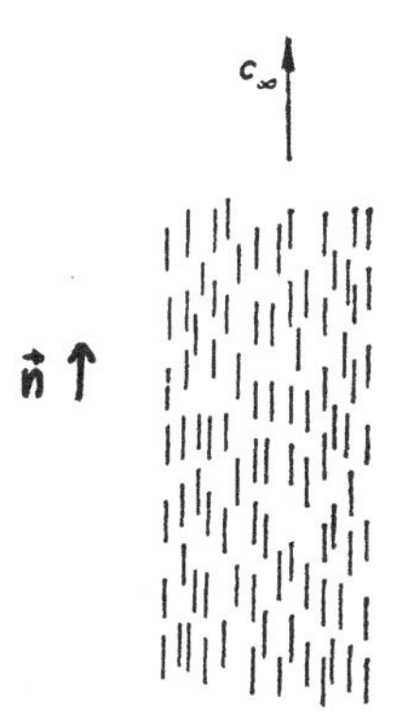

Fig.1 Schematic structure of nematic liquid crystals. The lines represent the molecules in their preferred orientations.

The main characteristic of nematic structures is their long range orientational order: the molecules tend to be parallel to one common axis, the nematic director, usually designated by the unit vector $\vec{n}$ (Fig.1). The existence of long range orientational order is reflected in all the macroscopic properties of the nematic mesophase.

Nematic liquid crystals flow as easily as organic liquids. The molecules, while preserving their average orientation along $\vec{n}$, move freely in all directions, and their centers of mass are distributed uniformly in the medium. Since the system is invariant with respect to the $\vec{n} \rightarrow -\vec{n}$ substitution, a nematic liquid crystal is an optically uniaxial medium with no polarity.

Although prevalently oriented along the nematic director, the molecules may execute Brownian reorientations about either the long or the short axes; in this case we speak of spinning or tumbling motions respectively.

In a non oriented nematic mesophase there are nematic domains whose director may adopt all possible orientations, with a smooth transition from one domain to the other. Observation of thin (some microns) layers with a polarizing microscope reveals thread-like structures. They are called disclinations, i.e. discontinuities in orientation. External perturbations, such as electric or magnetic fields, cell surface treatment with surfactants etc., orient the liquid crystalline molecules in the whole system, yielding a monodomain, or monocrystalline, sample.

Such alignement may be produced also by rubbing the glass surfaces which make up the cell walls in a thin sample. However, the alignement of the molecules along the direction of rubbing is far from being total, as depicted in the idealized structure. Thermal fluctuations induce large deviations from the perfect parallel alignement. Surface treatment such as coating with the appropriate surfactant improves the alignement; however in this case the molecules are oriented perpendicularly to the glass surface (homeotropic configuration). Surface effects and alignement procedures are discussed in detail in a recent review (1).

Thick nematic layers ($\phi \geq 1mm$) appear cloudy. At the nematic-isotropic (N-I) phase transition they clear up (hence "clearing point" for the transition temperature). The transition is usually reversible and shows no supercooling effects.

Upon cooling, a nematic phase will crystallize, eventually after passing through one or more smectic phases, or even another nematic (re-entrant) phase. Some of these transitions, and certainly the crystallization, may show supercooling effects. Sometimes a liquid crystalline mesophase may be obtained only by supercooling. In this case we speak of a monotropic phase, in order to differentiate it from the stable mesophases, enantiotropic liquid crystals.

The N-I transition is first order. However, it is weakly so; in fact the transition enthalpies are relatively small (varying from about 0.1 to about 4 kJ/mol). In a homologous series the transition energies increase with increasing molecular length. Although the viscosity coefficients of the anisotropic nematic phase will reduce to the isotropic viscosity of the normal liquid phase, the absolute change at the N-I transition turns out to be small.

The thermal volume expansion in nematics is not very different from that of normal liquids. Specific heats and compressibilities are also similar. Near the weakly first order N-I transition there are, however, pre-transitional effects which lead to an increase in these quantities. Also, at the N-I transition there is a small, discontinuous decrease in volume.

I.2.2 Cholesterics

The cholesteric phase may be regarded as a twisted nematic phase, and its thermodynamic properties are correspondingly similar. The molecular centers of mass are distributed randomly as in a nematic, and there are no Bragg peaks in the X-ray diffraction patterns. In a very thin layer the molecules are oriented predominantly in the n direction.

Cholesterics have the same orientational order as nematics but differ in the molecular arrangement: the molecular alignement in the planar texture is normal to the vertical axis and uniformly parallel in horizontal planes. The alignement direction rotates linearly with respect to the vertical axis, thus yielding a helical structure with a period equal to one half of the helical pitch (Fig.2)

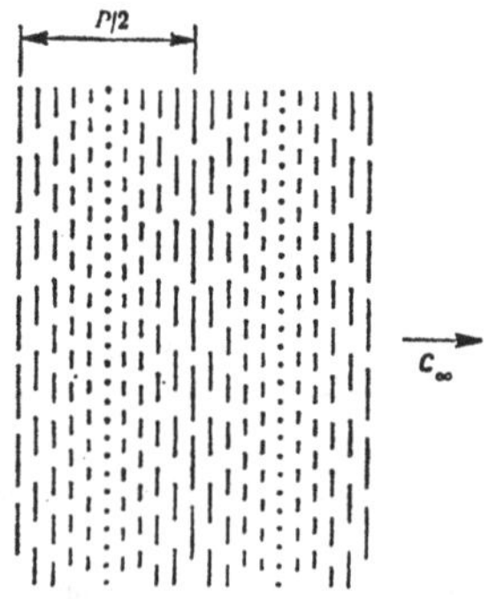

Fig.2 Schematic structure of cholesteric liquid crystals. The varying segment lengths measure the out-of-plane inclination of the molecules.

Many mesogens have only a cholesteric phase. Others may also show a lower temperature smectic phase. No compound known to us has both nematic and cholesteric phases.

Thus it would seem that, whereas the smectic phase may exist independently of the others, there is a sort of complementarity between the nematic and the cholesteric phases.

The enthalpies of the crystal-cholesteric transition (C-Ch) are approximately 20 to 50 kJ/mol; they are much lower for the Smectic A to Cholesteric (S_A-Ch) and Cholesteric to Isotropic (Ch-I) transitions, i.e. about 0.8 to 2.2 kJ/mol.

Hydrodynamic and optical properties of cholesterics depend on textures and may differ strongly from those of nematics. The most striking feature is the wavelength selective reflectivity of the cholesteric mesophase, which is due to a sort of Bragg-like scattering from the helical periodic structure, whose pitch turns out to be in the range of 200 to 2000 nm. Thus the sample appears beautifully chromatic, with the colour patterns changing with observation angle. Since temperature and perturbations such as mechanical stress or organic impurities affect the pitch, color is extremely sensitive to them.

The cholesteric phase is strongly optically active and rotates the plane of linearly polarized light by several thousands degrees/nm. Cholesterics are also circularly dichroic for some wavelengths. Therefore they will totally reflect, for example, right circularly polarized light by transmit left circular polarization. This effect is used to develop optical filters and display systems.

I.2.3 Smectics

All smectics are layered structures with well defined interlayer spacing, which may be determined by X-ray diffraction. Smectics are thus more ordered than nematics and cholesterics.

The smectics feature several substructural classes, which are currently labeled as smectic A through H. The best known ones are S_A, S_B and S_C.

Smectic structures may have ordered or disordered layers. In the latter group there is no long range order in the layer, and the molecular centers of mass are distributed randomly. The first group contains the most ordered mesophases. The molecules in the layers are arranged in a more or less regular two-dimensional lattice, which may be orientationally disordered to some extent due to molecular rotations about the long axis. X-ray studies on uniformly oriented samples confirm that the layers are all equally oriented and that they resemble a three dimensional lattice. The layers may glide against each other and this insures the fluidity of the mesophase.

In the S_A phase the molecules are parallel to each other and oriented perpendicularly to the layer (Fig.3a).

As a result, the layer thickness is close to the molecular length. The molecules may move within the layers and may rotate about their long axis. The system is optically uniaxial with the optic axis z normal to the layer plane. The z and -z directions are equivalent. The symmetry of S_A is $D_{\infty h}$ if formed by optically inactive compounds of racemic modifications; it reduces to D_∞ for optically active molecules.

When additional smectic phases are formed by a compound, S_A is always at the highest temperature. Upon heating, it leads to nematic, or cholesteric or directly to the isotropic phases. The transition enthalpy is about 6 to 8 kJ/mol, being higher for the transition to the isotropic phase since in this case all kinds of order disappear.

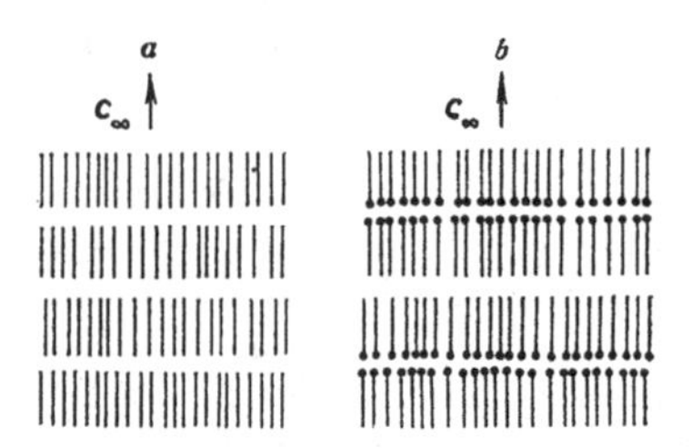

Fig.3 Schematic structures for smectics with $D_{\infty h}$ symmetry. The segments represent the molecules in their preferred orientation. a) S_A non-structured layers without polarity; b) non-structured layers with polar alternating orientational order.

There is another smectic phase with the same symmetry as the S_A phase. Here the molecules build double layers (fig.3b). Such a phase can be formed only by molecules without inversion symmetry, and the orientational order, in contrast to the normal S_A phase, is polar. In each layer the molecules are oriented in the same direction and the layers, therefore, are polar; however the polarization alternates in successive layers so that there is no net polarization; thus the double layer formation will affect only the dielectric properties.

Another unstructured smectic phase is S_C (fig.4), with C_{2h} symmetry, in which the layers are monomolecular and the molecules are tilted with respect to the normal to the plane. The layer thickness in this case is a little lower than the molecular length.

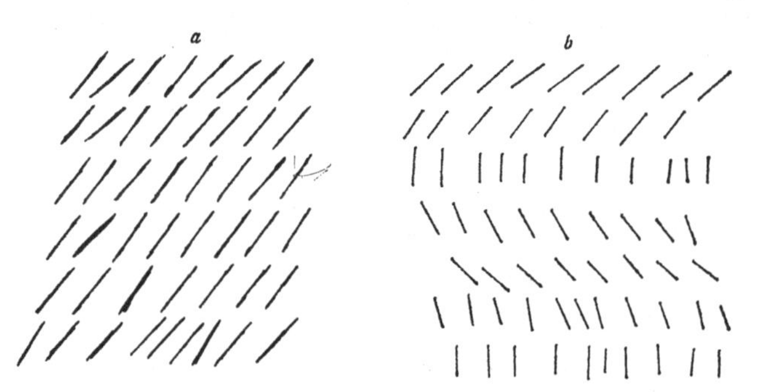

Fig. 4 Schematic structures of biaxial smectic liquid crystals with tilted non-structured layers. The segments show molecules and their preferred orientation. a) S_C single non-structured layers without polarity; b) twisted S_C, the twist axis being perpendicular to the molecular length.

The next higher temperature phase to S_C may be isotropic, nematic, S_A, or a recently discovered smectic phase which is optically isotropic. In the first two cases the transition enthalpies are similar to the corresponding transitions for the S_A phase. The transition enthalpy for the S_C-S_A case is instead very low, implying that such transition is very nearly second order. Experimental studies of the temperature dependence of the tilt angle show that, for a S_C-N transition, the angle has a large value (40-45°) which is temperature independent; for the S_C-S_A transition instead, the tilt angle varies from nearly zero at the transition temperature to about 25° at lower temperatures.

When tilted smectics are formed by optically active compounds, they form a twisted structure, similar to the cholesterics. In thin layers, a planar texture may be obtained which is similar to that shown by cholesterics.

The most common smectic with structured layers is S_B. The layer lattice symmetry is hexagonal. When the long molecular axis is perpendicular to the layer, the symmetry is D_{6h}. Tilted S_B has C_{2h} symmetry and is optically biaxial. S_B upon heating often tranforms to S_C or S_A. The transition enthalpies are considerably larger than for the transitions to a nematic phase (approximately twice as much). A direct S_B-N transition is rare.

I.3. Molecular structure of thermotropic liquid crystals

A wide variety of molecules may form mesomorphic phases. We have chosen the following examples to illustrate the diversity of structure which may be encountered:

Table I

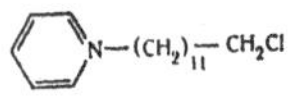

N-n-dodecylpiridinium chloride

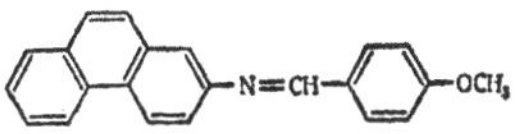

2-4-methoxybenzylideneaminophenantrene

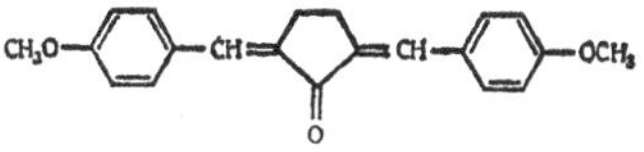

2,5 di-4-methoxybenzylidene cyclopentanone

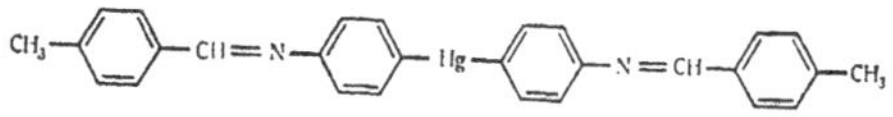

Bis- (4-p-methylbenzylidene aminophenyl) mercury(II)

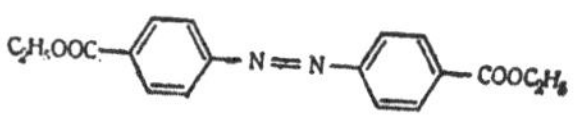

Cholesteryl bromide

Diethyl 4,4' -azoxydibenzoate

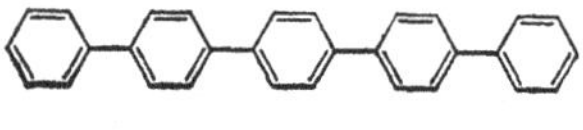

4-pentaphenyl

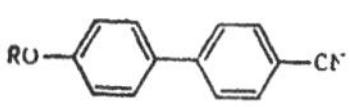

4-cyano 4', n-alkoxyphenyls

From the examples it is clear that, with the exception of cholesterol derivatives, mesomorphic molecules are rod-shaped and fairly rigid along their long axes. If the molecules lack rigidity, axial flexing may occur and this could prevent mesophase formation.

The parallel alignement of elongated molecules may be disturbed by several effects: thermal agitation will tend to disalign the molecules even though these will retain their rod-like shape; the elongated molecules may bend and flex, thus decreasing the possibility of parallel alignement. For instance, long chain n-alkanes can adopt elongated conformations but the flexible alkyl chain may coil and bend; thus hydrocarbons do not form liquid crystals. Normal alkanoic acids also do not yield mesophases.

Although a short rod-like central unit exists through hydrogen bonding between carboxyl groups, a high proportion of the dimeric molecule consists of potentially flexible alkyl chains. In both cases, intermolecular forces are very weak since the molecules are neither highly dipolar nor readily polarizable. Consequently, no strong constrains will be imposed on the potentially flexible alkyl chains to keep them in an extended conformation, and the system will not yield a liquid crystal.

The great majority of mesogen molecules have aromatic rings. Aromatic nuclei are polarizable, planar and rigid, and if suitable substituents are correctly positioned, elongated molecules are obtained, among which reasonably strong intermolecular forces should operate. For benzene nuclei, substituents must occupy para- positions and be such that they link up at least another benzene ring which also must carry a para-substituent. It is preferable if the central group which joins the rings is itself rigid: thus the bridge and rings would form the rigid core of the molecule. Central groups usually contain a multiple bond or a system of conjugated double bonds, or involve a ring formed by dimerization of carboxyl groups. Table II shows some common linkages in liquid crystal chemistry.

TABLE II

—CH=N—,	—N(→O)=N—,	—N=N—
—C(=O)—O—,	—C≡C—,	—CH=CH—
—C(=O)—NH,	$—O—CH_2—CH_2—O—$,	—CH=N(→)—,
—CH=CH—CH=N—,		—CH=CH—C(=O)—O—
CH=N—N=CH—	$(—C(=O\cdots)—O—H\cdots)_2$	$(—CH=CH—C(=O\cdots)—O—H\cdots)_2$

Terminal substituents may also influence considerably mesophase stability. If an unsubstituted compound forms a liquid crystal, then the substituted one will form a more stable mesophase. An exception are smectics, whose stability might be reduced by certain terminal substituents. They however do not simple lower the melting point or emphasize mesomorphic behavior of the parent system: actually, terminal substituents usually raise the melting points, but increase thermal stability even more. Some of the most common terminal substituents are shown in table III.

Table III

—O—R, —O—R', —R, —R', —C≡N, —Cl, —Br,

—C(=O)—O—R , —C(=O)—O—R' , —CH=CH—C(=O)—O—R ,

—O—C(=O)—R , —C(=O)—R' , —N(=O)=O , —N(R)—R

* R signifies saturated alkyl chain and R' — branched or nonsaturated hydrocarbon chain

I.4 The degree of order: Orientational order parameters

The mesomorphic phase symmetry is lower than that of the isotropic liquid. This is expressed qualitatively by the assertion that the mesophase is "more ordered". In order to define more quantitatively orientational order, it is useful to introduce quantities, the so-called order parameters, which may be calculated and determined from measurements of any anisotropic property of the liquid crystal.

If θ is the angle between the long molecular axis and the macroscopic symmetry axis of the mesophase, then the most fundamental parameter characterizing orientational order is the statistical average of the second order Legendre polynomial:

$$S_2 = 1/2 < 3cos^2\theta - 1 > \qquad (1)$$

For complete order (all molecules aligned parallel to the macroscopic axis), $cos^2\theta = 1$, so that $S_2 = 1$; for complete disorder, $cos^2\theta = 1/3$ so that $S_2 = 0$. Any intermediate degree of ordering will correspond to S_2 values between 0 and 1. If molecules are assumed to be long rigid cylinders, S_2 will depend only on temperature. For real molecules however S_2 will also depend on molecular structure; furthermore more parameters will be needed to describe the degree of order completely. Formally, it is convenient for this most general case to introduce the order parameter tensor:

$$S_{ij} = 1/2 < 3cos\theta_i \cdot cos\theta_j - \delta_{ij} > \qquad (2)$$

where i,j = x, y, z are the Cartesian coordinates of the mesogen molecule or of the test molecule dissolved in the liquid crystalline medium; $\theta_{i,j}$ is the angle between the molecular axes x, y, z and the director $\vec{n}$. As in eq.1 the angular brackets indicate averaging over all molecules. The second rank tensor S_{ij} is symmetric and traceless. In polar coordinates, the non-zero diagonal elements are:

$$S_{xx} = 1/2 < 3sin^2\theta \cdot cos^2\phi - 1 >$$

$$S_{yy} = 1/2 < 3sin^2\theta \cdot sin^2\phi - 1 > \tag{3}$$

$$S_{zz} = 1/2 < 3cos^2\theta - 1 >$$

where θ and ϕ are the director coordinates of the liquid crystalline monodomain sample in the coordinate system x, y, z. In eqs.3 the eventual hindered molecular group rotation is considered as a lowering of symmetry.

As stated, S_{ij} may be determined using any technique which probes some anisotropic property of the mesophase. For admixture molecules for instance, optical dichroic spectra or NMR from suitable proton couples may be used.

Bibliography

1. DeGennes P.G., The Physics of Liquid Crystals, Clarendon Press, Oxford 1974.
2. Blinov L.M., Electro- and Magnetooptics of Liquid Crystals (in Russian), Nauka, Moscow 1978.
3. Chandrasekhar S., Liquid Crystals, Cambridge University Press, Cambridge 1977.
4. Luckhurst G.R., Gray G.W., Molecular Physics of Liquid Crystals, Academic Press, New York 1979.

CHAPTER II

Vibrational Spectra

II.1 Introduction to the theory of vibrational spectra

II.1.1 Origin of vibrational spectra

The molecule is a dynamical system of atoms connected by the mutual interaction of their electrons. Molecular spectra appear as a result of energy exchanges δ E between the molecule and external fields such as electromagnetic radiation; in this case the energy exchange δ E may be due to absorption, emission, or scattering of light.

The total energy E of the molecule consists of several contributions: the motion of the molecular electronic cloud yields the electronic term, the intermolecular atomic oscillations and the molecular rotations yield the vibrational and rotational contributions respectively. There is finally a term due to translational motion, which for a free molecule is independent on the intramolecular dynamics.

II.1.2 Separation of vibrational and rotational motion

In order to separate molecular vibrations from rotations several cartesian coordinate systems are used:

1) The laboratory system X, Y, Z; 2) the molecular center of mass system X, Y, Z; 3) the rotating coordinate system x, y, z, in which the three axes coincide with the main axes of the inertia ellipsoid of the molecule in the equilibrium configuration. The following sets of nine coordinates are used : the cartesian coordinates of the molecular center of mass; the Euler angles defining the directions of x, y, z; the atomic coordinates in the rotating system.

A molecule with N atoms has 3N degrees of freedom, of which six correspond to rigid rotations and translations respectively. In order to eliminate them six conditions are necessary. Three conditions may be derived from the coincidence of the molecular center of mass with the origin of the moving coordinate system. From these the translational degrees of freedom may be eliminated. Three conditions which allow the elimination of the rotational degrees of freedom may be obtained by recalling that the vibrational angular momentum must be zero when the atoms pass through their equilibrium positions.

II.1.3 Normal vibrations

In term of the normal coordinates q_i (i=1....3N-6), the total vibrational kinetic energy of the molecule may be expressed as:

$$T = \frac{1}{2} \sum_{i=1}^{3N-6} q_i^2 \qquad (II.1)$$

The potential energy may be obtained as a Taylor expansion in terms of the small normal displacements q_i about the equilibrium configuration:

$$U = U_o + \sum_{i=1}^{3N-6} \frac{\partial U}{\partial q_i}_o q_i + \frac{1}{2} \sum_{i,j} \left(\frac{\partial^2 u}{\partial q_i \partial q_j} \right)_o q_i q_j + \ldots.. \qquad (II.2)$$

Since in the equilibrium configuration $\frac{\partial u}{\partial q_i} = 0$, neglecting the higher order terms and the inessential constant U_o, we may write:

$$U = \frac{1}{2} \sum_{i,j} \left(\frac{\partial^2 U}{\partial q_i \partial q_j} \right)_o q_i q_j \qquad (II.3)$$

which expresses the potential energy of an ensemble of 3N-6 harmonic oscillators. Using expressions (II.1) and (II.3) in Lagrange's equations for this set of oscillators, we obtain the equations of motion for the single normal coordinates q_i:

$$q_i + \sum_{i=1}^{3N-6} \left(\frac{\partial^2 U}{\partial q_i \partial q_j} \right)_o q_i = 0 \qquad (II.4)$$

A possible solution of this system of 3N-6 equations is:

$$q_i = A_i \cos(\lambda^{\frac{1}{2}} t + \epsilon) \qquad (II.5)$$

where, in the harmonic approximation, A_i, λ, ϵ are constants, which may be obtained by substitution of (II.5) in (II.4):

$$\sum_{i=1}^{3N-6} A_i [\left(\frac{\partial^2 U}{\partial q_i \partial q_j} \right)_o - \delta_{i,j} \lambda] = 0 \qquad (II.6)$$

Equations (II.11) form a system of 3N-6 linear, homogenous, algebraic equations in the 3N-6 unknown amplitudes A_i, which has non-zero solutions only for the values of λ which are solutions (eigenvalues) of the secular equation:

$$\begin{pmatrix} \frac{\partial^2 U}{\partial q_1 \partial q_1} - \lambda & \frac{\partial^2 U}{\partial q_1 \partial q_2} & \cdots & \frac{\partial^2 U}{\partial q_1 \partial q_{3N}} \\ \frac{\partial^2 U}{\partial q_2 \partial q_1} & \frac{\partial^2 U}{\partial q_2 \partial q_2} - \lambda & \cdots & \frac{\partial^2 U}{\partial q_2 \partial q_{3N}} \\ \vdots & \vdots & \ddots & \vdots \\ \frac{\partial^2 U}{\partial q_{3N} \partial q_1} & \frac{\partial^2 U}{\partial q_{3N} \partial q_2} & \cdots & \frac{\partial^2 U}{\partial q_{3N} \partial q_{3N}} - \lambda \end{pmatrix} = 0 \qquad (II.7)$$

From eq. II.7 we obtain 3N-6 (3N-5 for linear molecules) solutions λ_k, such that

$$q_i^k = A_i^k \cos(2\pi\nu_k t + \epsilon_k) \qquad (II.8)$$

where $\nu_k = \frac{1}{2\pi} \cdot \sqrt{\lambda_k}$. Thus every vibrational degree of freedom is seen to correspond to a simple harmonic oscillator with amplitude A_i^k, frequency ν_k and phase ϵ_k. This means that the frequency and phase of vibration are the same for a given normal coordinate, whereas the amplitudes A_i^k may vary from atom to atom, and, in particular, they may be zero. Also, all the atoms in the molecule reach their maximum displacement and pass through the equilibrium position simultaneously. Vibrational modes of this kind are called normal vibrations and the corresponding frequencies normal frequencies.

Although eqs. (II.7) have 3N-6 solutions, they need not be all distinct. In fact, for a given molecular symmetry there will be m equal roots of (II.7), *ie* an m-fold degeneracy in the frequency spectrum. The normal vibrations of the tetrahedral CH_4 molecule are presented in fig. 5 as an example: ν_2 corresponds to two-fold degenerate normal modes, and ν_3, ν_4 are three-fold degenerate.

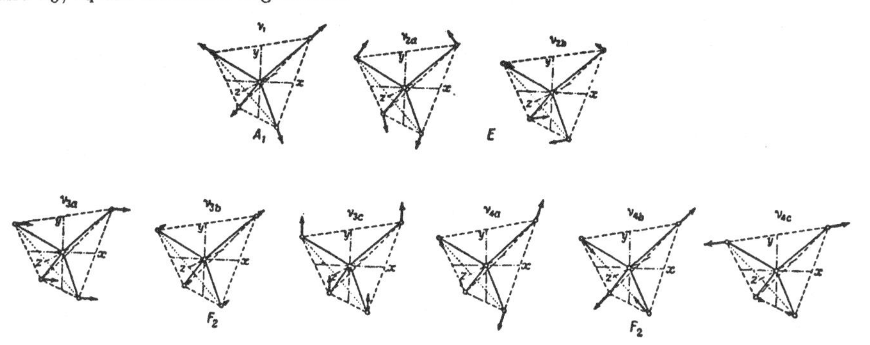

Fig. 5 Normal vibrations of the CH_4 molecule.

When passing from classical to quantum mechanics, the normal coordinate transformation must be changed to take into account the non-commutability of cartesian coordinates. We now have:

$$Q_k = \sum_{i=1}^{3N-6} \ell_{ki} q_i \qquad (II.9)$$

where the coefficients ℓ_{ki} are chosen so that the kinetic and potential energies T and U are diagonal.

The total vibrational wavefunction for the 3N-6 degrees of freedom can be put into the product form:

$$\Psi_k = \prod_{k=1}^{3N-6} \psi_v^k(Q_k) \qquad (II.10)$$

to which will correspond the energy eigenvalues $E_v = \sum_k E_v^k$, where ψ_v^k and E_v^k are the wavefunctions and energies of the single vibrational normal modes respectively, with

$$E_v^k = (n_k + \frac{1}{2}) \cdot h\nu_k \qquad (II.11)$$

where $n_k = 1, 2, ...$ is the vibrational quantum number. Thus the total vibrational energy of an N-atom molecule is:

$$E_v = (n_1 + \frac{1}{2})h\nu_1 + (n_2 + \frac{1}{2})h\nu_2 + ...(n_{3N-6} + \frac{1}{2})h\nu_{3N-6} \qquad (II.12)$$

of which

$$E_v^o = \frac{1}{2}h \sum_{k=1}^{3N-6} \nu_k \qquad (II.13)$$

is the zero-point ground state energy which, for a polyatomic molecule may consitute a considerable fraction of the total vibrational energy.

All the previous theory was in the framework of the harmonic approximation, according to which only transitions such that $\Delta n_k = \pm 1$ are allowed. In general however real systems will be anharmonic to some extent, and there will be deviations from the parabolic dependence of energy on vibrational displacement. This will lead to a decrease of the energy level spacing as the dissociation energy is approached; an example of this is shown in fig. 6 for the case of the hydrogen molecule.

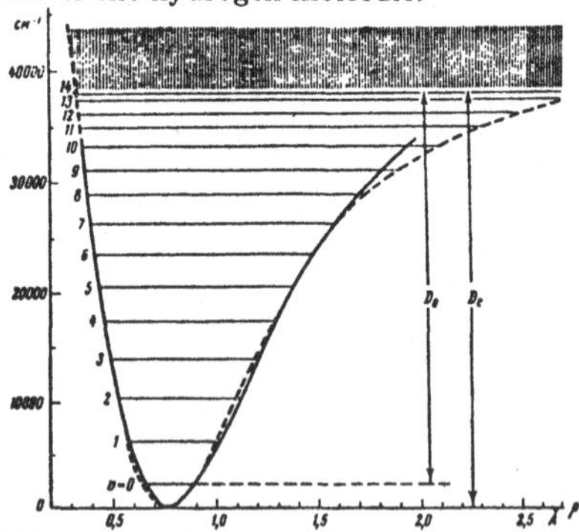

Fig. 6 -Vibrational Energy level spacing for the hydrogen molecule.

Anharmonicity can be taken into account by including higher order terms in the expansion of the potential energy U in terms of the normal coordinates; in practice the cubic and quartic terms will be sufficient. These additional terms will scramble the normal coordinates, so that the total vibrational energy cannot be expressed as a sum of independent oscillator terms any longer. The anharmonic coupling between modes will lead, to a good approximation, to the following form of the energy level distribution:

$$E_{n_1 n_2 ..} = h \sum_i \nu_i(\nu_i + \frac{1}{2}) + h \sum_{i,j} d_{i,j}(n_i + \frac{1}{2})(n_j + \frac{1}{2}) \qquad (II.14)$$

where the anharmonic coupling shows through the non-diagonal coefficients $d_{i,j}$, which are in general small relatively to the frequencies ν_j.

Another effect of anharmonicity is to allow transitions for which $|\Delta n_i| > 1$, corresponding to the so-called overtone and combination frequencies.

II.2 Molecular Symmetry

II.2.1 Molecular symmetry in the equilibrium configuration

The shape and size of molecules may be quite different depending on chemical structure and composition. The molecule may be linear, planar, it may contain branched chains, ring structures etc. Simple molecules, as for instance water, benzene, methane etc., in their equilibrium configuration may have a definite symmetry. The calculation of the vibrational spectrum may be simplified considerably by taking into account the molecular symmetry. Consider for example the case of the NH_3 molecule, which has the symmetry of a triangular pyramid (Fig. 7).

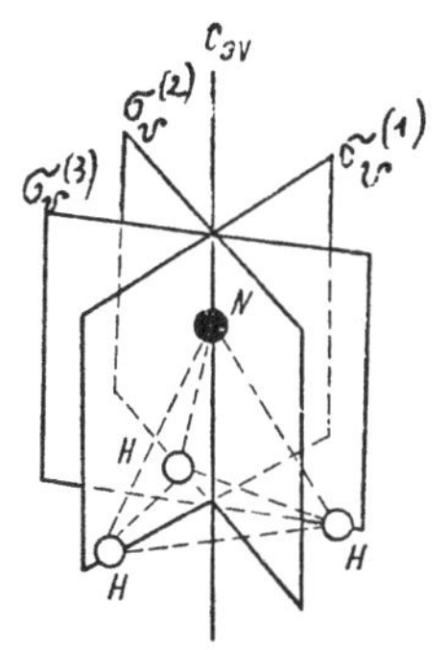

Fig. 7 -Symmetries of the ammonia molecule.

It has a three-fold rotational symmetry axis, C_3, passing through the N atom and the center of the pyramidal base. A second symmetry element (reflection, σ) is the plane passing through the N atom and one of the H atoms. The third symmetry element is the center of inversion i. The fourth element corresponds to an n-fold rotation-reflection axis (S_n). For n=2, it would reduce to the element i.

In a similar way, all molecules may be characterized according to their symmetry characteristics - formally, according to their point symmetry groups. We list here some representative point groups:

C_n: This is a class of groups with an n-fold rotational symmetry axis;

D_n: similar to C_n, with n additional two-fold rotational symmetry axes perpendicular to C_n;

D_{nh}: now a symmetry plane perpendicular to C_n is added; examples are the p-dichlorobenzene (D_{2h}) or benzene (D_{6h}) molecules;

C_{nv}: to the C_n axis we add n σ_v symmetry planes which contain it; examples are the already discussed NH_3 molecule (C_{3v}) and the water molecule (C_{2v}).

At higher symmetries we find the so-called "cubic" groups, corresponding to regular polyhedra: tetrahedron (T, T_h, T_d); octahedron (O, O_h) and icosahedron (I, I_d, I_h) (see fig. 8).

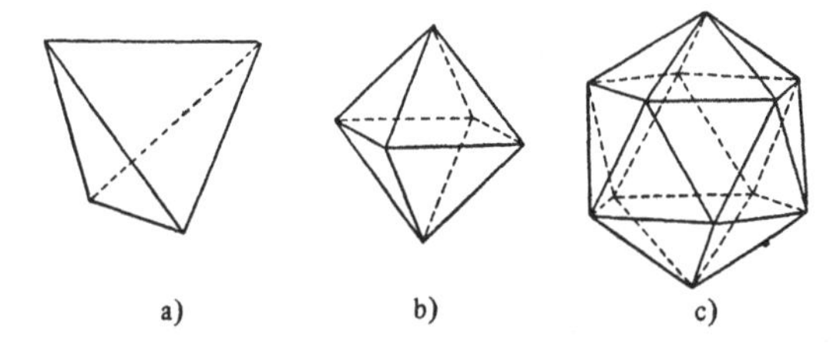

Fig. 8 Regular polyhedra

To O_h symmetry belong for instance the SF_6 and UF_6 molecules. No icosahedral symmetry molecules are known to exist.

II.2.2 The symmetry of normal vibrations

Under the action of a symmetry operation the molecular potential energy does not change, and so do the force constants and the corresponding vibrational frequencies. However the displacements of the atoms will deform the equilibrium molecular symmetry. Three things may happen: the configuration of the deformed molecule is transformed into itself by a symmetry operation of the molecular point group; the new configuration may differ from the preceding one only because the displacement vectors change sign; finally, there may be also an orientational shift. For molecules with up to two-fold symmetry axes only non-degenerate modes are possible and therefore the number of normal frequencies equals the number of vibrational degrees of freedom. For molecules with a C_3 axis, there are two-fold degenerate modes and therefore the total number of observable frequencies appears to be less than 3N-6. As molecular symmetry increases, so do the number and the level of degeneracy of vibrational modes.

A non-degenerate vibration may be symmetric or antisymmetric relatively to a specific symmetry operation, if it does not, or does, change sign in its displacement vectors when acted upon by such operation. For instance, consider the vibrations of a planar molecule XYZ_2 (fig. 9a-f) under the action of σ_v which yields the configurations shown in fig. 9g-u .

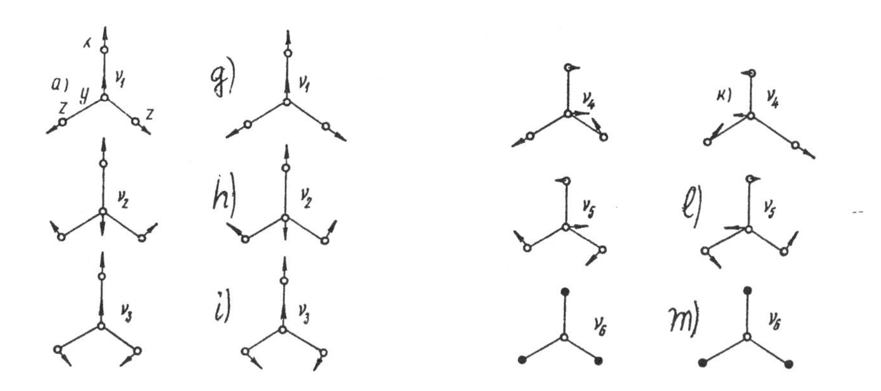

Fig. 9 Normal modes of a planar molecule and effect of the application of symmetry operation σ_v.

The ν_1, ν_2, ν_3 and ν_6 modes are not affected, and are thus symmetric, whereas the displacement patterns of ν_4 and ν_5 reverse sign, making these modes anti-symmetric. Similarly we may observe that ν_6 is anti-symmetric relatively to σ_h.

For non-degenerate vibrations there are some simple and useful rules; for instance:

a. an atom located (at equilibrium) in a symmetry plane can only move within such plane if the vibration is symmetric, and only perpendicularly to it if the vibration is anti-symmetric;

b. an atom on a symmetry axis can move only along such axis for a symmetric mode, and perpendicularly to it for an anti-symmetric one.

Vibrations which are symmetric with respect to all symmetry operations of the molecule are called totally symmetric. For further details on the application of point groups theory to molecular symmetry and vibrational mode classification, see the appropriate literature.

II.2.3 Characteristic normal vibrations in polyatomic molecules

The problem of vibrational dynamics of complex molecules is considerably simplified by the possibility of isolating vibrational modes of specific sub-constituents. If the fre quency of the mode belonging to the molecular subgroup is the same (or nearly so) as that which the same sub-group would have if it were isolated, then such mode acts as a "signature" of the presence of the given sub-group in the more complex overall molecular structure. Thus the existence of these "characteristic" frequencies is of great help in elucidating the structure and dynamics of complex molecules.

Let us consider for instance the C-H bond. If the distorsion due to the presence of other molecular groups acts only perpendicularly to the displacement vectors of the C and H atoms (as f.i. for the stretching mode), the C-H bond force constant will not be significantly altered. For the double C-H bond in C_2H_4 and H_2CO it is equal to $5.28 \cdot 10^2$ N/m; in C_2H_6 and similar molecules with single C-H bond it is $4.97 \cdot 10^2$ N/m etc. The same results apply to modes which involve a modulation of the bond angles.

The preceding considerations apply, with some modification, to more complex situations such as those of atoms vibrating in the middle of molecular groups. In such cases the single oscillator model which works well for the atom vibrating at the end of a group is not sufficient, and coupled oscillator models must be used. This gives rise to resonances which may delocalize the mode over the whole molecule. This is the case f.i. of skeletal modes: altough the vibrational amplitude is not localized on a single bond any more, the mode can still be considered as "characteristic".

An internal group will still have a characteristic frequency if it is very different from the frequency of the other coupled oscillator. This is the case for instance of the groups -C=C-, -C=N, >C=O, >C=C<.

II.3 Intensity of vibrational spectra of polyatomic molecules

II.3.1 Infrared absorption

Any atomic motion which modulates the molecular dipole moment will cause absorption (or emission) of electromagnetic radiation, mainly in the infrared spectral range. Since the motion of any atom of the molecule may be viewed as due to the superposition of the motional patterns of all the normal modes, each with its specific amplitude and frequency, it follows that electromagnetic radiation will be absorbed (emitted) at the normal mode frequencies.

We have in fact, expanding the molecular dipole moment $M(q_1, q_2, \ldots, q_{3N})$ in terms of the normal coordinates q_i:

$$\vec{M} = \vec{M}_o + \sum_{i=1}^{3N} \left(\frac{\partial \vec{M}}{\partial q_i} \right)_o q_i + \frac{1}{2} \sum_{i,j=1}^{3N} \left(\frac{\partial^2 \vec{M}}{\partial q_i \partial q_j} \right)_o q_i \cdot q_j + \ldots \qquad (II.15)$$

where $\vec{M}_o$ is the dipole moment at equilibrium. Substituting in (II.15) expresssions (II.8) for q_i we obtain:

$$\vec{M} = \vec{M}_o + \sum_{i=1}^{3N} K_i \left(\frac{\partial \vec{M}}{\partial Q_i} \right)_o \cos(2\pi\nu_i t + \epsilon_i) + \frac{1}{4} \sum_{i,j=1}^{3N} K_i K_j \left(\frac{\partial^2 \vec{M}}{\partial Q_i \partial Q_j} \right)_o \cdot$$
$$\cdot \left\{ \cos\left[2\pi(\nu_i + \nu_j)t + (\epsilon_i + \epsilon_j)\right] + \cos\left[2\pi(\nu_i - \nu_j)t + (\epsilon_i - \epsilon_j)\right] \right\} + \ldots \qquad (II.16)$$

The second term on the right side leads to absorption of radiation at the fundamental frequencies ν_i, whereas the third term leads to the so-called overtone or combination bands at frequencies $\nu_i \pm \nu_j$.

The intensity of a fundamental absorption band will be proportional to

$$\left[\left(\frac{\partial \vec{M}}{\partial Q_i} \right)_o \right]^2$$

The quantum mechanical treatment leads to essentially the same result, with the intensity now being proportional to the square of the transition matrix element:

$$[\vec{M}]_{v'v''} = \int \psi^*_{v'}(Q_k)\, \vec{M}\, \psi_{v''}(Q_k)\, dQ_k \qquad (II.17)$$

where v', v'' are the vibrational quantum numbers of the upper and lower levels respectively.

If the product $\psi^*_{v'} \cdot \psi_{v''}$ does not transform according to the same symmetry representation of the molecular point group as the dipole moment operator, integral (II.17) will be zero: the corresponding vibrational mode is inactive, or not allowed in infrared absorption. A simple and important consequence can be seen for molecules with a center of inversion symmetry (such as, f.i., CO_2): all wavefunctions may then be classified as even (g) or odd (u) relatively to the inversion of coordinates. A g-type vibration would lead to a wavefunction product which would be even and therefore the integral (II.17) would vanish. Conversely a u-type vibration would be active. In a polyatomic molecule, where the symmetry of a local sub-group might be slighly distorted and possibly lose inversion symmetry, the selection rule we just mentioned would not hold any more, in principle, and all vibrations should become active. However, if the distortion is not too large, vibrational modes which in the undistorted configuration would have been even will have a much smaller intensity than the previously odd modes. Thus intensity ratios may be a very useful tool in studying local symmetry changes in complex molecules.

The absolute intensity of a fundamental infrared absorption band is given by:

$$A_i = \frac{\pi N}{3c} \left(\frac{\partial \vec{M}}{\partial Q_i} \right)_o^2 = \int_{-\infty}^{+\infty} \alpha(\nu)\, d\nu \qquad (II.18)$$

where N is the number of molecules per unit volume, c is the velocity of light and $\alpha(\nu)$ is the absorption coefficient, which is experimentally defined as:

$$\alpha(\nu) = \left[\log \frac{1}{T(\nu)}\right] \cdot d^{-1} \qquad (II.19)$$

where d is the sample thickness and $T(\nu)$ is the transmittance.

II.3.2 Raman scattering

A normal mode will be active in Raman scattering if it modulates in time the amplitude of the molecular dipole induced by the electric field of the incident radiation. Such field may be written as:

$$\vec{E} = \vec{E}_o \cos(2\pi\nu_o + \varphi_o) \qquad (II.20)$$

where E_o is the field amplitude, ν_o the frequency and φ_o the phase. Under the action of such field a dipole moment density $\vec{P}$ is induced in the molecule:

$$\vec{P} = \{\alpha\} \cdot \vec{E} \qquad (II.21)$$

where $\{\alpha\}$ is the molecular polarizability tensor. From eq. II.21 and II.20 we have:

$$\vec{P} = \{\alpha\} \cdot \vec{E}_o \cos(2\pi\nu_o t + \varphi_o) \qquad (II.22)$$

As the atoms vibrate the electronic distribution in the molecule will be deformed, and therefore the polarizability will depend on the vibrational normal modes. In the small displacement approximation $\{\alpha\}$ may be expanded in a Taylor series. Stopping at the linear term and substituting the expansion in eq. II.22 we have:

$$\vec{P} = \{\alpha_o\}\, \vec{E}_o \cdot \cos(2\pi\nu_o t + \varphi_o) + \frac{1}{4}\sum_{i=1}^{3N}\left(\frac{\partial\alpha}{\partial Q_i}\right)_o \vec{E}_o\, Q_i \{\cos[2\pi(\nu_o + \nu_i)t + (\varphi_o + \varphi_i)] +$$
$$+ \cos[2\pi(\nu_o - \nu_i)t + (\varphi_o - \varphi_i)]\} \qquad (II.23)$$

The first term in eq. II.23 yields the elastic (Rayleigh) contribution to the scattering. The second term yields Raman scattering at scattered light frequencies $\nu_s = \nu_o \pm \nu_i$.

Generally the molecular polarizability is a symmetric tensor: $\{\alpha\}_{i,j} = \{\alpha\}_{j,i}$, for i,j = x, y, z, with trace

$$b_i = \alpha_{xx} + \alpha_{yy} + \alpha_{zz} \qquad (II.24)$$

and anisotropy factor

$$g_i^2 = \frac{1}{2}\left[\left(\alpha_{xx} - \alpha_{yy}\right)^2 + \left(\alpha_{yy} - \alpha_{zz}\right)^2 + \left(\alpha_{zz} - \alpha_{xx}\right)^2\right] \qquad (II.25)$$

From eqs. II.20 through II.26 we see that a normal mode of frequency ν_i may be active in Raman scattering only if at least one of the six components $\left(\partial\alpha_{i,j} \,/\, \partial Q_k\right)_o$ is non-zero.

The absolute intensity of an allowed Stokes $(\nu_o = \nu_s + \nu_k)$ Raman line is given by:

$$I_s = \frac{64\pi^3 h(\nu_o - \nu_k)^4}{9c^4\,\mu_k\,\nu_k} \cdot \left(5b(b_{k'})^2 + 13(g_{k'})^2\right) \cdot \frac{1}{1 - exp\left(-\frac{h\nu_k}{k_B T}\right)} \qquad (II.26)$$

where μ_k is the reduced mass of the oscillator with characteristic frequency ν_k, and $b_{k'}$ and $g_{k'}$ are the first order derivatives of b_k and g_k relative to the normal coordinate Q_k; T is the absolute temperature and k_B is Boltzmann's constant.

The intensity of a given Raman line will depend on the quantity

$$5(b_{k'})^2 + 13(g_{k'})^2 \qquad (II.27)$$

which in general is difficult to determine experimentally. For this reason in Raman spectroscopy relative intensities are used; thus well behaved sharp and isolated lines are particularly useful as standards.

Finally, we must recall that the above treatment is valid only in off-resonance conditions and for linear processes only. Much more information may be obtained by non-linear, coherent Raman processes and, for liquid crystals with chromophoric groups, by resonant Raman scattering.

II.3.3 Polarization of Raman lines

In general the polarization of a Raman line will depend on the angle between the directions of the incident and scattered light respectively, on molecular symmetry and on the specific symmetry of the related normal mode. The usual scattering geometry is shown in fig.10 (the so-called 90° geometry).

The measured intensities are therefore $I_{||}$ and $I_{|}$ (polarizers parallel or perpendicular to each other respectively), and we may define the **degree of depolarization** ρ as:

$$\rho = \frac{I_{|}}{I_{||}} \qquad (II.28)$$

In the case of a totally symmetric vibration in an isotropic molecule, it is easy to see that the Raman line will be totally polarized, i.e. $\rho = 0$. If the molecular polarizability is anisotropic, the induced dipole moment may have, besides the P_z and P_x (this last one will not be observed in the chosen scattering geometry) components, also P_y. As a result the scattered light will be depolarized, i.e. $\rho \neq 0$.

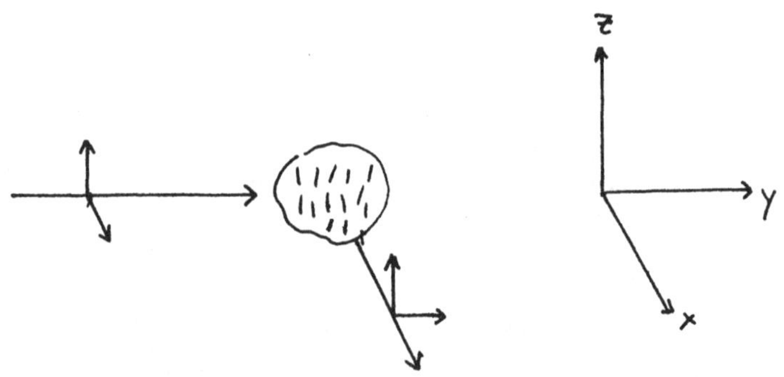

Fig. 10 Experimental Raman scattering geometry: the small arrows indicate the possible directions of the polarizer and analyzer axes.

If the incoming light is linearly polarized, the measured degree of depolarization will differ from the case of natural incident light. For instance, total depolarization would yield $\rho = 6/7$ for natural light and $\rho = 3/4$ for linearly polarized light.

The value of ρ yields information on molecular symmetry; it is thus a very useful quantity to measure, especially for simple molecular systems. On a more qualitative level, it is also useful in the case of more complex systems. Some simple rules are always valid:

- Non-totally symmetric vibrations always yield $\rho = 3/4$ (from now on we shall only consider excitation with linearly polarized light.)
- Very low (say, ≤ 0.1) values for ρ are always associated with totally symmetric vibrations in highly symmetric molecules (nearly spherical symmetry)
- As molecular anisotropy increases, so does ρ
- For a given molecular group in a complex system, ρ may feature different values for the **same** normal mode. This is an indication of different local environments: thus a measure of ρ and its variations gives important information on local symmetry in complex molecules.

Bibliography

Wilson E.B., Decius, J.C., Cross, P.C., *Molecular Vibrations*, Mc Graw-Hill Publ. Co., New York.

Sverdlov, L.M., Kovner, M.A., Krainov, E.P., *Vibrational Spectra of Polyatomic Molecules*, (in Russian), Moskow, Nauka, 1970.

Steele, D., *Theory of Vibrational Spectroscopy*, W.B.Saunders, Philadelphia, 1971.

Volkenstein, M.V., Gribov, L.A., Eljashevich, M.A., Stepanov, B.I., *Molecular Vibrations*, (in Russian), Moskow, Nauka, 1972.

Chapter III

VIBRATIONAL SPECTRA OF THERMOTROPIC LIQUID CRYSTALS

III.1 Introduction

The standard method for assigning complex molecules vibrational spectra, including those of liquid crystals, is to study the spectra of progressively simpler compounds, assuming the validity of the characteristic frequency approach. In some cases, additional information may be obtained from polarized spectra. A full normal coordinate calculation from the complex mesogen molecules is extremely difficult, so that alternative, more approximate methods must be used.

There have been basically two approaches to the problem. One uses the fact that most compounds which form nematic or smectic phases derive from the basic para-substituted structures such as:

The D_{2h} and C_{2v} structures have been analyzed in detail by many authors. Data on IR and Raman spectra of several parasubstituted benzenes can be found in many papers in the literature (see e.g. L.M. Sverdlov, M.A.Kovner, E.P.Krainov, **Vibrational Spectra of Polyatomic Molecules**, Moscow, Nauka, 1970).

In the case of C_{2v} symmetry, all vibrations will be Raman-active and three modes will also be IR-active. In D_{2h} symmetry, IR and Raman spectroscopy will be mutually exclusive, because of inversion symmetry: therefore both techniques are required for a complete vibrational study.

In addition to the fundamental motions of the aromatic rings, there will be vibrations associated with para-substituted and bridging groups such as azo-, azoxy-, etc. These groups will contribute substantially to the complexity of the vibtrational spectra, especially for molecules with longer tail structures.

Another approach to the problem is to consider the molecule as a whole, without privileging its aromatic core. In this case, since most mesomorphic molecules do not have any symmetry elements, all vibrational modes are allowed both in IR and Raman spectra.

III.2 Internal vibrations assignements

Except for the CH stretching vibrations in the $2800 cm^{-1}$ - $3100 cm^{-1}$ region, most modes are concentrated below $1800 cm^{-1}$. Only the intense C=N stretching mode is found at about 2220cm^{-1} in both IR and Raman spectra. At about 2215cm^{-1} we find the C=C stretch vibration, which is strong in Raman scattering and very weak in IR absorption.

The free carbonyl in mesomorphic esthers and ethers is centered at 1720-1730 cm^{-1} in both IR and Raman spectra. In 4-substituted benzoic and cynnamic acids the C=O group absorbs or scatters at about 1685cm^{-1}, as a consequence of hydrogen bonding.

The characteristic skeletal stretching vibrations of the semi-unsaturated C-C bonds of the benzene rings cause the appearance of four bands between 1600 and 1400 cm^{-1}. Those near 1600-1500cm^{-1} are characteristic of the aromatic ring itself and, together with the CH stretching bands near 3000cm^{-1} they provide a good signature of the aromatic structure. The actual positions of these bands are influenced to some extent by the nature and arrangement of the substituent groups around the ring. The frequency positions of the two most prominent benzene ring stretching modes are 1610-1590cm^{-1} and 1525-1490cm^{-1} respectively. Generally in mesomorphic molecules the higher wavenumber band is somewhat weaker than the other one. A weak ring band appears at 1600-1560cm^{-1}, mainly as a shoulder to the stronger 1600cm^{-1} band in non-conjugated rings. In most mesogens with conjugated structure this third band is usually observed between 1580 and 1565cm^{-1}. In 4-substituted alkylcyano biphenyls and tolanes this band is very weak. With HC=N and N=N bridge groups the intensity of this band is considerably enhanced.

The fourth skeletal C-C ring frequency for all investigated liquid crystals is in the range 1470-1430cm^{-1} and is usually of moderate intensity. This band frequently overlaps with the strong CH_2 deformation modes and therefore its utility is reduced.

In most cases, the three bands connected with the aromatic structure of mesogens can be recognized without difficulty, and provide sufficient identification of the structure itself. However, heterocyclic aromatic compounds such as those containing pyridine and pyramidine rings also yield a similar set of bands about 1600cm^{-1}.

Other vibrations, besides those due to the benzene rings, are also present in the 1600-1400cm^{-1} region. In all Schiff base liquid crystals the HC=N stretching mode yields intense IR and Raman bands localized in the 1625-1620cm^{-1} range. This band is thus very characteristic of such mesogens. When the C=C double bond is conjugated with an aromatic ring, the corresponding wavenumber shifs to lower values relatively to compounds containing the isolated group. The main C=C absorption (IR) for 4-substituted cynnamic acids is centered near 1625cm^{-1}.

In-plane OH deformation coupled with C=O stretching yields a mode usually observed at 1420cm^{-1}. Hydrogen bending in CH_3 and CH_2 groups yields in all mesogens moderate bands in the region 1450 ± 20 and $1450 \pm 30 cm^{-1}$ respectively. All IR absorption bands in the 1600-1400cm^{-1} range are strongly longitudinally polarized.

There are two other CH_3 and CH_2 deformation modes giving rise to usually intense bands in the 1395-1380 and 1390-1370cm^{-1} regions. CH wagging and twisting modes on the other hand, as well as CH_3 rocking, yield weak bands which cannot be easily identified in the complex mesogenic spectra. However, the Raman active CH_2 twisting mode is easily recognizable at about 1300cm^{-1} and is not sensitive to the specific molecule. The azoxy linkage absorbs in the 1310-1250cm^{-1} region. It is associated with the $N \rightarrow O$ link; we must point out however that the C-N stretching mode is active in the same spectral region. Thus the specific assignement of these bands is generally difficult. The strongly longitudinally polarized absorption due to the phenyl ring-oxygen stretching mode, as well as the less strongly polarized alkyl chain-oxygen stretching mode, also appear in the $1300cm^{-1}$ region as intense and broad bands.

The C-C stretching mode yields weak absorption bands in the 1070 to $950cm^{-1}$ range in normal paraffins. In liquid crystalline molecules the contribution of polar groups, such as f.i. the C=O group, to the skeletal vibrations causes a marked increase in the intensity of these bands. In the 1220-$1170cm^{-1}$ and 1130-$1090cm^{-1}$ ranges we find sharp and longitudinally polarized bands of moderate intensity (IR), which can be identified as due to aromatic CH in-plane deformation modes.

Strong bands appear in the region 1000-$600cm^{-1}$ due to out-of-plane deformations of hydrogen atoms in the rings. Their frequency is determined almost totally by the position rather than the nature of the substituents; thus they provide a very sensitive method to detect the type of substitution. The strong out-of-plane bending mode of the aromatic CH group yields a band at 860-$800cm^{-1}$ which is characteristic of 4-substituted systems and is usually a doublet for mesogens which do not contain a symmetrical central link group. The CH out-of-plane mode is transversally polarized. The other CH out-of-plane deformation modes appear at 780-$720cm^{-1}$. Typical in this region is the CH_2 rocking mode at 730-$720cm^{-1}$ (only active in IR). In the crystalline phase of some mesogens with long alkyl chains (i.e. $n \approx 7$ or higher), this band is usually a doublet.

It is very difficult to interpret the region below $600cm^{-1}$. However it is very useful to try since even small changes in the molecular environment and/or intermolecular forces (as f.i. at phase transitions) may result in substantial changes in the spectrum.

In Raman scattering, molecular tail vibrations contribute strongly below $\approx 400cm^{-1}$. The so-called accordion mode of the end chain of 4-azoxyanisole (PAA) and its homologues appear at about $270cm^{-1}$. Actually the band is a band envelope which contains a number of modes depending on the actual chain conformation. The existence of the accordion mode in the alkane series and its behavior has implications on the conformational changes which accompany the solid-fluid phase transitions. For the alkane solids, one, or at most two, strong bands have been observed; this implies the existence of one or two dominant conformations. New bands, which appear in the fluid phase, may be correlated with kinks in the alkyl chain, which would produce many gauche-conformers.

The series of bands between 350 and $150cm^{-1}$ depends on the type of substituent. Molecular vibrations causing IR absorption in this region may be described as frustrated rotations of the C_6 ring against the essentially stationary substituents.

The modes below $220cm^{-1}$, which are due for 20 to 60 % to C bonds, show coupling to flipping/twisting ring deformations, which usually appear at 110-$80cm^{-1}$. Frustrated ring deformations occur in the 160-$130cm^{-1}$ region.

Several torsional modes may also be observed in the low frequency range. Accurate assignement of the torsional frequency is often very difficult for complex mesogens. In the crystalline phase the assignement is often complicated by the presence of lattice modes in the same range. Mass or positional changes (other than in the group under study) may affect the position of low frequency bending modes but will have little or no effect on the torsional modes.

In general, IR spectra are more complicated than Raman spectra below $600cm^{-1}$.

Typical Schiff bases such as azo- and azoxy- compounds show two moderate or strong bands at 580 and $520cm^{-1}$. In substitued biphenyls and tolanes this structure is more complex: four bands appear in this region. Mesomorphic molecules with esther structure yield only one band at $540cm^{-1}$. These bands are usually trasversally polarized (IR).

III.3 Mode assignements for specific compounds

In Table III.1 we give the assignements for the well studied mesogen PAA and some of its higher homologues. Another well studied mesogen molecule is MBBA. In table III.2 we give the vibrational assignements. In table III.3 we give the Raman assignements for N-(4-hexyloxybenzilidene) 4,n-propylanyline and in tables III.4 and III.5 the IR assignements for EBBA and anisylidene-4-aminophenylacetate (APAPA).

Many liquid crystal molecules contain a central ring system formed by dimerization of carboxylic groups by hydrogen bonding, as f.i. para-substituted benzoic and cynnamic acids. The hydrogen bond is an example of a weak bond (energy $\approx 15 - 30kJ \cdot mol^{-1}$); it is however stronger than the ordinary Van der Waals interactions. In many cases it may be partially or totally broken by a moderate increase in the kinetic energy, or by simple dilution.

Usually the hydrogen bond perturbs the stretching and bending vibrations of OH groups in IR and Raman spectra, i.e. in the regions $3500\text{-}3000cm^{-1}$ and $1730\text{-}1680cm^{-1}$. As an example we show in table III.6 the IR bands and assignements for 4,n-heptyloxybenzoic acid. In table III.7 we show the possible assignements for Raman bands in 5CB and its deutero-derivative $5CBd_{11}$.

4,4'-disubstituted biphenyls are interesting compounds: they have a strong tendency to form low temperature liquid crystalline phases and are chemically and photochemically stable; they have strong dielectric anisotropy and are very suitable for displays and other industrial applications. In table III.8 we summarize vibrational assignements for 8OCB, which we selected as typical of vibrational modes in alkyl and alkoxy substituted biphenyls.

III.4 Spectral changes at phase transitions

All investigations show that there are considerable spectral changes at the crystal-nematic phase transition for all molecules studied. This is not surprizing, since in this strongly first order transition both ΔH and ΔS are large, implying that most of the changes in the motional freedom and intermolecular forces occur here, rather than at the nematic-isotropic point. Most of the Raman and IR bands in the crystalline phase are sharp, including bending and stretching vibrations of the paraffin chain. In the nematic phase many IR and Raman bands shift and broaden and change intensity as the molecules recover rotational and translational freedom. Broadening and intensity decrease are typical of modes of the polymethilene chain, such as CH_2 bending at $1450cm^{-1}$, wagging and twisting modes between 1350 and $1190cm^{-1}$. The C-C stretching modes at $1060\text{-}1020cm^{-1}$ (IR) and $1000\text{-}950cm^{-1}$ (Raman), the CH_2 and CH_3 rocking vibrations etc., also become broader and lose considerable intensity in the mesomorphic state. For example, at the crystal-mesophase transition, all IR-active CH_2 bending modes in the region $1300\text{-}1200cm^{-1}$ of the spectra of 4,n-alkoxybenzoic acids merge into two broad bands. This implies that many rotational isomers exist in the mesophase, including gauche forms, while only one stable trans form exists in the solid.

Thus the Raman line at $271cm^{-1}$ in 4,4'-di-n-heptyloxyazoxybenzene strongly decreases in intensity on going into the smectic phase, and virtually disappears in the nematic and isotropic phases. The accordion mode in the $300\text{-}200cm^{-1}$ region of the Raman spectrum of PAA does not disappear, however. This is in agreement with the idea that this mode is connected with the "melting" of alkyl tails, similarly to the melting of alkane solids.

Strong changes in the band structure at very low frequencies (say below $150cm^{-1}$) also occur at several of the mesomorphic phase transitions. These changes however are due also to changes in the lattice vibrational spectrum; thus their quantitative analysis and use is very difficult.

Much work has been devoted also to the nematic-isotropic phase transition; the spectra in the two phases have been found to be nearly identical. All observations lead to the conclusion that the short range intermolecular interactions in the nematic phase are essentially the same as in the isotropic liquid.

Essential bibliography

Sverdlov, L. M., Kovner M.A., Krainov E.P., Vibrational spectra of polyatomic molecules (in Russian), Moskow, Nauka, 1970

Bulkin B.J., in Liquid crystals and ordered fluids, J.F.Johnson, R.S.Porter eds., New York, Plenum Press, 1974

Kirov N., Simova P., Vibrational Spectroscopy of liquid crystals, Sofia, Publishing House of the Bulgarian Academy of Sciences, 1984.

Table III. 1

Possible Assignment of IR and Raman Bands of Some 4,4'- Di-n-alkoxyazoxybenzenes

PAA			PAF*		PAB**		HAB***		Assignment
[+]	[++]	[+++]	[+]	[+++]	[+]	[+++]	[+]	[+++]	[+++]
21									
					37				
40									
52			54						
72									
210	213								
234									
317	313		318		317				
360	365								
417									
474			467						
494					494				
536									
611									
629			628		628		629		
670		667		669	671	665		663 }	Γ_1
		717		718		718		718 }	Γ_3

Table III. 1 (continued)

PAA			PAF*		PAB**		HAB***		Assignment	
[+]	[++]	[+++]	[+]	[+++]	[+]	[+++]	[+]	[+++]	[+++]	
725		723	724	726	724	724		?		
				813		731		728	$\rho(CH_2)$	
								753		
						778				
		753								
797		756		770		787		788	ω_{11}	
		805				800		799		$\rho\ (CH_2)$
				837		810		807		
832		836		844	833	831		835		
								843	γ	
848					850	853				
						892		897	$\rho(CH_3)$?	
911	914	908	911	914	910	910	910	910	ω_1	
				922		942			ω_c	
		943		938		948		935	γ	
		959		966		958		?		
						978				ω_c
								988		

Table III. 1 (continued)

PAA			PAF*		PAB**		HAB***		Assignment	
[+]	[++]	[+++]	[+]	[+++]	[+]	[+++]	[+]	[+++]	[+++]	
		1006		1008					δ_5	
						1007		1008		
		1020				1017		1028	ω_{c-o}	
				1045				1058		
						1060		1068		
						1088				
1095	1094	1091	1095	1099	1095	1097	1095	1096		
		1109		1110				1111	$?\delta_4$	
1114	1112	1115	1114	1122	1115	1118	1113	1118		
			1140		1142	1129	1141	1129	$\tau(CH_2)$	
1157		1154	1159	1160	1159	1157	1156	1157	δ_3	
			1167							
1171	1174	1181	1178	1174	1178	1171	1176	?	$\gamma(CH_3)$?	$\tau(CH_2)$
1186					1188		1192	1203		
1220		1218	1220	1217	1221	1216	1219	1221		
1246	1247	1250	1246			1253	1246	1251	ω_4	
1252		1260	1253	1258			1261			
1261			1262		1260		1268			

Table III. 1 (continued)

PAA			PAF*		PAB**		HAB***		Assignment	
[+]	[++]	[+++]	[+]	[+++]	[+]	[+++]	[+]	[+++]	[+++]	
			1271							
1276	1276	1277	1280	1278	1271	1270	1278	1281		
					1281					$\gamma(CH_2)$
1301		1300	1300	1300	1300	1302	1300	1296	ω_3	
								1003		
1319		1311	1319	1320	1320	1320	1320	1318	ω_9 δ_1?	
1333	1334	1331	1332	1332	1332	1336	1333	1336	ω(NO/N)	
						1367		1348		$\gamma(CH_2)$
		1370				1380		1382		
			1393			1394		1397		
1410	1408	1414	1411	1414	1411	1417	1410	1417	ω_9 or δ_1?	
		1425		1422		1427		1425	δ (CH_3)	
1438		1446	1439	1449		1440		1438	ω_5	
1454		1456	1454	1457	1454		1454		ω(N(O)N)	
1465	1460	1464	1464	1463	1466	1464	1466	1462	δ (CH_2)	
		1473				1472		1470		
				1477		1481				
1501	1496	1500	1499	1502	1500	1503	1500	1502	ω_6	

Table III. 1 (continued)

PAA			PAF*		PAB**		HAB***		Assignment
[+]	[++]	[+++]	[+]	[+++]	[+]	[+++]	[+]	[+++]	[+++]
				1547		1540			
1570	1572	1567	1570	1564		1566		1564	
1582			1580				1580		
1596			1595		1595	1595	1595	1594 }	ω_7, ω_8
1604	1603	1600	1604	1603	1604	1603	1605	1607 }	ω_7, ω_8
						1626		1629	
		1648		1650		1640			

The accuracy of the frequency determination is ±2 cm^{-1};
* 4-azoxyphenotele; ** 4,4'-n-di-n-pentyloxyazoxybenzene;
*** 4,4'-n-dii-n-hexyloxyazoxybenzene

[+] Amer, N. M., Y. R. Shen, J. Chem. Phys., 56, 2654 (1972)

[++] Freyman, R., R. Servant, Ann. Phys., 20, 131 (1945)

[+++] Maier, W., G. Englert, Z. Elektrochem., 62, 1020 (1958)

Table III. 2
Possible Assignment of IR Bands of MBBA[+] and BA[++]

MBBA			BA
Frequency cm^{-1}	Relative Intensity	Assignments (Wilson's notation)	Observed Frequency
180	6		
315	1		250
340	2		
410	4	16a	405
			540
630	5	6b	620
			650
720	2		
760	2		750
775	4		770
790	4		
825	3		825
886	3	δ(C-H)	875
934	1	17b	920
975	7	5	970
			1000
1014	1	18a (I and II)	1025
1105	4		
1164	70	9a (I and II)	1165
1182	37	(Φ-N)	1190
1245	5	(Φ-C)	1240
1305	5		1320
1370	4	(C-H)	1370
1422	12	19b (I and II)	1450
1503	9	19a (I and II)	1480
1575	53	8a (II)	1575
1596	100	8a ( )	1590
1626	41	γ (C=N)	1630

[+]N-(4-metoxybenzylidene) 4`,n-butilaniline

[++]benzylideneaniline

Table III. 3

Tentative assignment of Raman Bands of N-(4-hexy)loxylbenzylidene) 4',n-propylaniline (S_H and S_m phases) *

S_H phase 300 K	S_m phase 262 K	Tentative assignments	S_H phase 300 K	S_m phase 262 K	Tentative assignments
103			778	777	
114		Lib. C_6H_4	800	803vw	
	138	$\tau(CH_2CH_4)$	828	830vw	
	170			842	γ (=CH)
205	203	$\tau(CH_3)$		852vw	
210			860	862	
	220 sh		890	882vw	$\rho(CH_3)$ $\rho(CH_2)$
	240vw		940	940vw	
270	260vw		965	962vw	γ(=CH)
300	294vw	δ (hc) Accordian	977	977	
	318vw	modes	1013	1012	
330	333 sh	(C-C-C)	1032	1034	β(=CH) ω_c
355	353	Lib. CH=N bond		1065	
	387vw			1073	
412	414		1085	1086vw	β(=CH)
427sh				1094vw	
443	441		1108	1116	β(CH)

Table III. 3 (continued)

S_H phase 300 K	S_m phase 262 K	Tentative assignments	S_H phase 300 K	S_m phase 262 K	Tentative assignments	
	458vw			1136	τ (CH_2)	
480vw	471vw	(C-C-C) bond	1167	1165	γ (=CH), ω_e	
497vw	491		1192	1193		τ (CH_2) ↓
513vw	513vw		1250	1250	Phenyl-O stretch	
441	536vw		1282	1289		
563	559vw		1308	1308	$\omega(CH_2)$, Ring Vib.	γ (CH_2) ↓
592vw			1370	1365		
611sh	608	α(C-C-C)		1393		
620sh	625		1421	1422	δ	
636	638	α(C-C-C)	1501	1500	δ (CH_3)	
670vw		δ ring	1571	1571	C=C stretch	
710	720		1594	1595		
730	732	ρ (CH_2)	1625	1423	C=N stretch	

vw - very weak; sh - shoulder; S_m - long-lived metastable solid

* Deniz, K. U., A. J. Mehta, U. R. K. Rao, P. S. Parvathanathan, A. S. Paranjpe, Phys. Lett., 63A, 105 (1977)

Table III.4

Possible Assignment of IR Bands of $EBBA^+$(Crystal Phase)

3185 w	CH aromatic stretch vibration
3140 w	CH aromatic stretch vibration
3040 w	CH aromatic stretch vibration
3020 m	CH aromatic stretch vibration
2955 s	CH_3 asymmetric stretch vibration
2922 s	CH_2 asymmetric stretch vibration
2860 w	CH_3 symmetric stretch vibration
2845 s	CH_2 symmetric stretch vibration
1622 s	CH=N symmetric stretch vibration
1602 s	CC aromatic stretch vibration
1590 m	CC aromatic stretch vibration
1568 m	CC aromatic stretch vibration
1510 m	CC aromatic stretch vibration
1470 w	CC aromatic stretch vibration
1455 s	CH_3 asymmetric deformation
1420 m	CH_3 asymmetric deformation
1390 m	CH_3 asymmetric deformation
1378 m	CH_3 symmetric deformation
1360 w	CH_3 symmetric deformation
1308 m	CH aromatic in-plane deformation
1302 w	CH aromatic in-plane deformation
1288 w	CH aromatic in-plane deformation
1252 vs	benzene ring - O stretch vibration
1205 w	benzene ring - C stretch vibration
1192 m	benzene ring - N stretch vibration
1160 m	CH aromatic in-plane deformation
1152 s	CH aromatic in-plane deformation
1112 vs	CH aromatic in-plane deformation
1042 vs	O-C stretch vibration
1015 w	CH_3 rocking
978 s	CH aromatic out-of-plane deformation
888 s	benzene ring - N deformation

Table III.4 (continued)

840 vs	CH aromatic out-of-plane deformation
824 vs	CH aromatic out-of-plane deformation
780 w	CH aromatic out-of-plane deformation
728 m	CH_2 rocking
642 m	benzene ring out-of-plane deformation
632 w	
622 m	benzene ring in-plane deformation
577 s	
532 s	benzene ring out-of-plane deformation
519 w	benzene ring out-of-plane deformation
492 m	benzene ring - O deformation
415	benzene ring out-of-plane deformation

vs - very strong; s - strong; m - middle; w - weak

+ N-(4-ethoxybenzylidene) 4`,n-butylaniline

Table III.5

Possible Assignment of IR Bands of APAPA$^+$ (Crystal Phase)

3080 w	CH aromatic stretching vibration
3060 vw	CH aromatic stretching vibration
3025 w	CH aromatic stretching vibration
3004 m	CH aromatic stretching vibration
2964	CH_3 asymmetric stretch vibration
2925	CH_2 asymmetric stretch vibration
2875	CH_3 symmetric stretch vibration
2843	CH_2 symmetric stretch vibration
1748 vs	C=O stretch vibration
1624 vs	CH=N stretch vibration
1600 m	benzene ring stretch vibration
1540 m	benzene ring stretch vibration
1570 m	benzene ring stretch vibration
1510 m	benzene ring stretch vibration
1490 m	benzene ring stretch vibration
1455 m	CH_3 asymmetric deformation
1430 m	benzene ring stretch vibration
1373 s	CH_3 symmetric deformation
1308 m	CH_3 symmetric deformation
1298 m	CH_3 symmetric deformation
1255 s	benzene ring - O stretch vibration
1240 s	benzene ring - O stretch vibration
1215 s	benzene ring - N stretch vibration
1285 vs	CH aromatic in-plane deformation
1160 s	CH aromatic in-plane deformation
1108 m	CH aromatic in-plane deformation
1098 s	CH aromatic in-plane deformation
1020 vs	O-C stretch vibration
974 m	CH aromatic out-of-plane vibration
912 s	CH aromatic out-of-plane vibration
883 m	benzene ring - N deformation

Table III.5 (continued)

851	CH aromatic out-of plane deformation
830 vs	CH aromatic out-of-plane deformation
783 vs	CH aromatic out-of-plane deformation
765 w	benzene ring in-plane deformation
722 w	benzene ring in-plane deformation
650 m	benzene ring in-plane deformation
632 m	benzene ring in-plane deformation
590 m	
570 m	
540 vs	benzene ring out-of-plane deformation
508 m	benzene ring - O deformation
478 m	
462 m	
416 m	benzene ring out-of-plane deformation
403 m	
358	
348 m	
305 m	

vs - very strong; s - strong; m - middle; w - weak; vw - very weak

+N-(4-etoxybenzylidene) 4'-aminophenylacetate

Table III. 6

Possible Assignment of IR Bands of 4, n-heptyloxybenzoic Acid (Crystal phase)*

Band	Assignment
2965 m	CH_3 asymmetric stretch vibration
2922w	CH_2 asymmetric stretch vibration
2868 w	CH_3 symmetric stretch vibration
2850 m	CH_2 symmetric stretch vibration
2670 m	OH...O stretch vibration(?)
2565 m	
1685 s	C=O stretch vibration
1612 s	C-C aromatic stretch vibration
1580 m	C-C aromatic stretch vibration
1518 m	C-C aromatic stretch vibration
1472 m	CH_2 symmetric deformation
1438 m	C-C aromatic stretch vibration
1402 m	C-O stretch + O-H deformation
1380 m	CH_3 symmetric deformation
1338	
1320 m	CH_2 deformation
1310 s	C-O stretch + OH deformation
1300 s	CH_2 deformation
1283 m	CH_2 deformation
1260 s	benzene ring - O stretch vibration
1210 m	CH aromatic in-plane deformation
1190 m	CH_2 deformation
1172 s	C-O-C stretch vibration
1150 m	CH_2 deformation
1130 m	
1110 w	
1068 m	O-C-C stretch vibration
1042 m	C-C stretch vibration
1010 m	CH aromatic in-plane vibration
972 s	O-C-C stretch vibration

Table III. 6 (continued)

Band	Assignment
950 s	OH in-plane deformation
885 m	CC stretch vibration
855 w	
848 s	CH aromatic out-of-plane deformation
830 w	
815 w	
797 w	
773 s	CH aromatic out-of-plane deformation
740 w	CH_2 rocking
721 w	CH_2 rocking

vs - very strong; s - strong; m - middle;
vw - very weak in intensity

*Hodjaeva, B. L., Izv. AN SSSR, ser. chimicheskaya, 1, 2409 (1969)

Table III. 7
Possible Assignments for tha Raman Spectra of the Isotropic Liquid Phases of 4-cyano-,4' n-pentylbiphenyl (5CB) and 4-cyano-4',n-pentil d_{11}-biphenyl (5CB-d_{11})*

5 CB		5 CB-d_{11}		Possible Assignment
Fre-quency	Inten-sity	Fre-quency	Inten-sity	
179	3	178	3	
216	3	213	3	
410	23	410	25	Ring vibration and/or phenyl-C wagging
		468	2	
474	3			
509	3			
		498	3	
551	3	541	2	
639	23	552	3	
651	5	637	18	Ring vibration
		650	5	
		697	1	
		713	1	
738	1	732	3	
780	10			
		767	1	
		795	40	Ring breathing vibration
806	28	801	sh	
826	25			Ring breathing vibration Alkyl chain vibrations
862	1	831	sh	Alkyl chain vibrations (862–955)
886	1			
899	2			
967	3	955	5	

Table III. 7 (continued)

5 CB		5 CB-d_{11}		
Frequency	Intensity	Frequency	Intensity	Possible Assignment
976	sh	971	4	
1018	10	1007	2	
		1018	9	Aromatic C-H in-plane deformation
1062	3	1052	1	
				Alkyl chain vibration
1112	8	1070	15	C-D deformation
		1112	9	
1179	75	1123	8	C-D deformation
1185	60	1181	69	Aromatic C-D in-plane deformation
1235	3	1187	sh	
		1235	vw	Ring vibration
1285	92	1248	12	$-CD_2$ stretch
1436	4	1287	90	C-C stretch of biphenyl link
1445	3			C-H deformation of pentyl chain
1452	2			
1527	10			
1606	100	1525	110	C-C stretch of aromatic rings
1613	sh	1607	00	
		1613	sh	
		2067	15	
		2098	15	
		2116	22	C-D stretching vibrations
		2150	5	
		2175	3	
2224	90			
		2224	80	C≡N stretch

* Gray. G. W., A. Mosley, Mol. Cryst. Liq. Cryst., 35, 71 (1976)

Table III. 8

Possible Assignment of IR Bands of 8OCB$^+$ (Crystal Phase)

Band	Assignment
3060 w	CH aromatic stretch vibration
3040 w	CH aromatic stretch vibration
2945 s	CH_3 asymmetric stretch vibration
2930 s	CH_2 asymmetric stretch vibration
2870 s	CH_3 symmetric stretch vibration
2855 s	CH_2 symmetric stretch vubration
2225 s	C≡N stretch vibration
1602 vs	C-C aromatic stretch vibration
1577 m	C-C aromatic stretch vibration
1510 m	C-C aromatic stretch vibration
1492 vs	C-C aromatic stretch vibration
1470 s	CH_3 asymmetric deformation
1460 s	
1390 m	CH_3 symmetric deformation
1310 m	CH aromatic in-plane deformation
1290 s	CH aromatic in-plane deformation
1260 s	
1247 s	Benzene ring - O stretch vibration
1175 s	CH aromatic in-plane deformation
1122 s	
1110 w	C=C stretch vibration
1060 s	
1030 m	C-C stretch vibration
997 w	
965 w	
845 m	
826 vs	CH aromatic out-of-plane deformation
820 s	CH aromatic out-of-plane deformation
755 w	Benzene ring in-plane deformation
732 w	
725 w	CH_2 rocking

Table III. 8 (continued)

Band	Assignment
718	
658	
558	
550	
545	benzene ring out-of-plane deformation
538	

+4,n-octyloxy 4`-cyanobiphenyle

FAR INFRARED SPECTRA

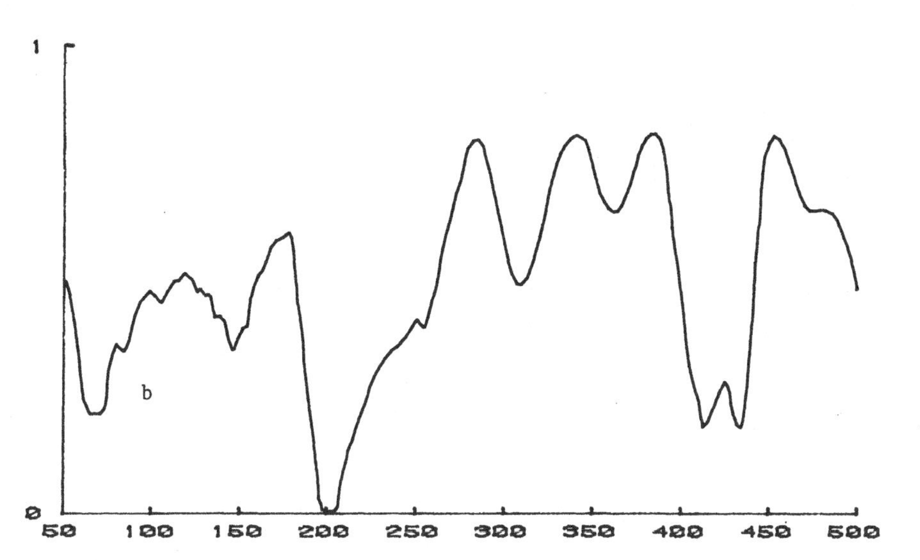

4,4'-di-methoxyazoxybenzene: a-nematic phase 125°C; b-isotropic liquid 145°C.

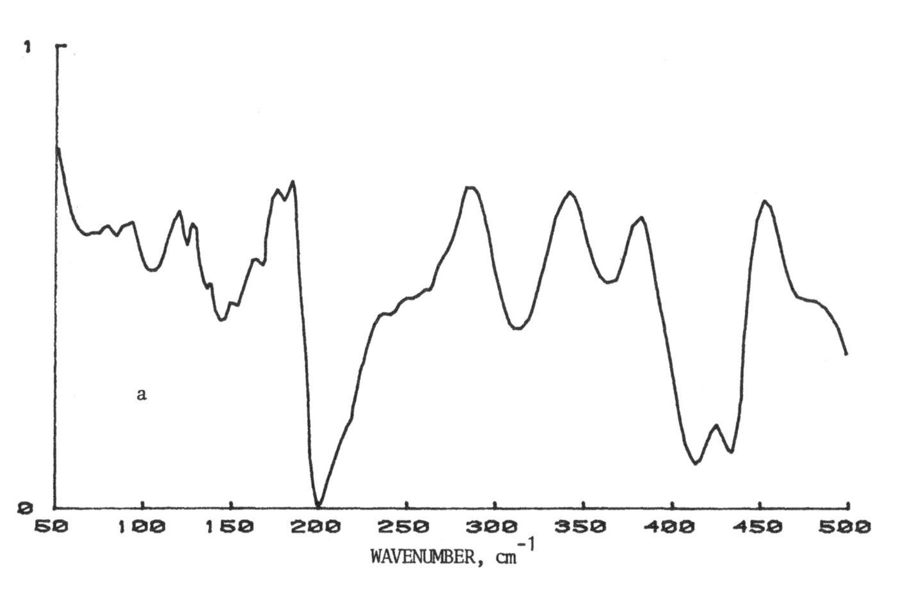

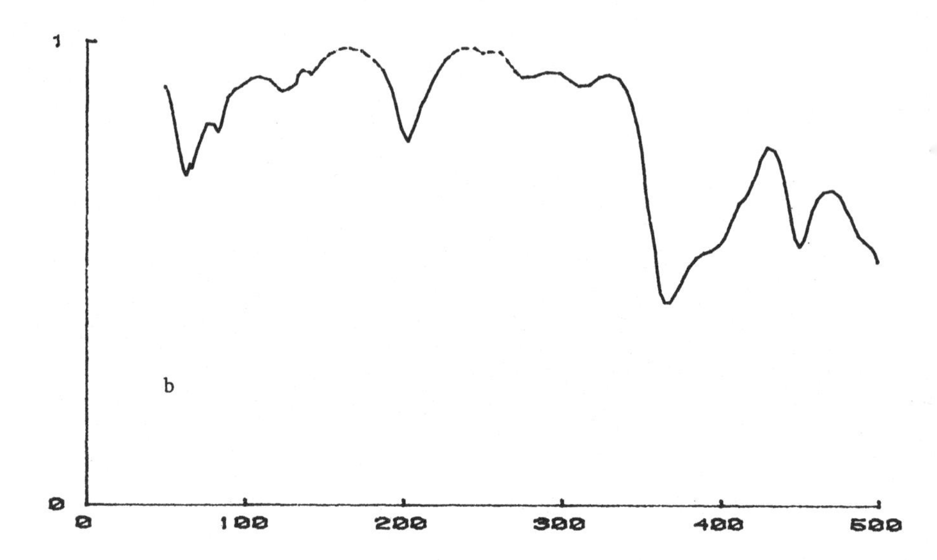

4,4'-di-ethoxyazoxybenzene: a-crystal phase 40°C; b-nematic phase 145°C.

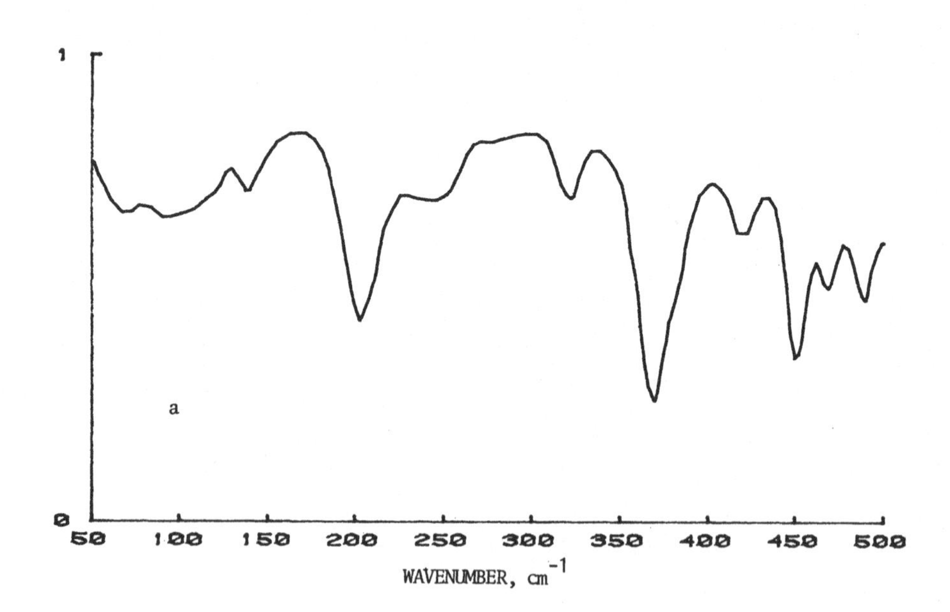

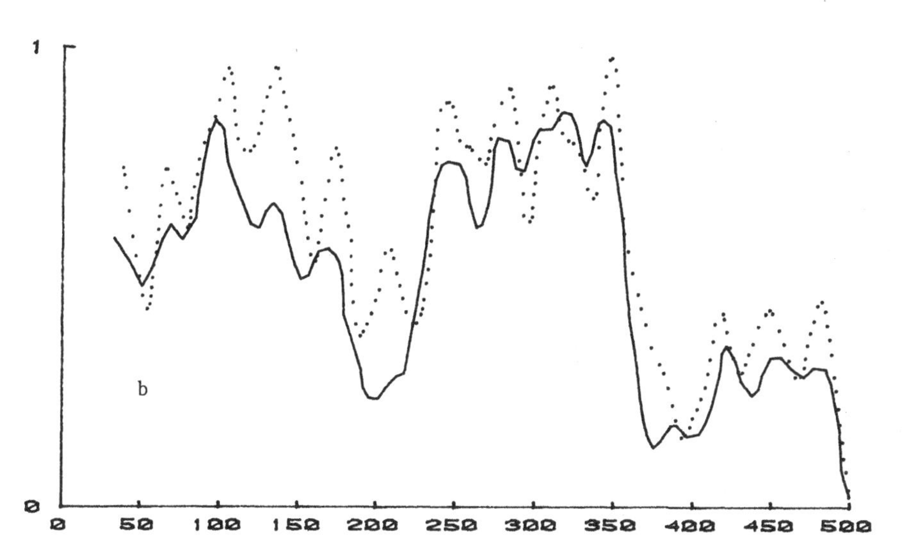

4,4'-di-n-propyloxyazoxybenzene: a-crystal phase 20°C; b- full line -nematic phase 125°C, dotted line-isotropic liquid 140°C.

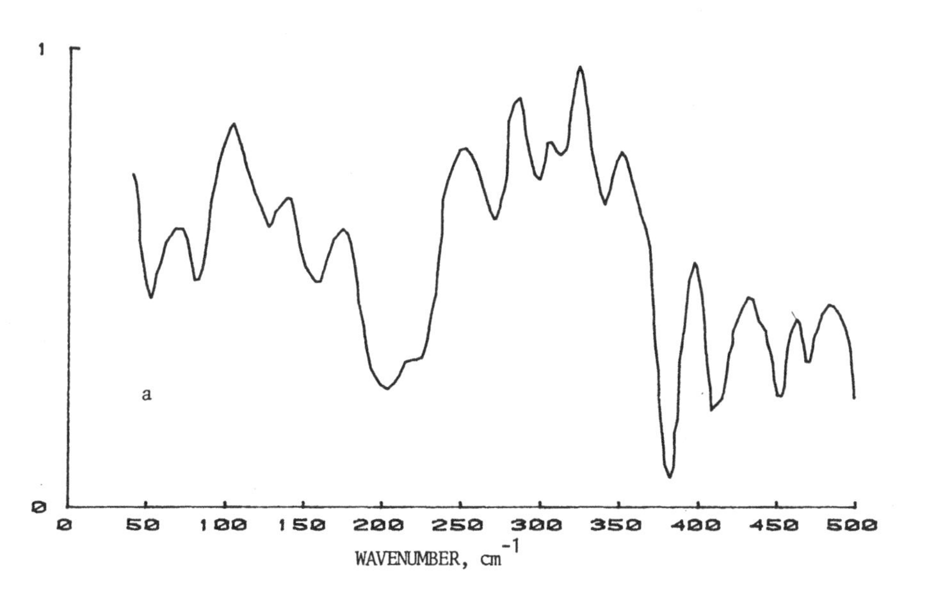

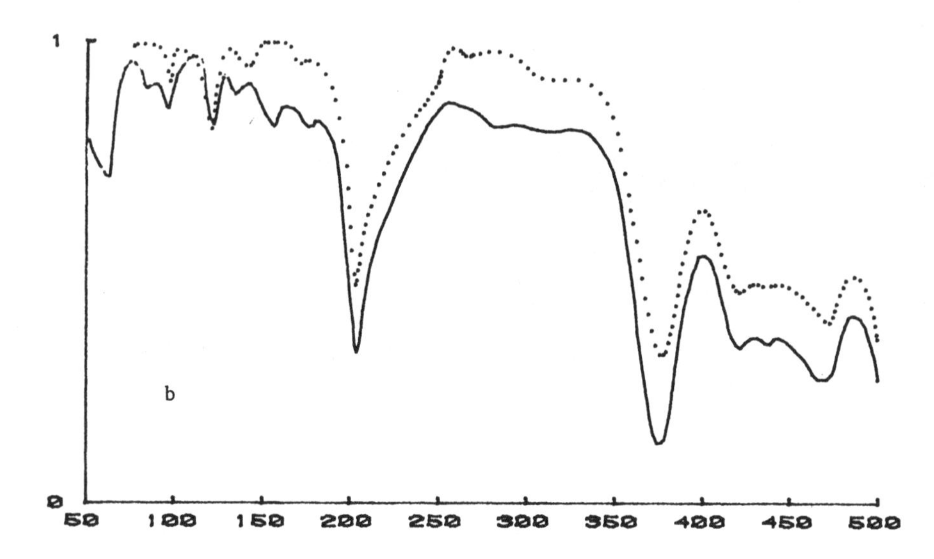

4,4'-di-butyloxyazoxybenzene: a-crystal phase 20°C; b-full line-nematic phase 115°C, dotted line- isotropic liquid 135°C.

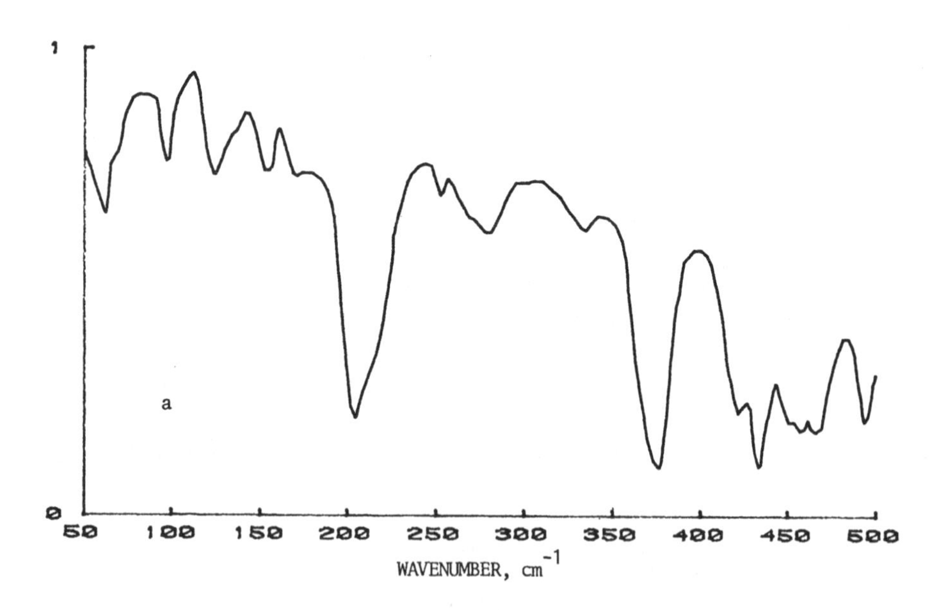

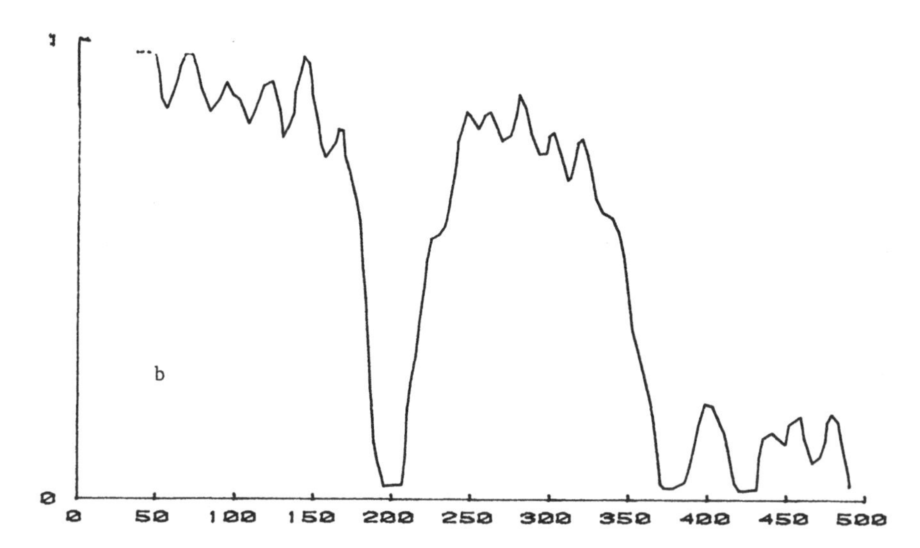

4,4'-di-n-pentyloxyazoxybenzene: a-crystal phase 20°C; b-nematic phase 105°C.

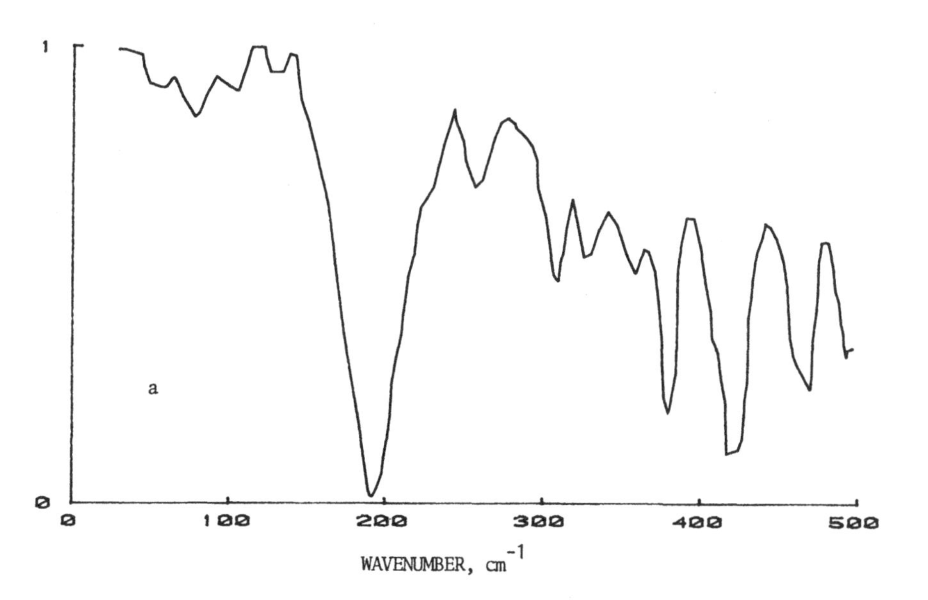

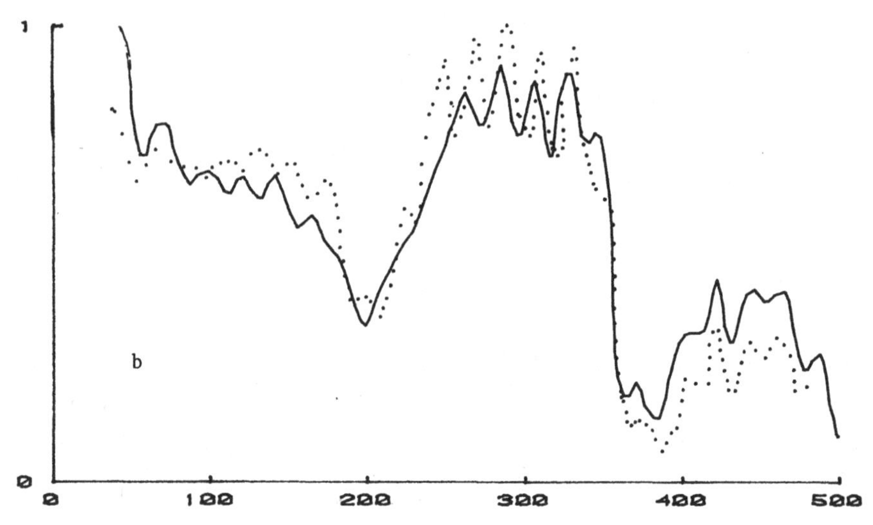

4,4'-di-hexyloxyazoxybenzene: a-crystal phase 20°C; b-full line-nematic phase 105°C, dotted line-isotropic liquid-135°C.

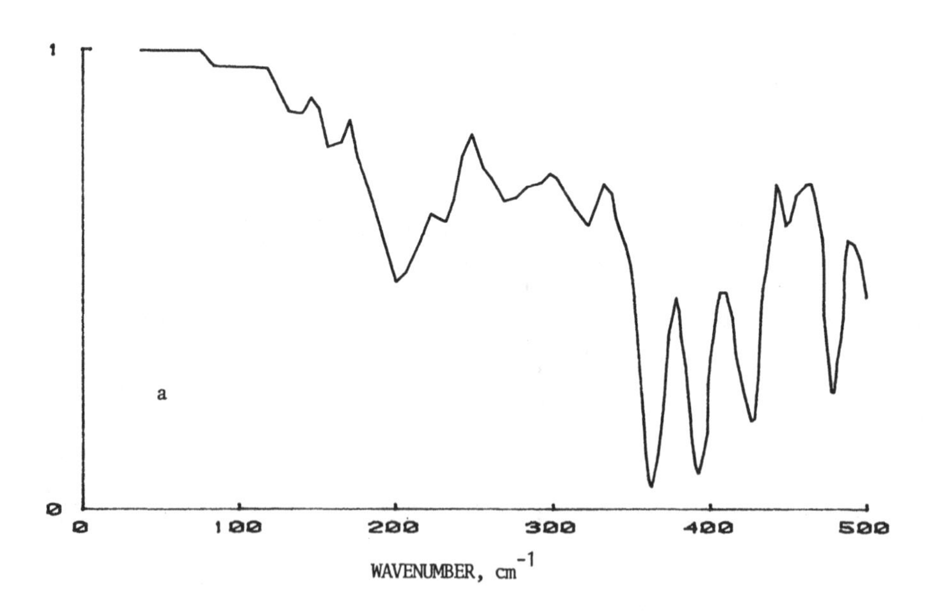

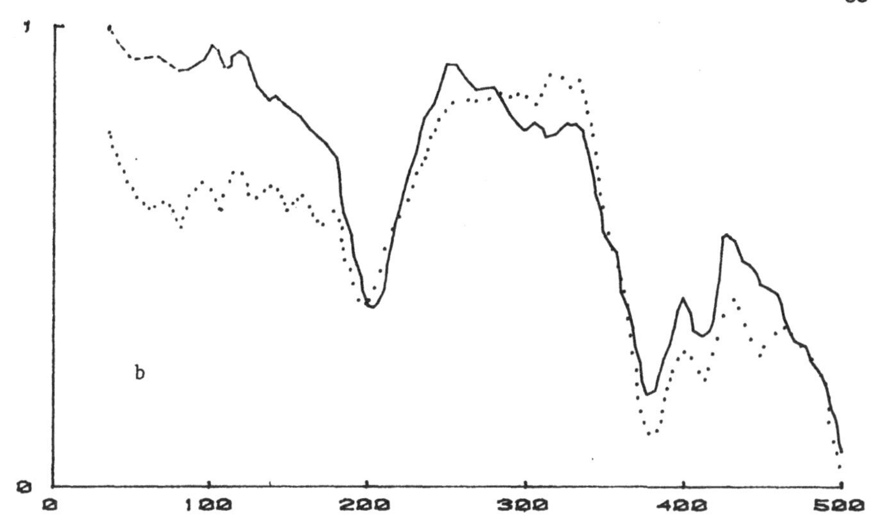

4,4'-di-n-heptyloxyazoxybenzene: a-crystal phase 20°C; b-full line-smectic C phase 85°C, dotted line-isotropic liquid 135°C.

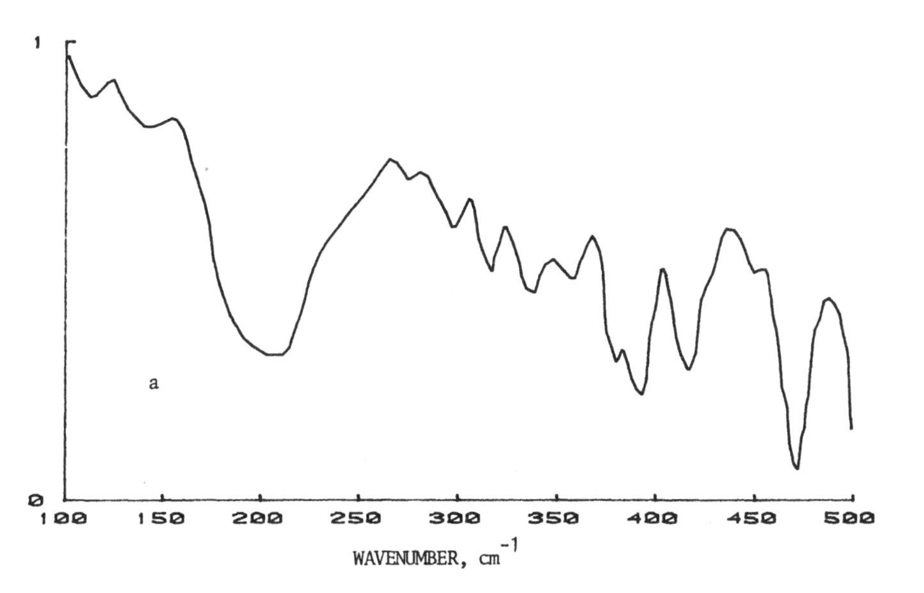

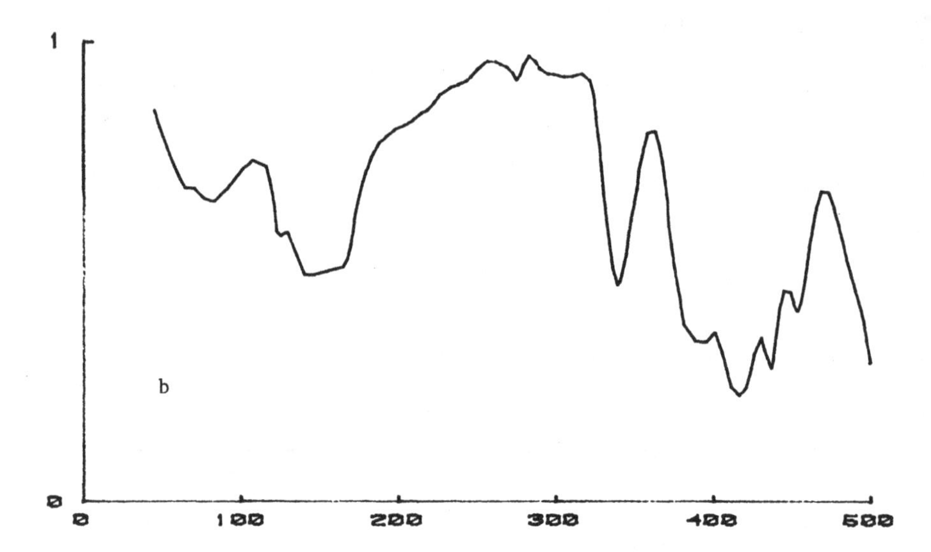

4,n-butyl 4'-methoxyazoxybenzene: a-nematic phase 30°C; b-isotropic liquid 85°C.

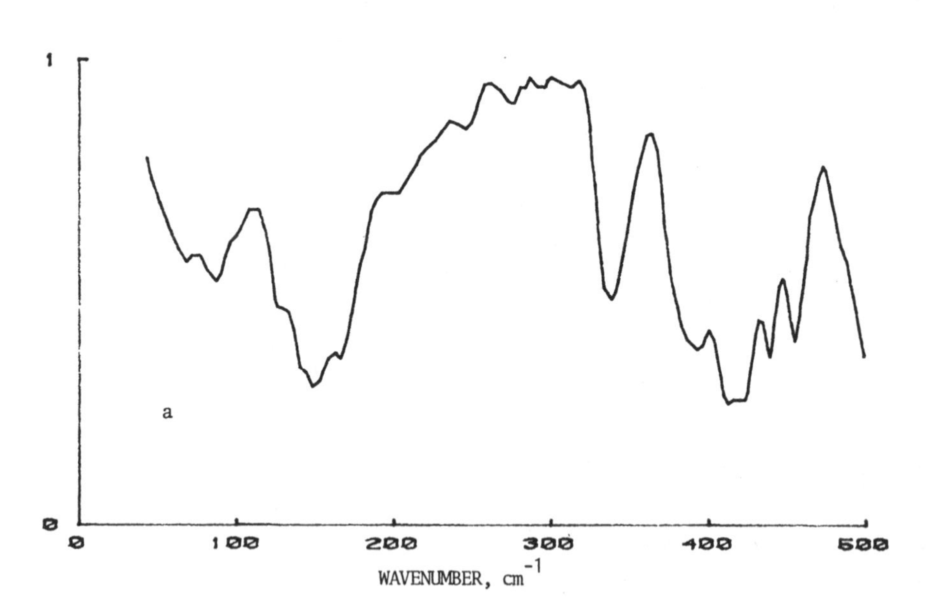

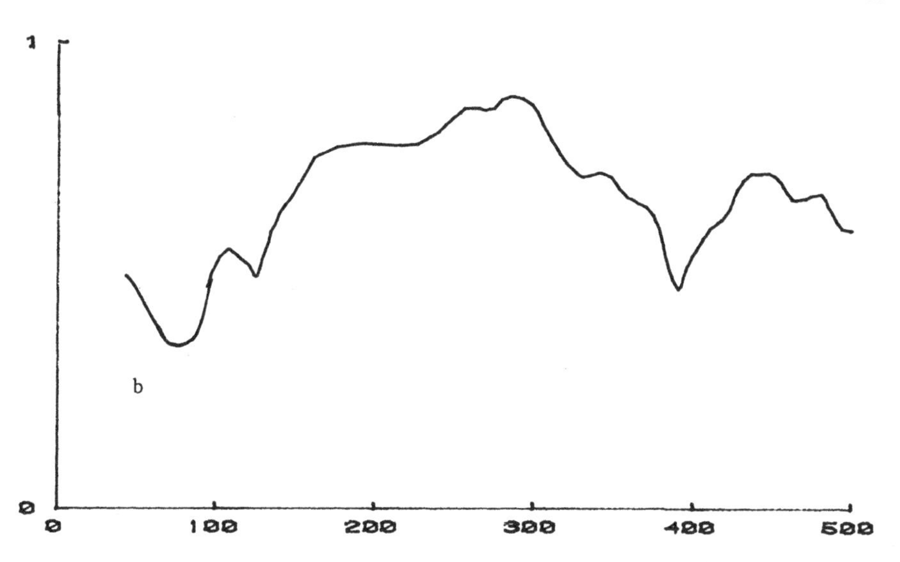

4-methoxy 4',n-butylazoxybenzene: a-nematic phase 20°C; b- isotropic liquid 80°C.

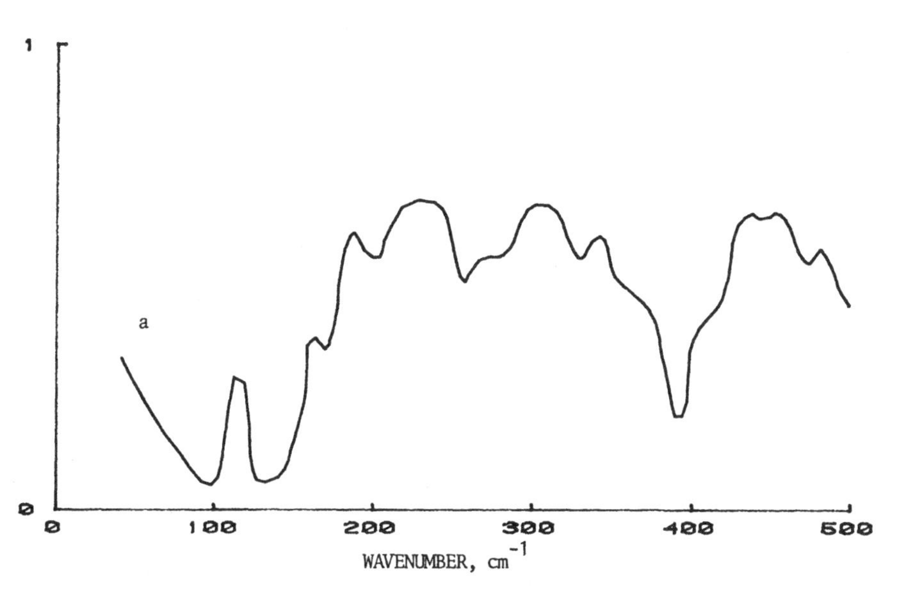

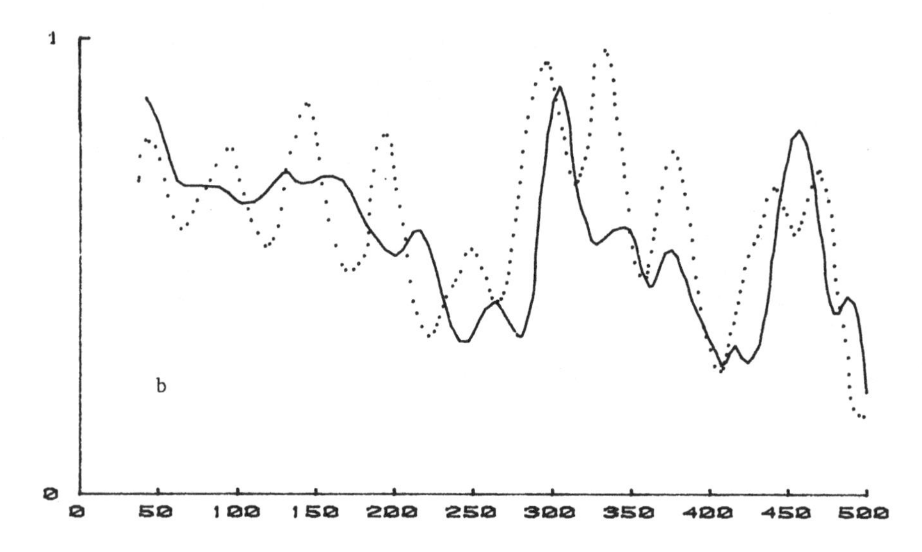

4-ethoxy 4',n-hexanoyloxyazoxybenzene: a-crystal phase 20°C; b-full line-nematic phase 90°C, dotted line-isotropic liquid 120°C.

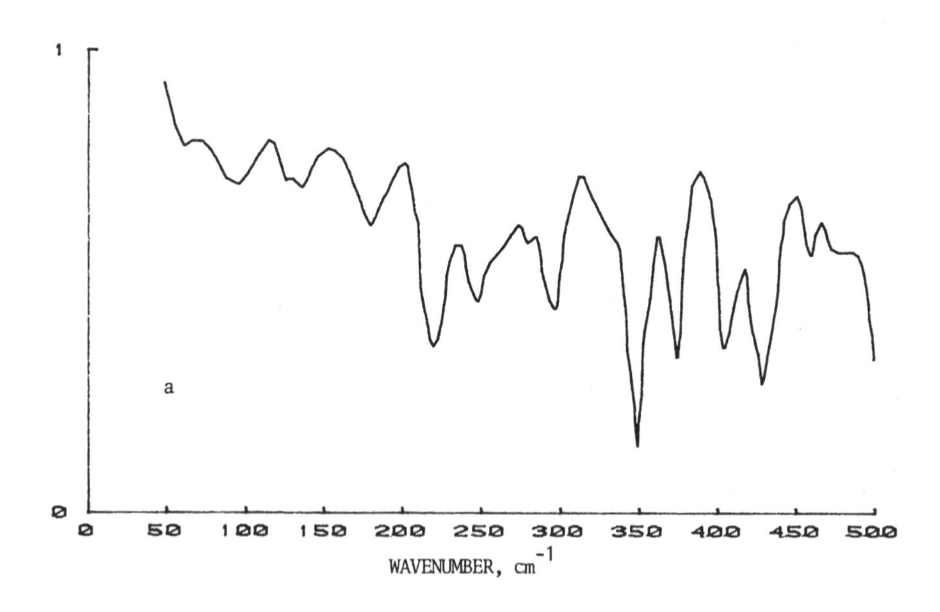

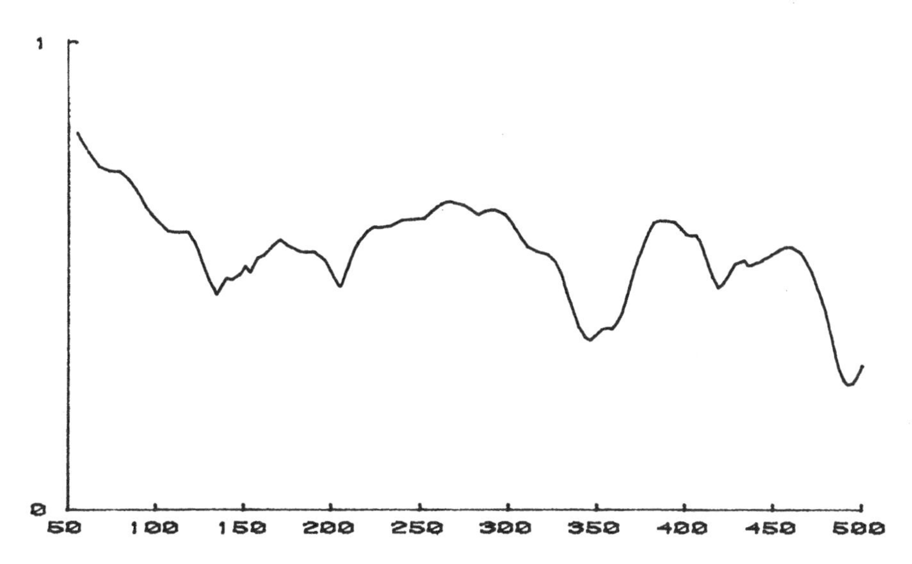

N-(4-methoxybenzylidene) 4',n-butylaniline (MBBA): nematic phase 25°C.

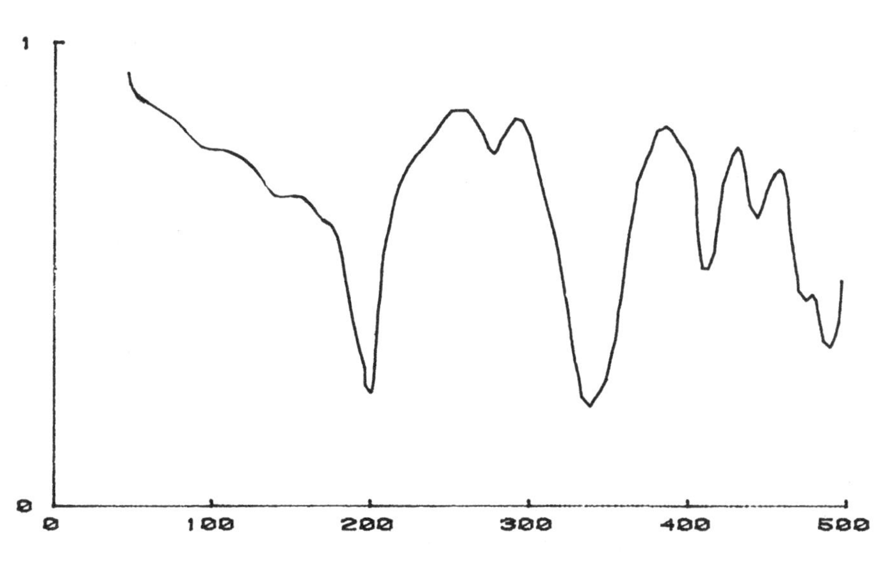

N-(4-propyloxybenzylidene) 4',n-butylaniline: nematic phase 50°C.

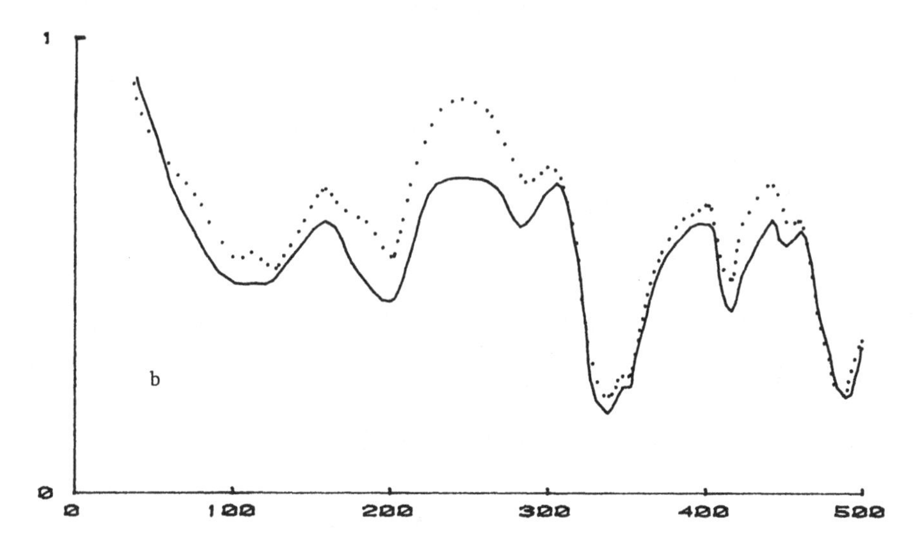

N-(4-ethoxybenzylidene) 4',n-butylaniline: a-crystal phase 20°C; b-full line nematic phase 55°C, dotted line-isotropic liquid 85°C.

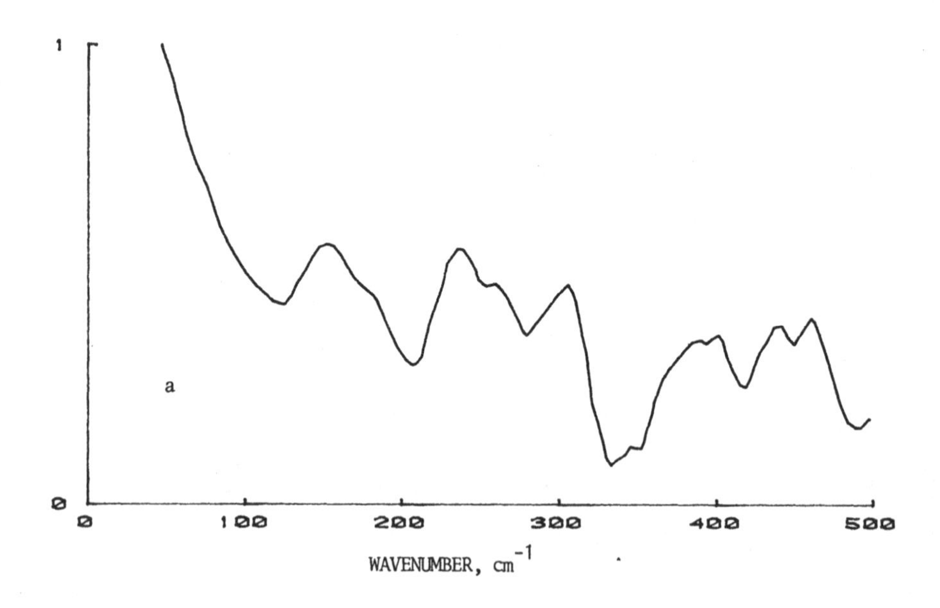

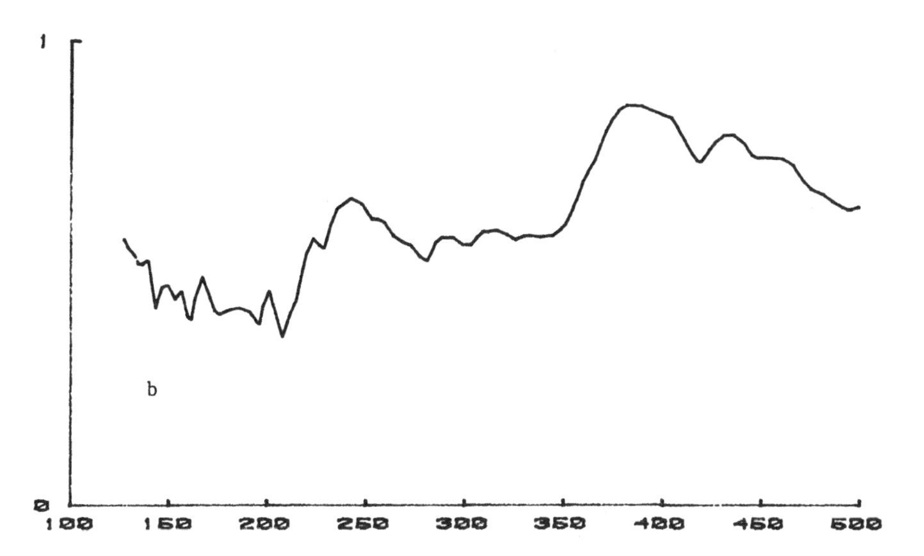

N-(4-propyloxybenzylidene) 4',n-pentylaniline: a-full line-crystal phase 20°C, dotted line-nematic phase 45°C; b-isotropic liquid 80°C.

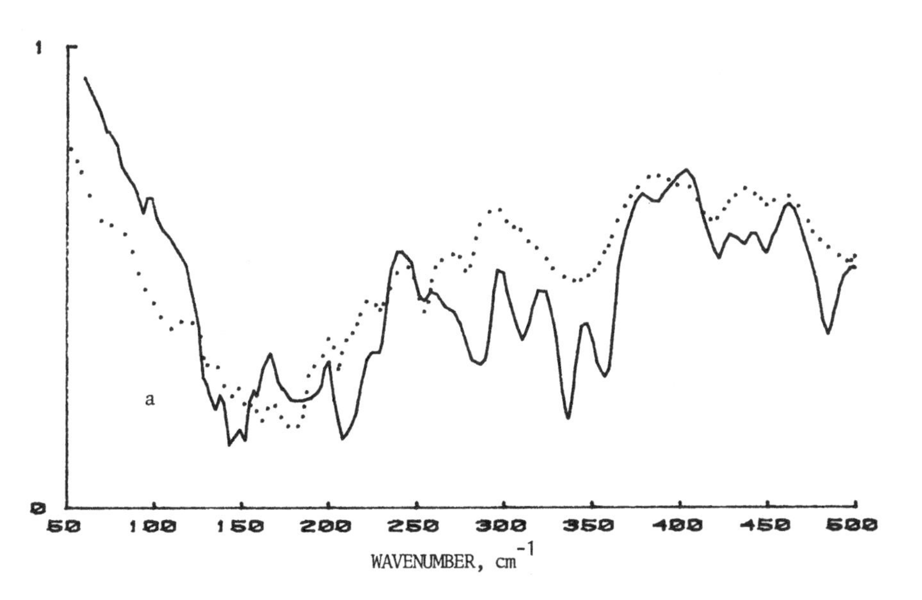

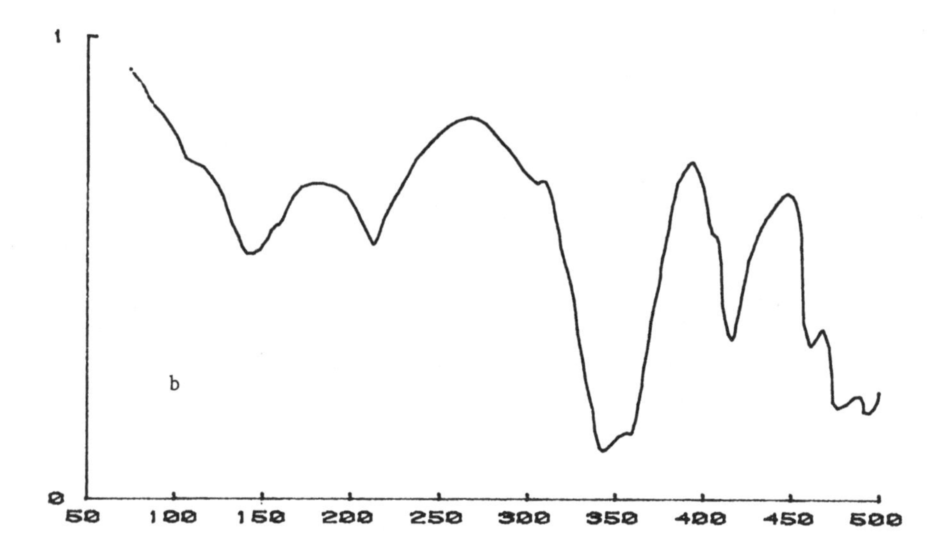

N-(4-butyloxybenzylidene) 4',n-pentylaniline: a-crystal phase 20°C; b-nematic phase 45°C.

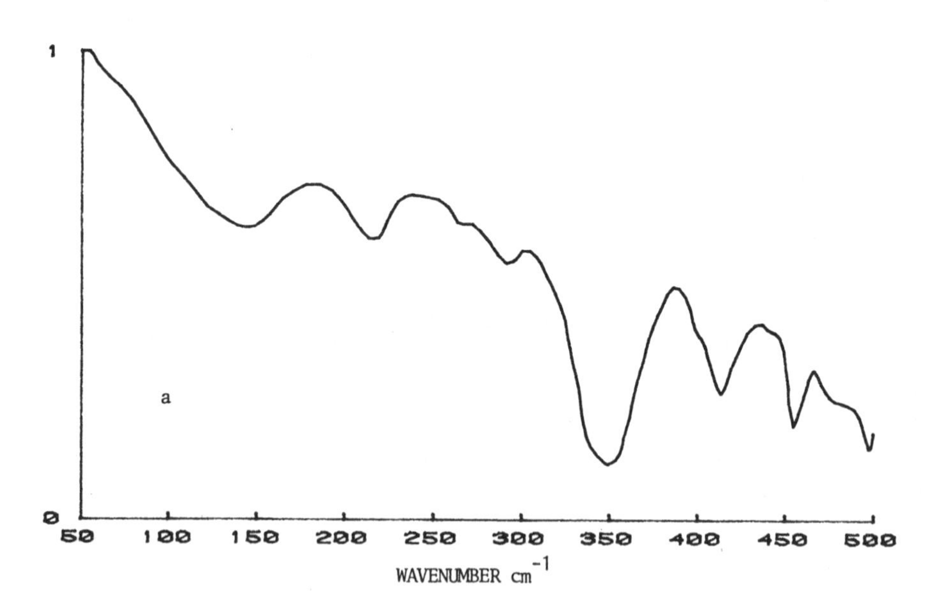

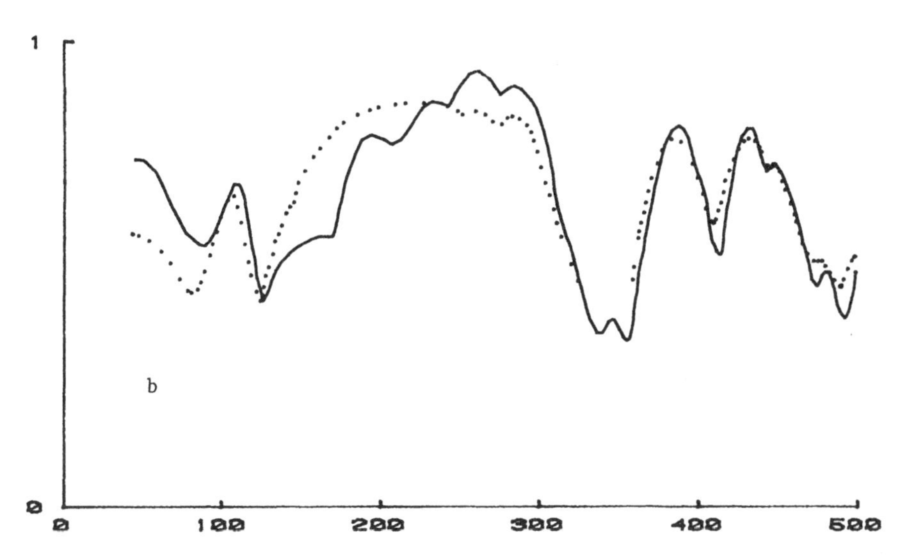

N-(4-pentyloxybenzylidene) 4',n-pentylaniline: a-crystal phase 20°C; b-full line-nematic phase 50°C, dotted line-isotropic liquid 75°C.

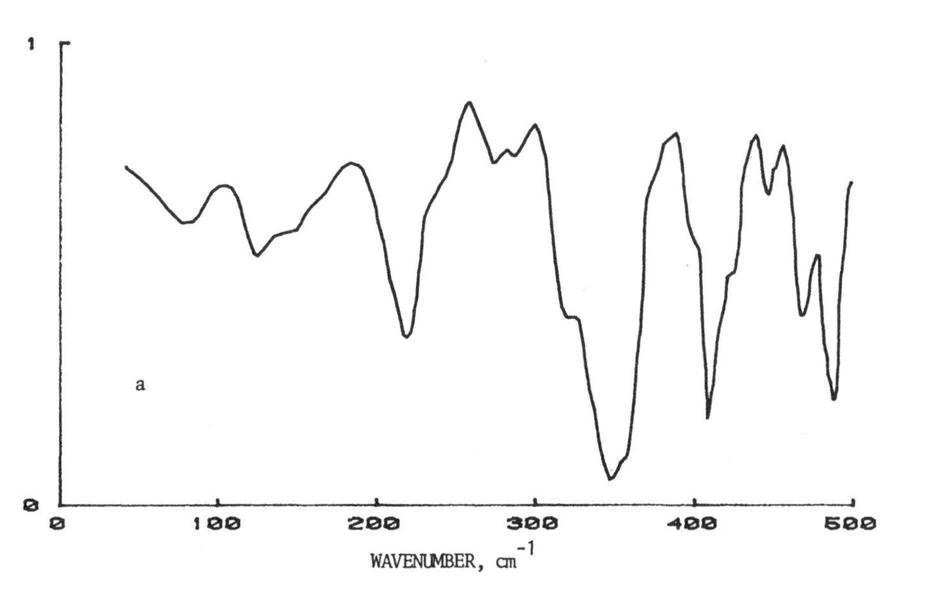

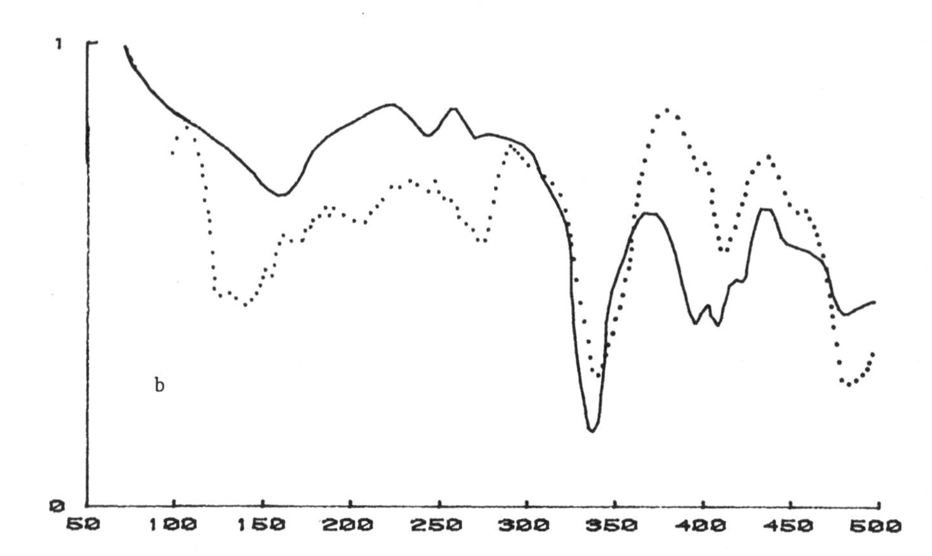

N-(4-pentyloxybenzylidene) 4'-chloraniline: a-full line-crystal phase 20°C, dotted line-nematic phase 90°C; b-full line-smectic A phase 60°C, dotted line-isotropic liquid 95°C.

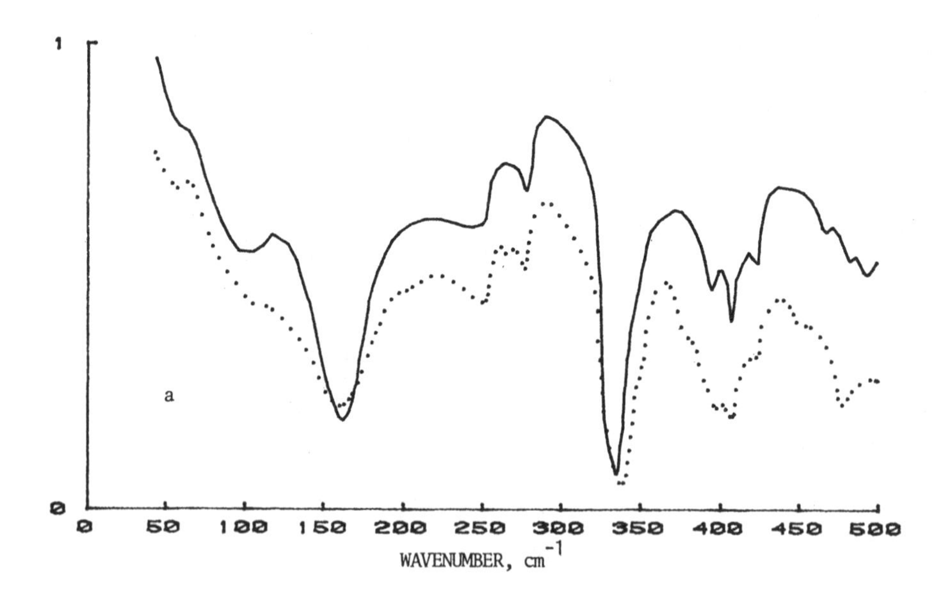

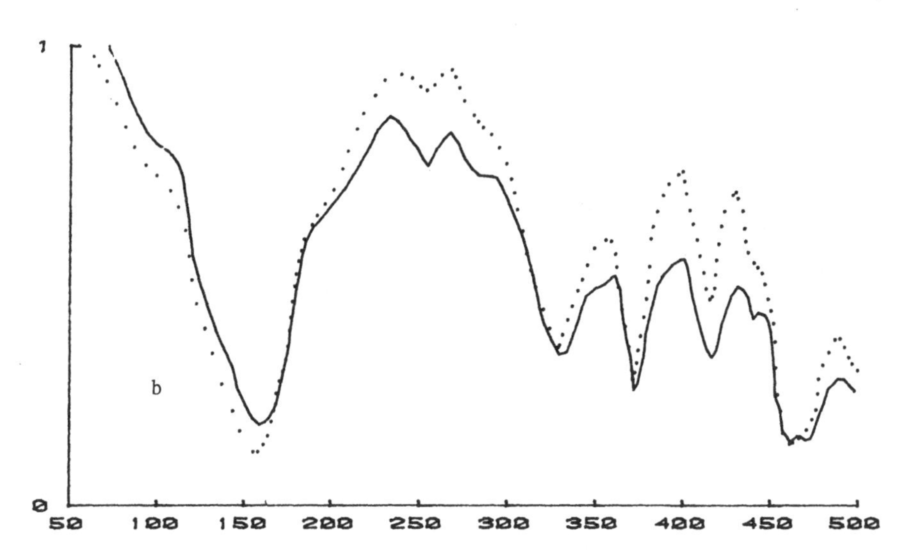

N-(4-hexyloxybenzylidene)4'-cyanoaniline(HBCA): a-crystal phase 20°C; b-nematic phase 80°C (full line), dotted line-isotropic liquid 110°C.

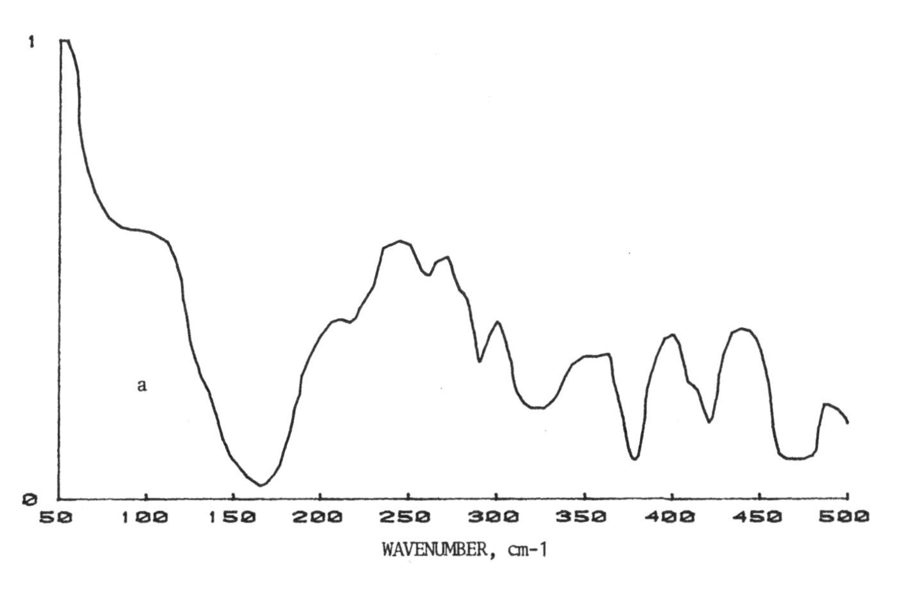

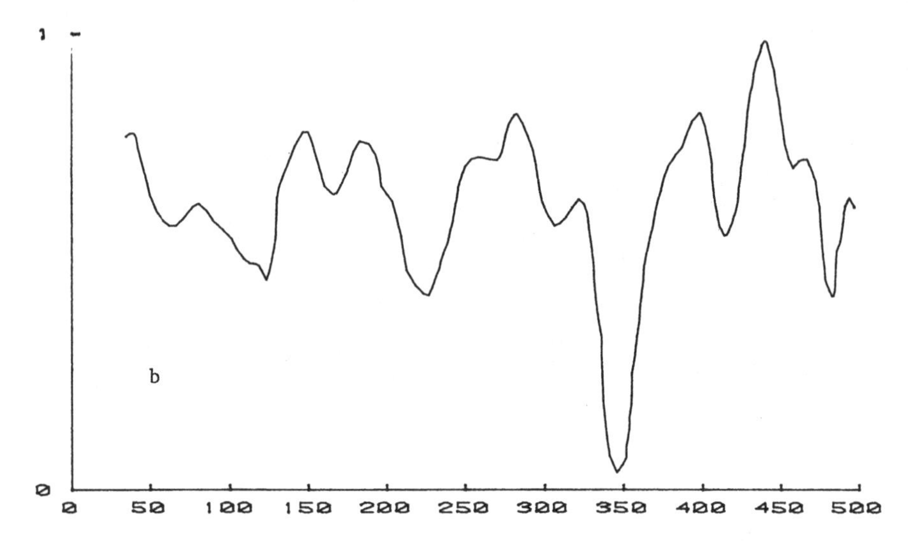

N-(4-ethoxybenzylidene)4',n-phenylacetate:a-nematic phase 90°C, b-isotropic liquid 120°C.

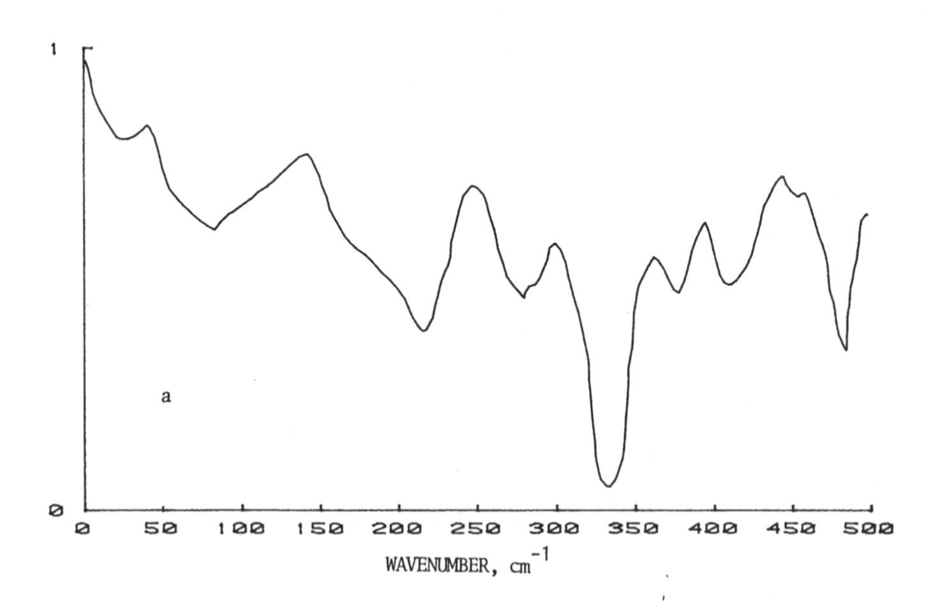

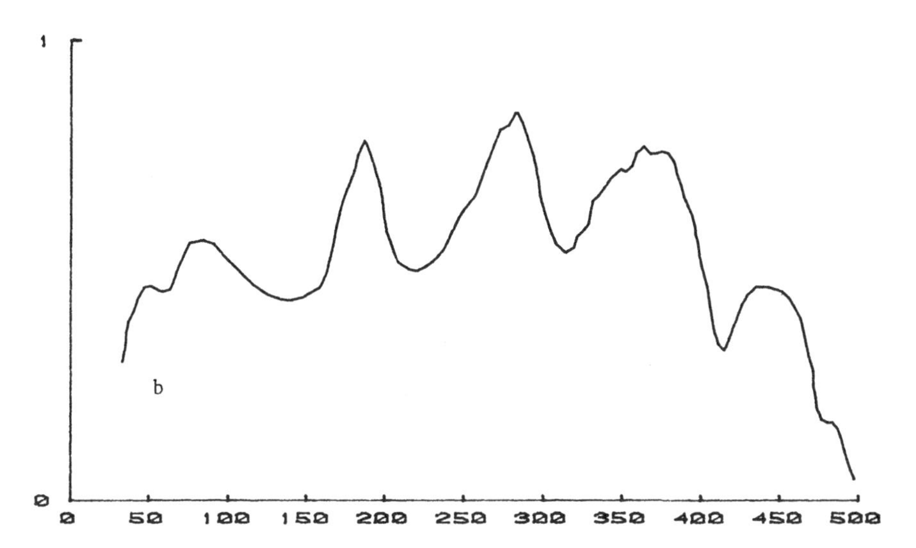

2,6-bis (4,n-butyloxybenzylidene) cyclohexane: a-nematic phase 125°C; b-isotropic liquid 160°C.

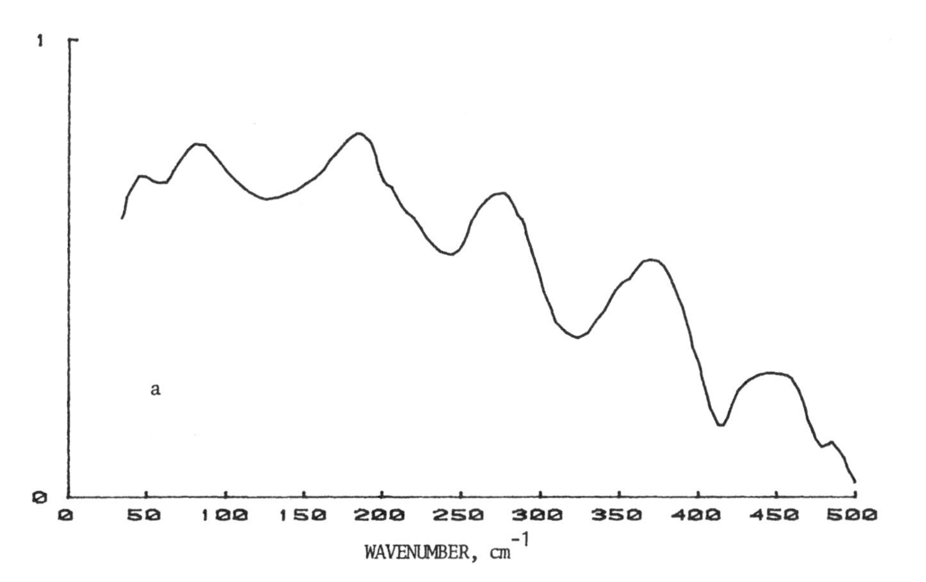

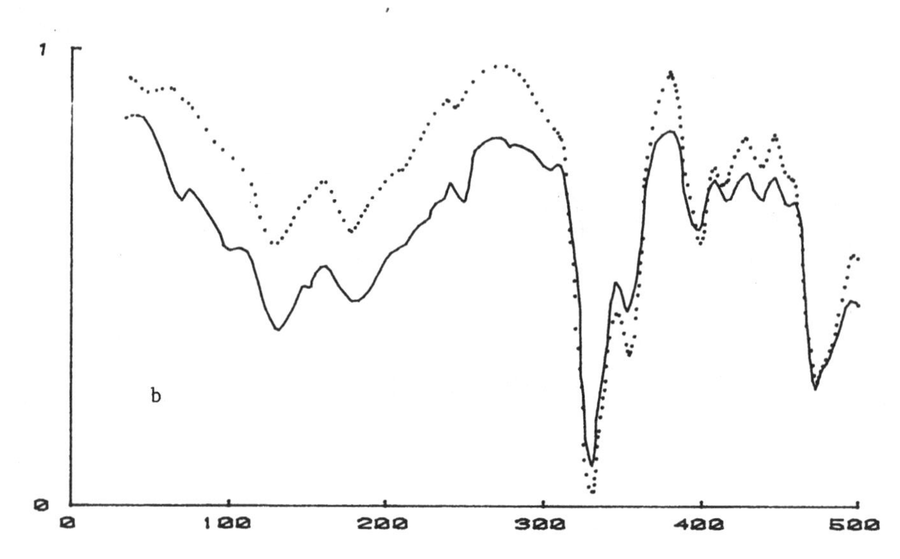

N-(4-aminooxybenzylidene) 4'-toluidine:a-crystal phase 20°C, b-full line - nematic phase 60°C, dotted line-isotropic liquid 75°C.

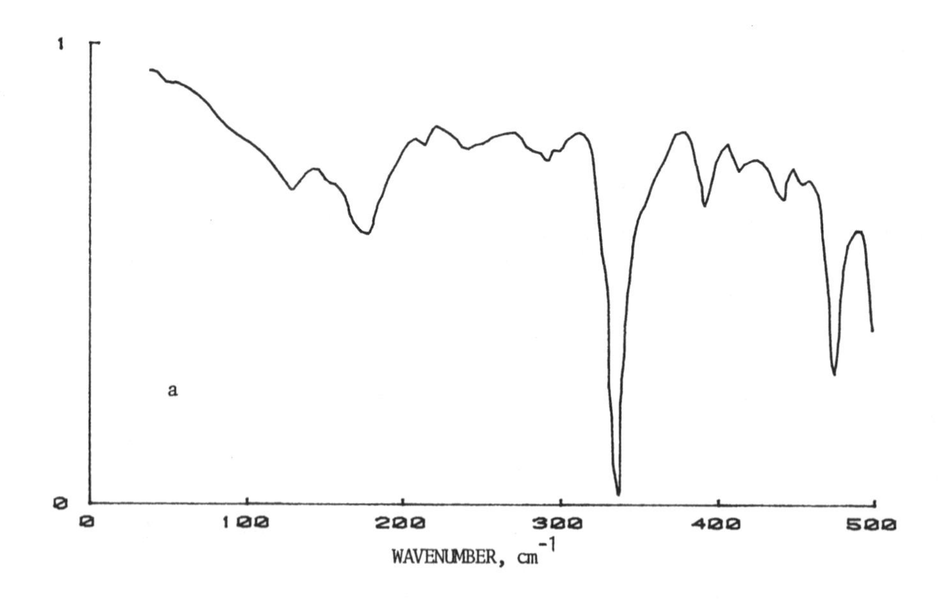

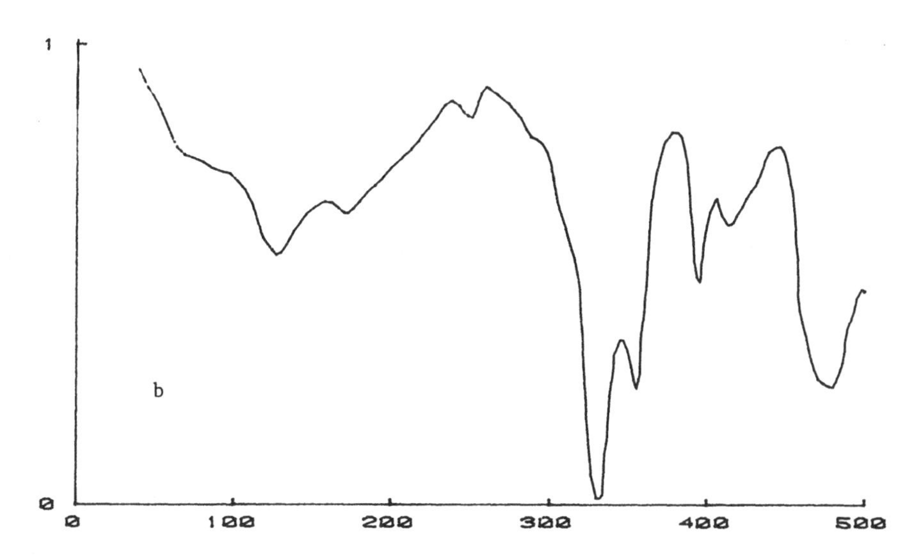

N-(4-butyloxybenzylidene) 4'-toluidine: a-nematic phase60°C, b-isotropic liquid 75°C.

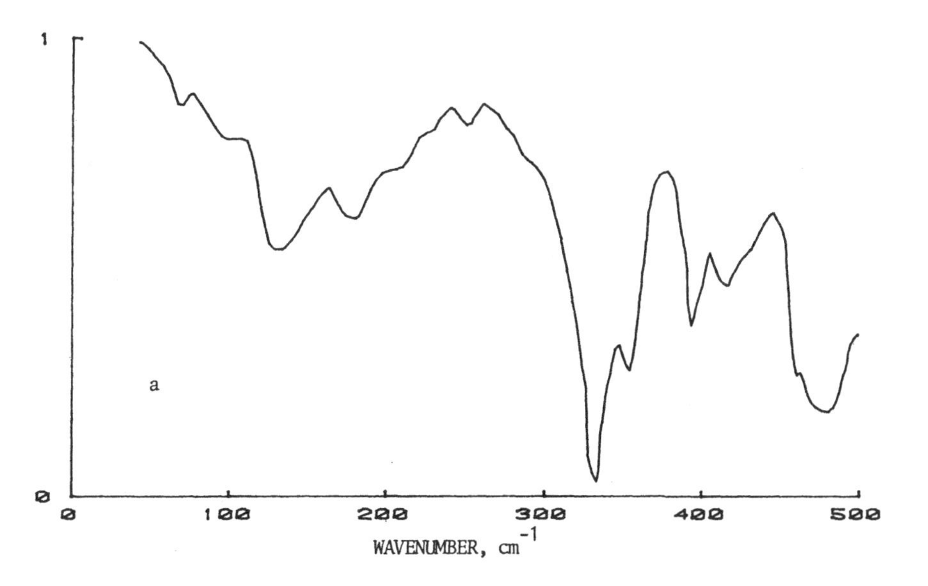

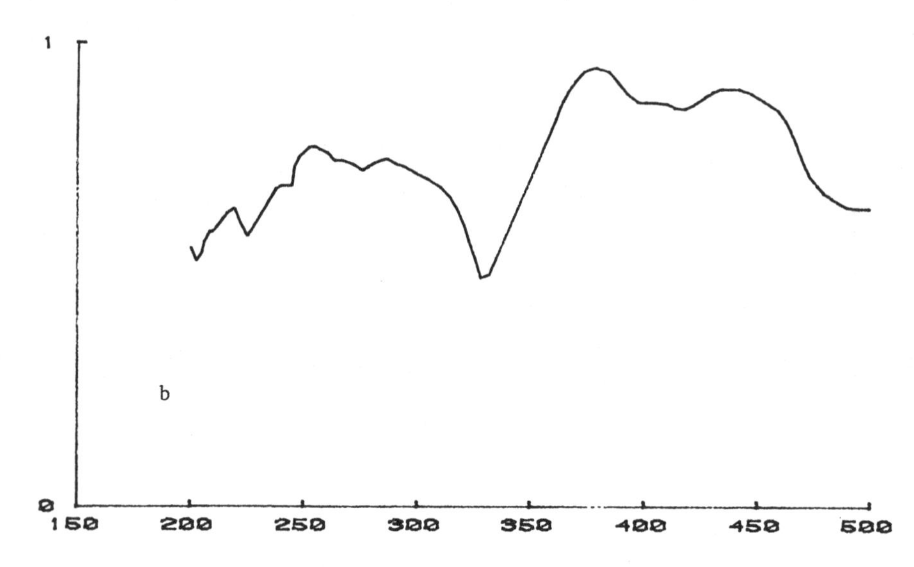

N-(4-hexyloxybenzylidene) 4'-toluidine: a-full line-crystal phase 20°C,dotted line-nematic phase 60°C; b-isotropic phase 90°C.

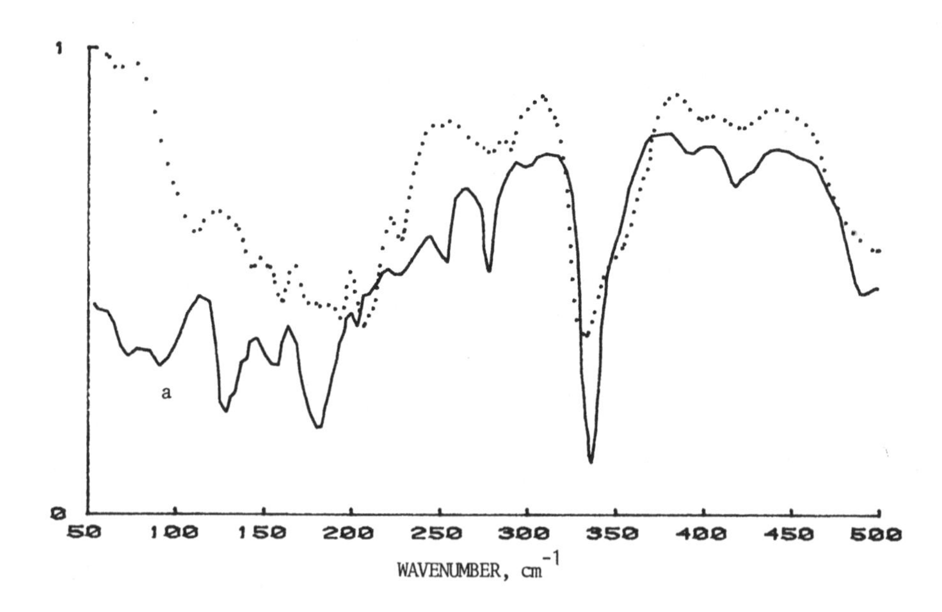

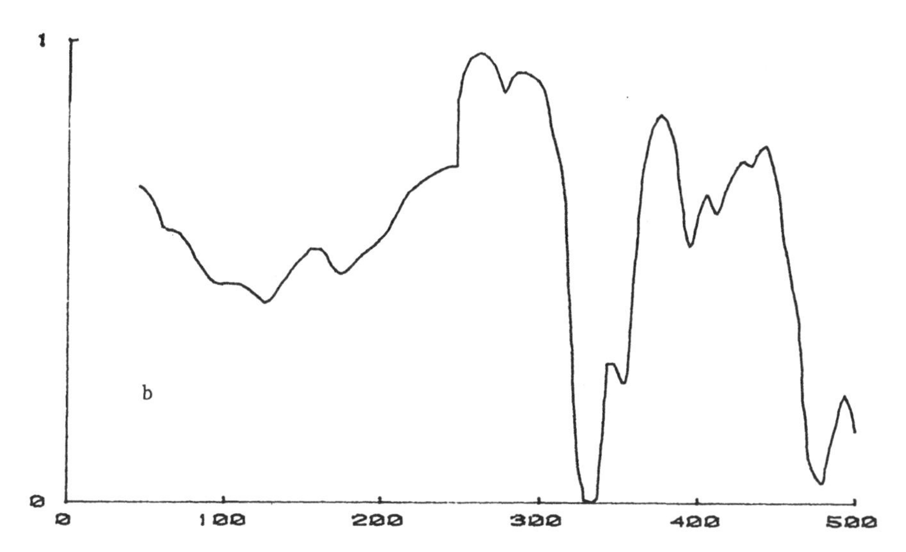

N-(4-nonyloxybenzylidene) 4'-toluidine: a-full line-crystal phase 20°C, dotted line-isotropic phase; b-nematic phase.

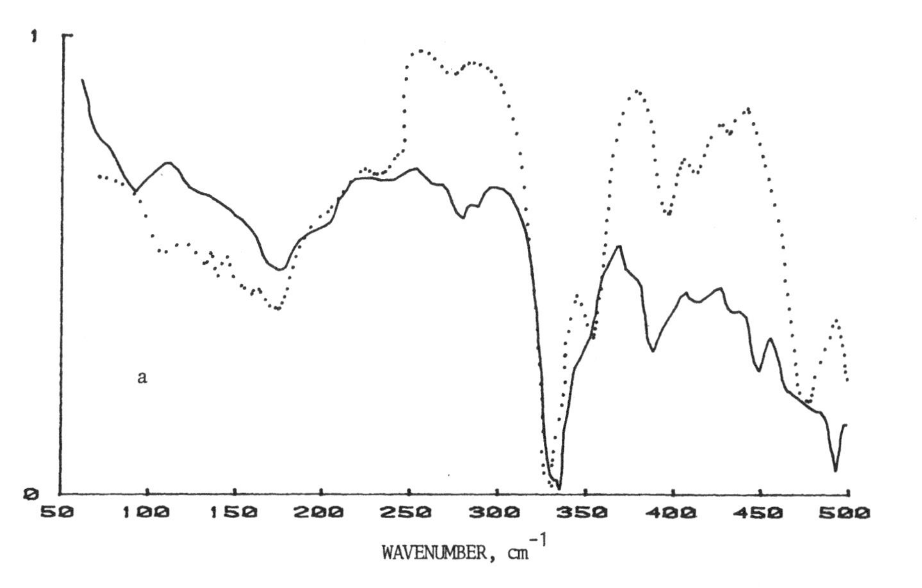

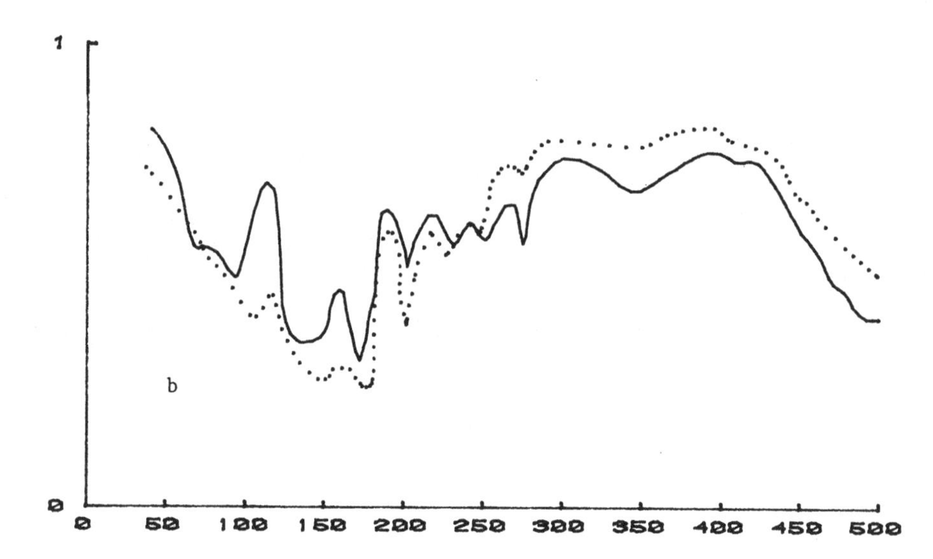

4,n-pentylacetophenonoxim 4',n-heptylbenzoate: a-crystal phase 20°C; b-full line-nematic phase 55°C, dotted line-isotropic liquid 75°C.

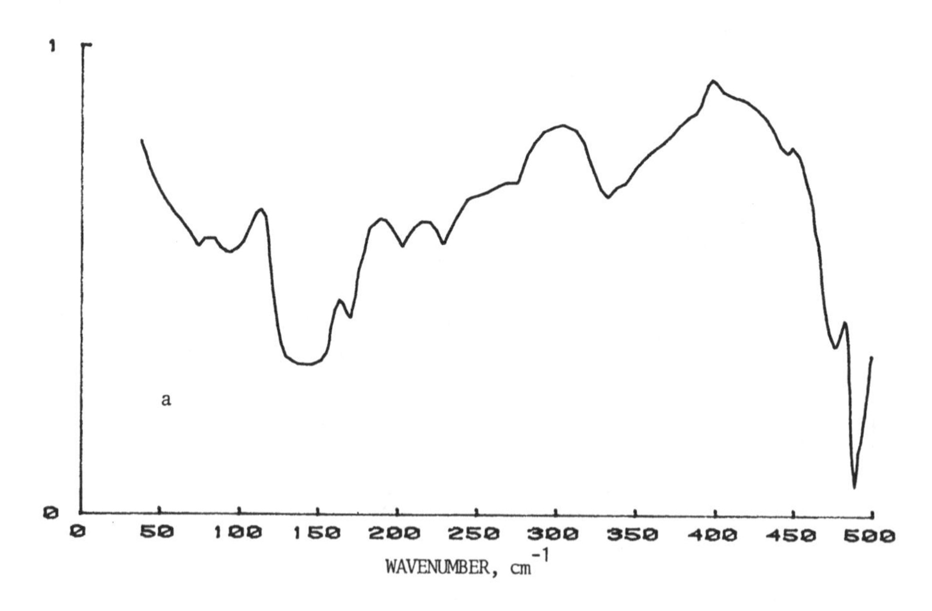

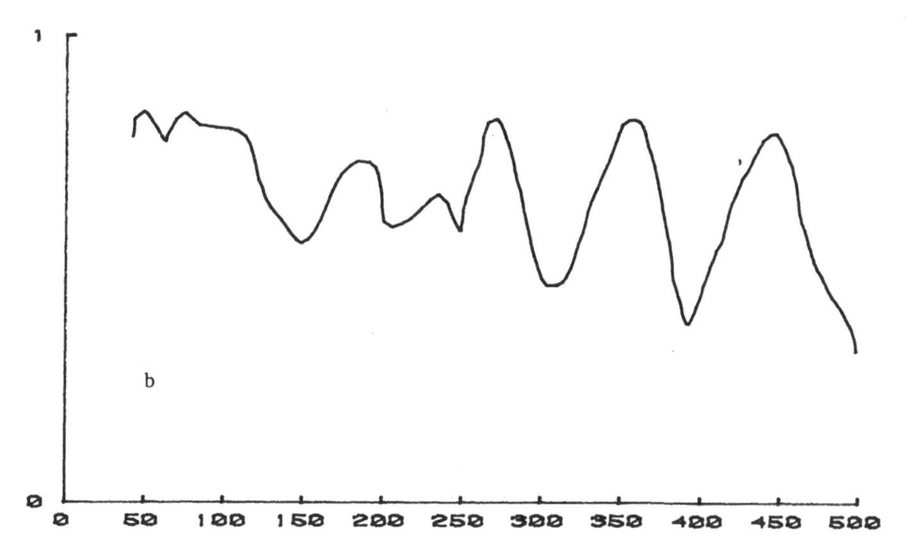

hidrochinone bis-4,n-heptyloxybenzoate: a-crystal phase 20°C; b-isotropic liquid 140°C.

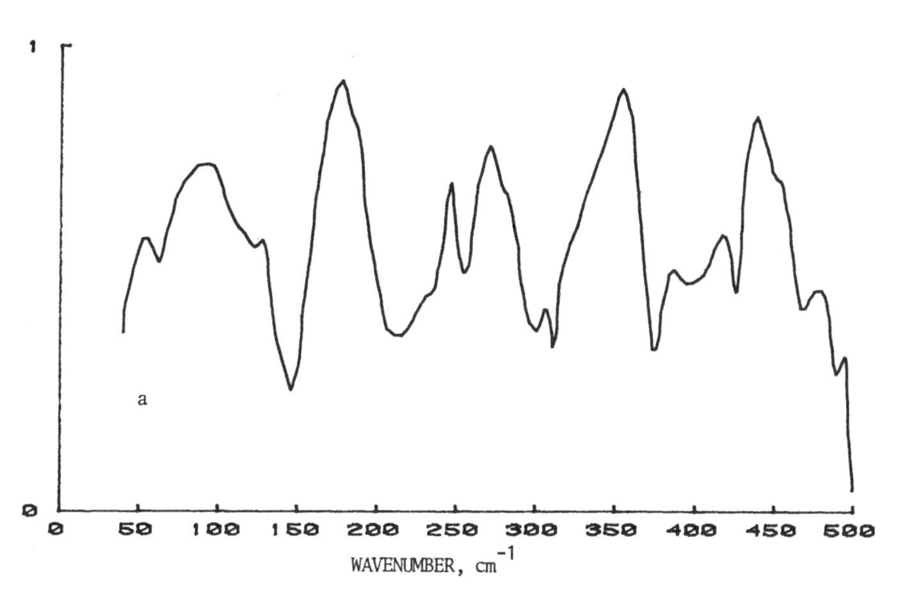

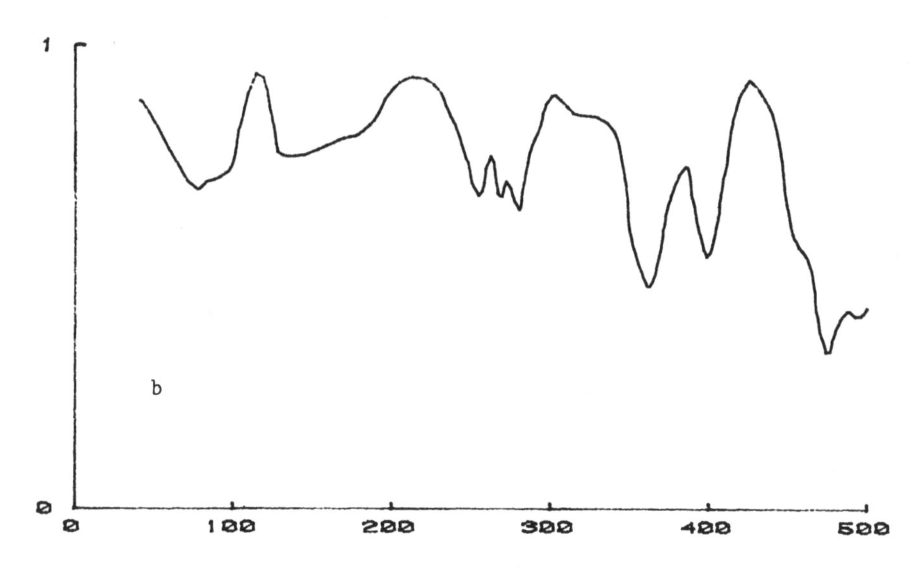

2-ethylhidrihinone 4,n-hexylbenzoate: a-crystal phase 20°C; b-nematic phase 60°C.

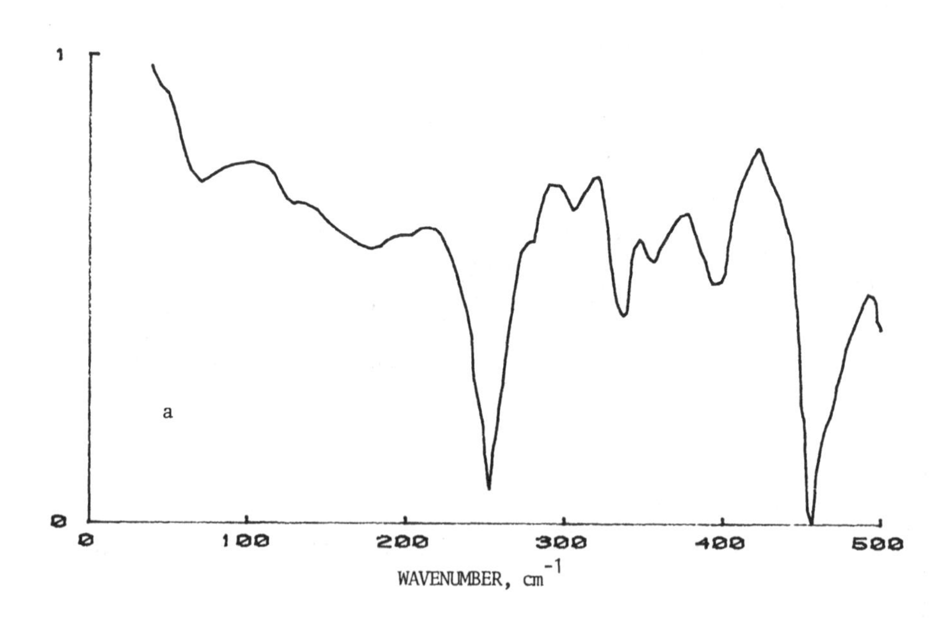

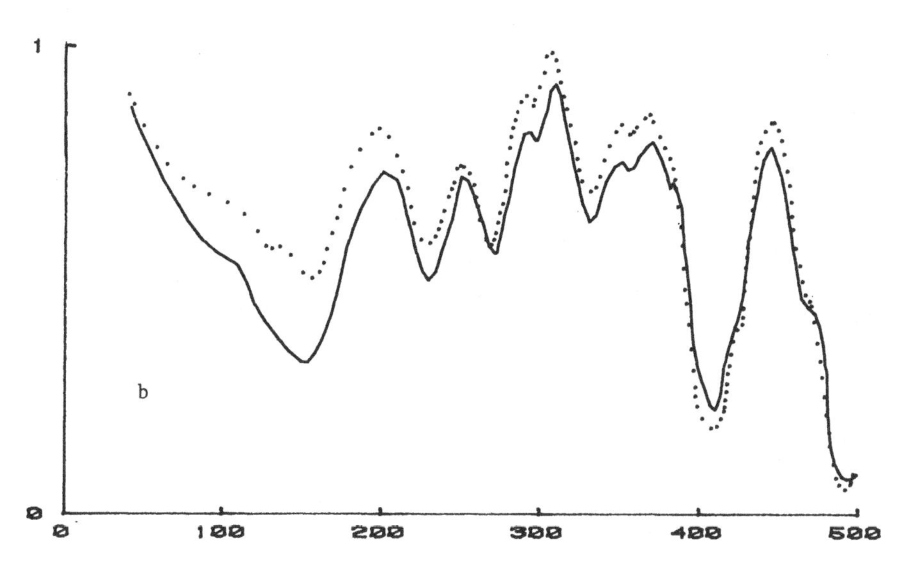

4-nitrophenol 4',n-octyloxybenzoate: a-crystal phase 20°C; b-full line-smectic A phase 55°C, dotted line-nematic phase 70°C.

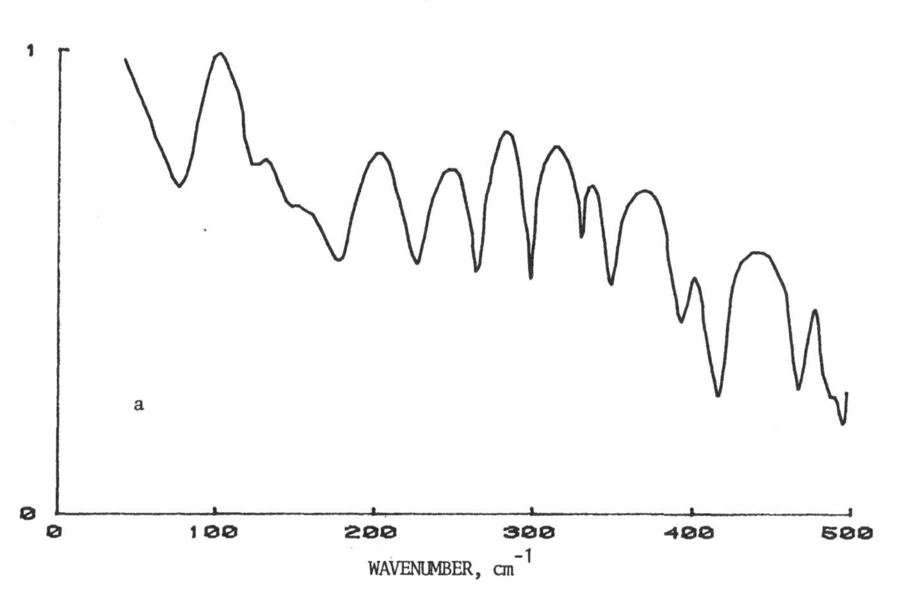

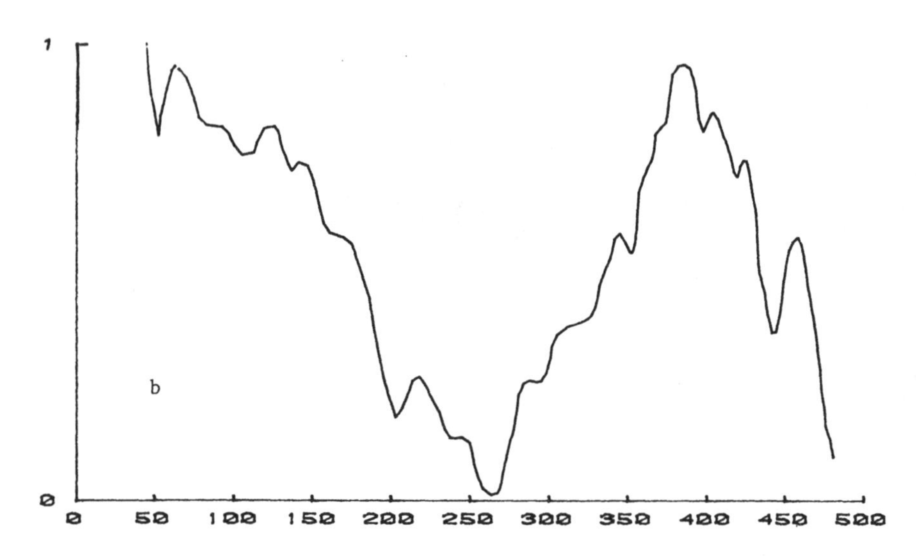

4,n-propyloxybenzoic acid: a-crystal phase 20°C; b-nematic phase 150°C.

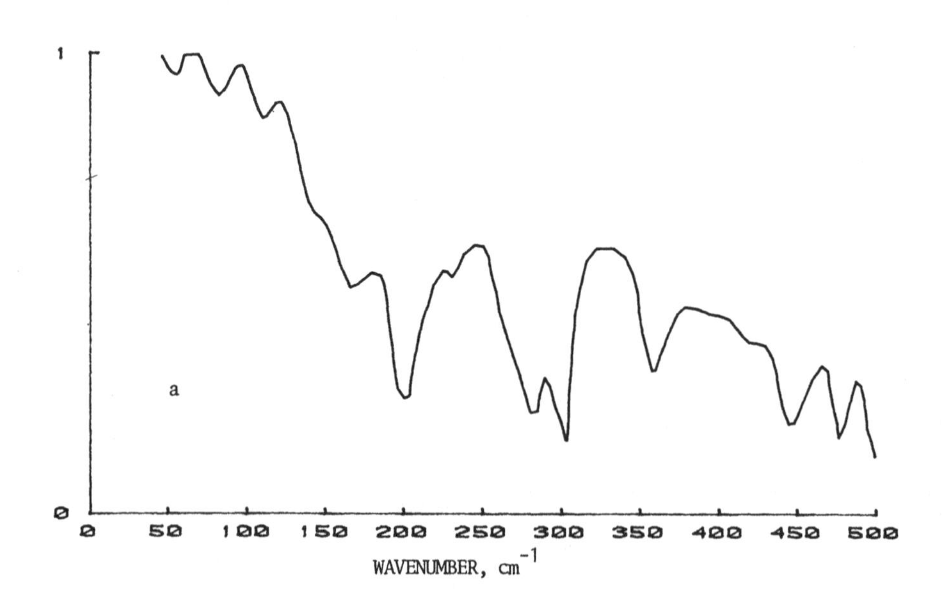

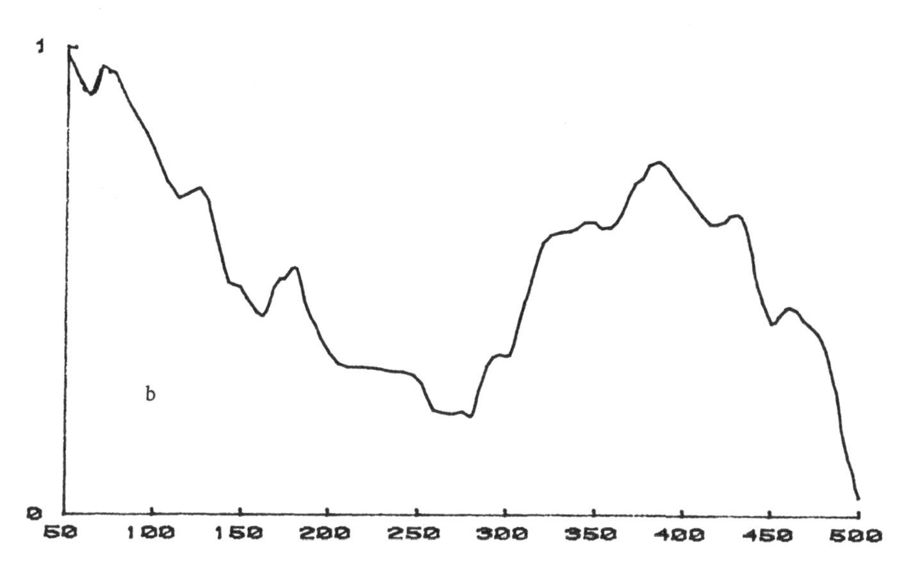

4,n-pentyloxybenzoic acid: a-crystal phase 20°C; b-nematic phase 130°C.

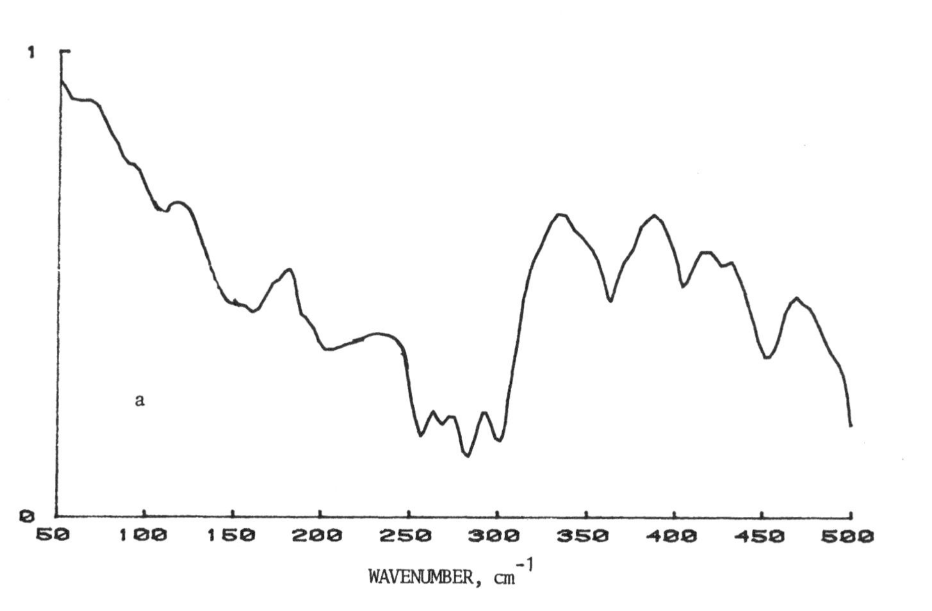

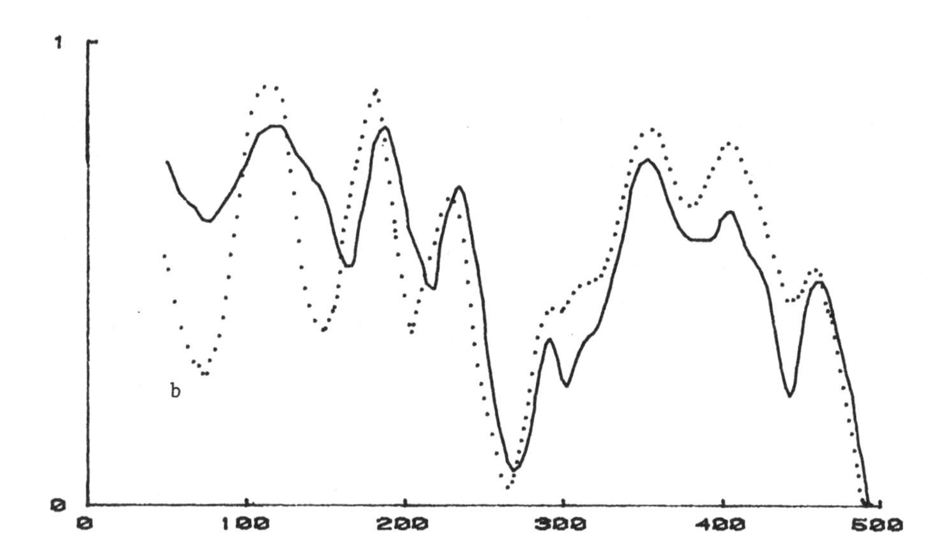

4,n-octyloxybenzoic acid: a-full line-crystal phase 20°C, dotted line-nematic phase 125°C; b-full line-smectic C phase 105°C, dotted line- isotropic liquid 160°C.

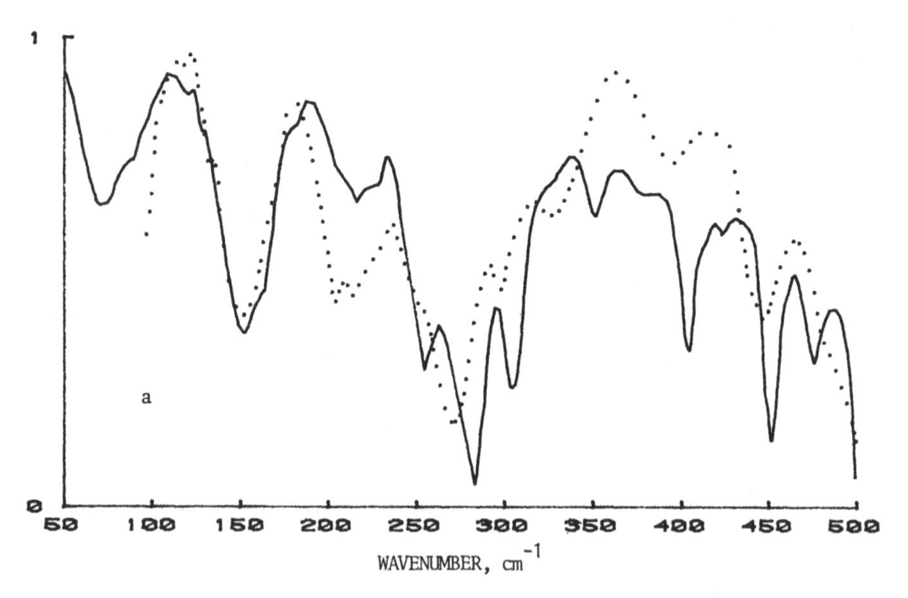

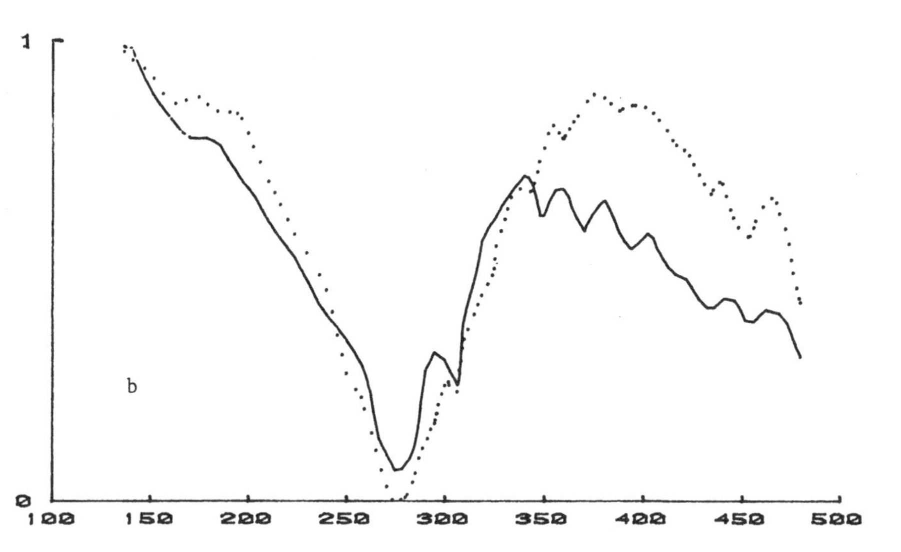

4,n-decyloxybenzoic acid: a-crystal phase 20°C; b-full line-smectic C phase 110°C, dotted line-nematic phase 130°C.

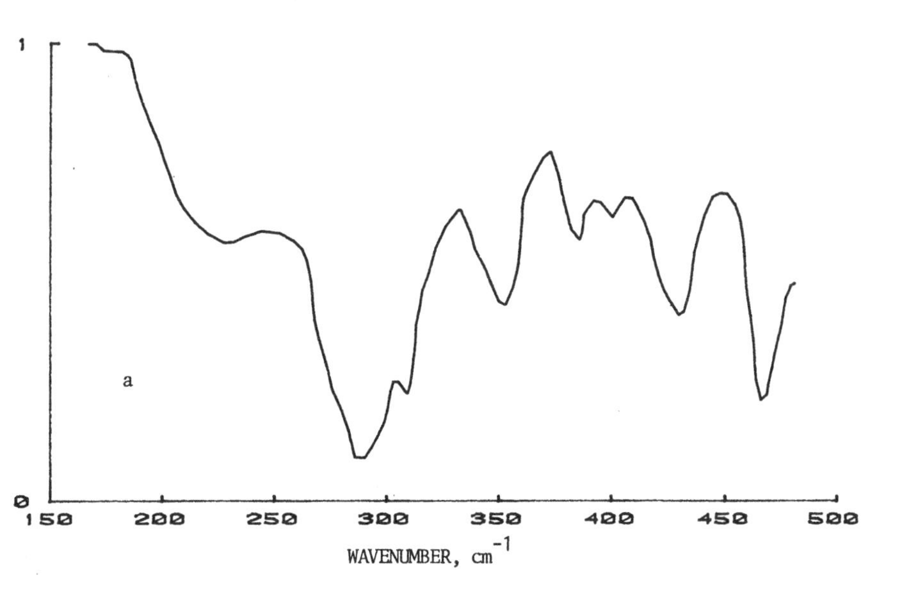

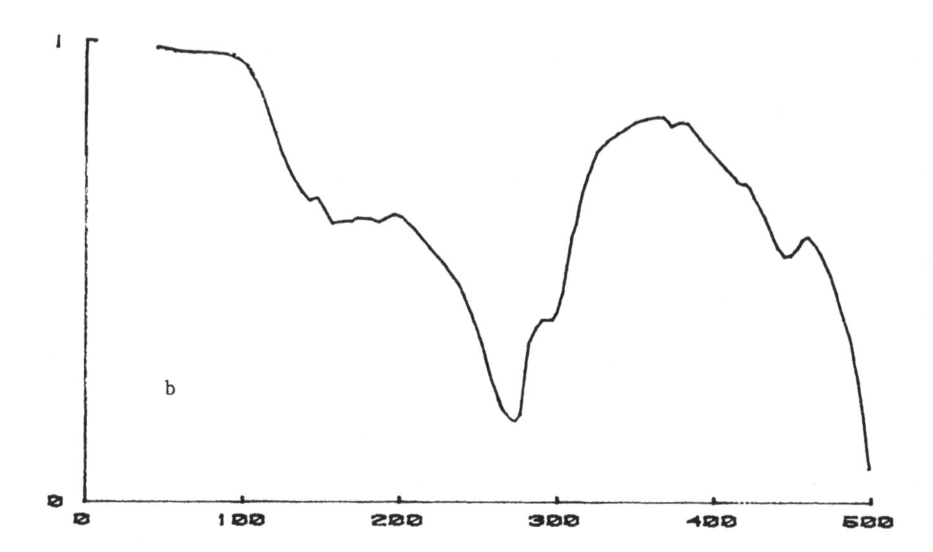

4,n-dodecyloxybenzoic acid: a-crystal phase 20°C; b-smectic C phase 110°C.

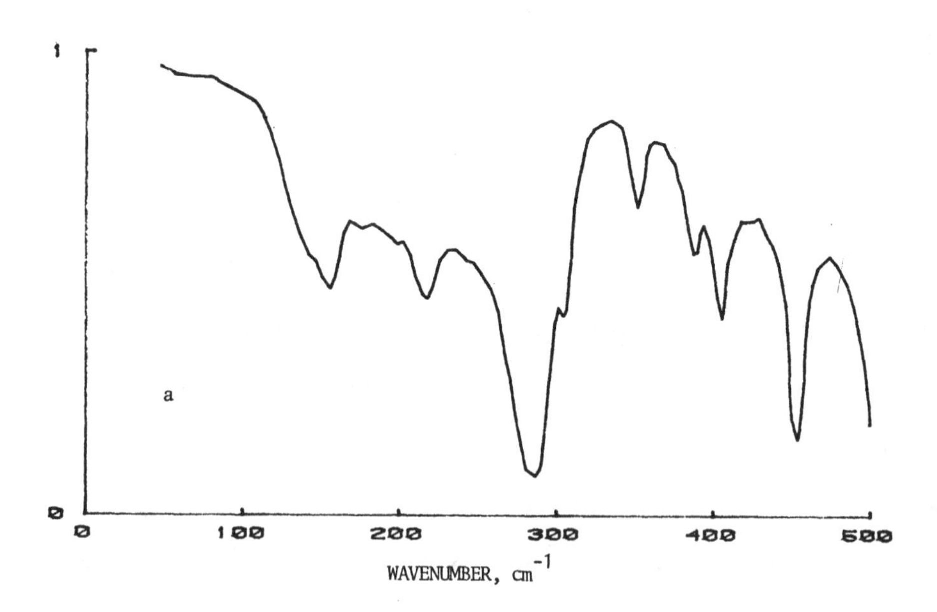

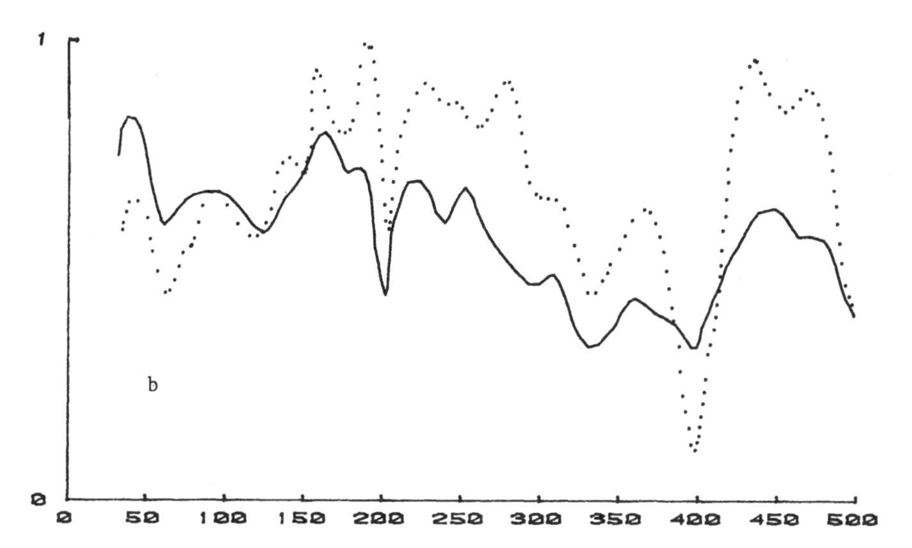

4,4'-diethylester of azoxybenzoic acid: a-crystal phase 20°C; b-full line-smectic A phase 115°C, dotted line-isotropic liquid 130°C.

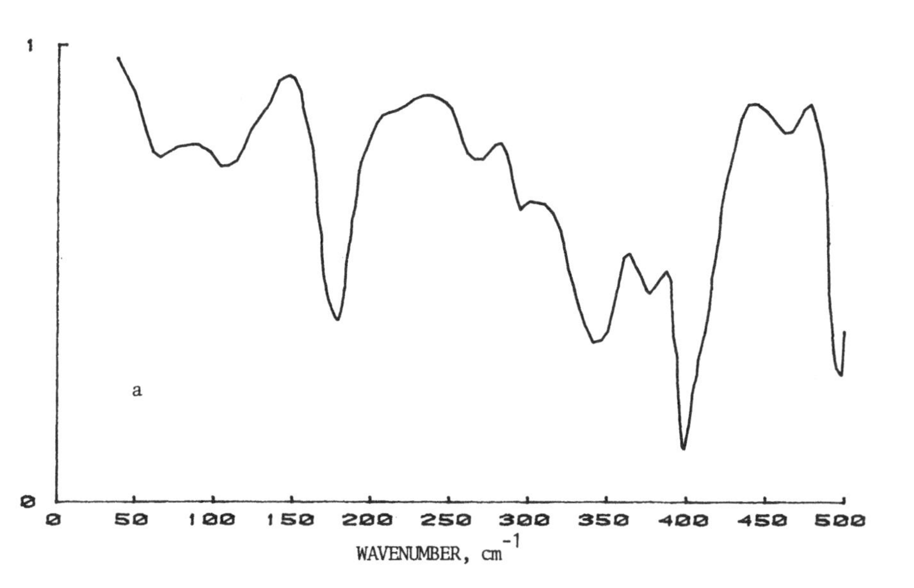

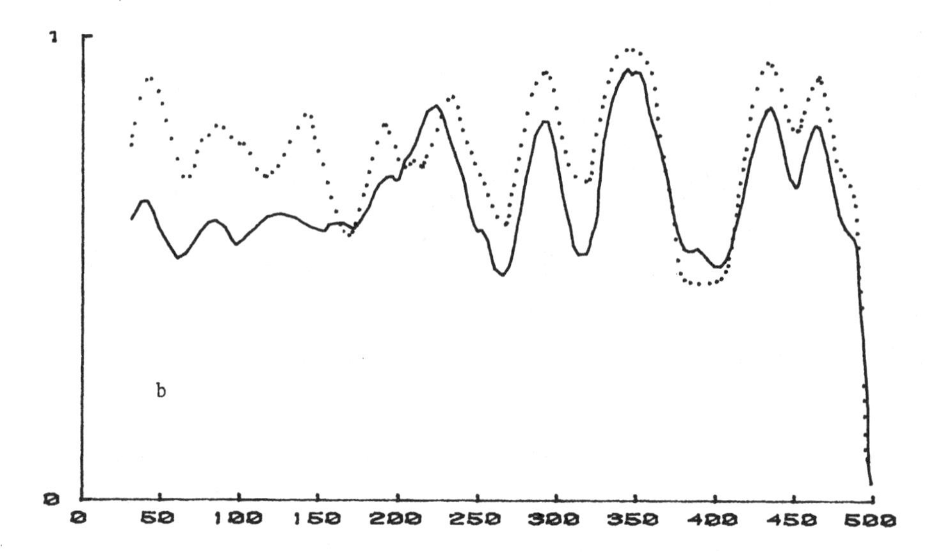

4-methoxyphenylester of 4,n-butyloxybenzoic acid: a-crystal phase 20°C; b-full line-nematic phase 85°C, dotted line-isotropic liquid 100°C.

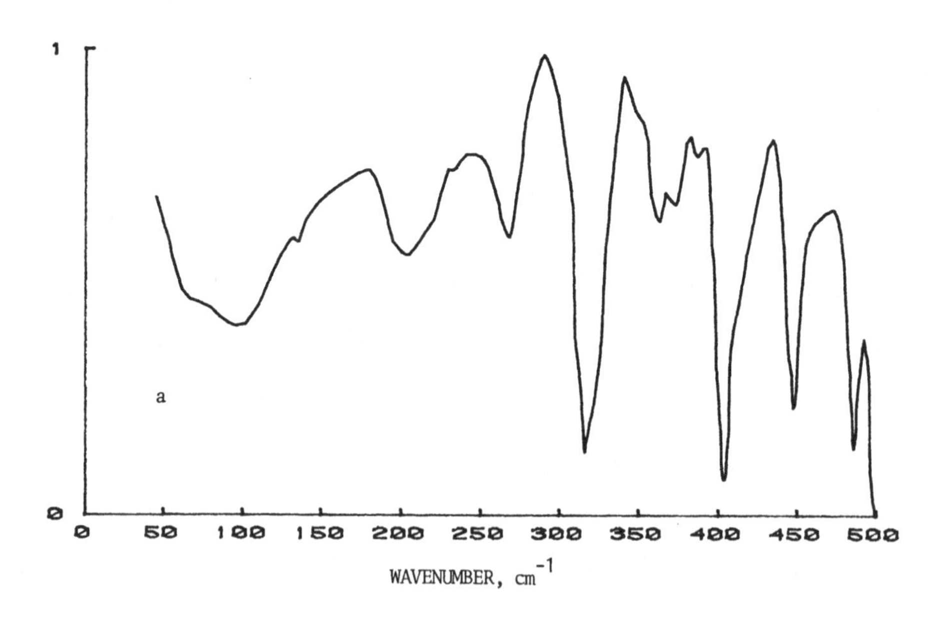

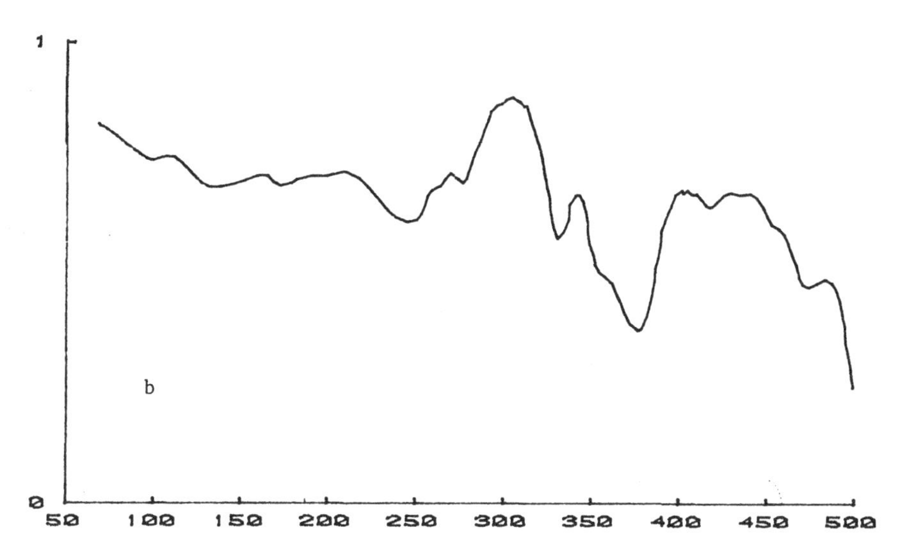

4,n-hexylphenylester of 4,n-butylbenzoic acid: a-nematic phase 35°C; b-isotropic liquid 60°C.

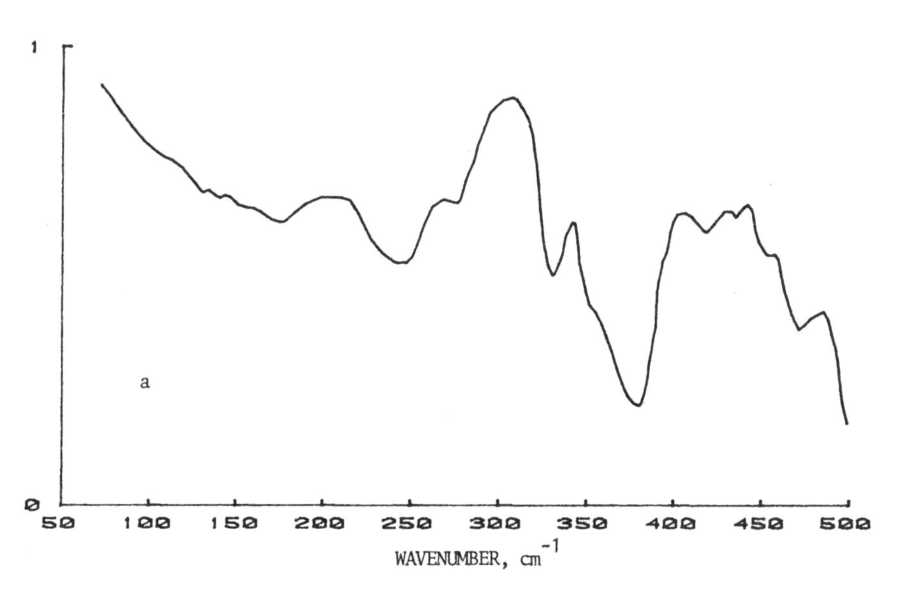

4,n-hexyloxyphenylester of 4,n-butyloxybenzoic acid: crystal phase 20°C.

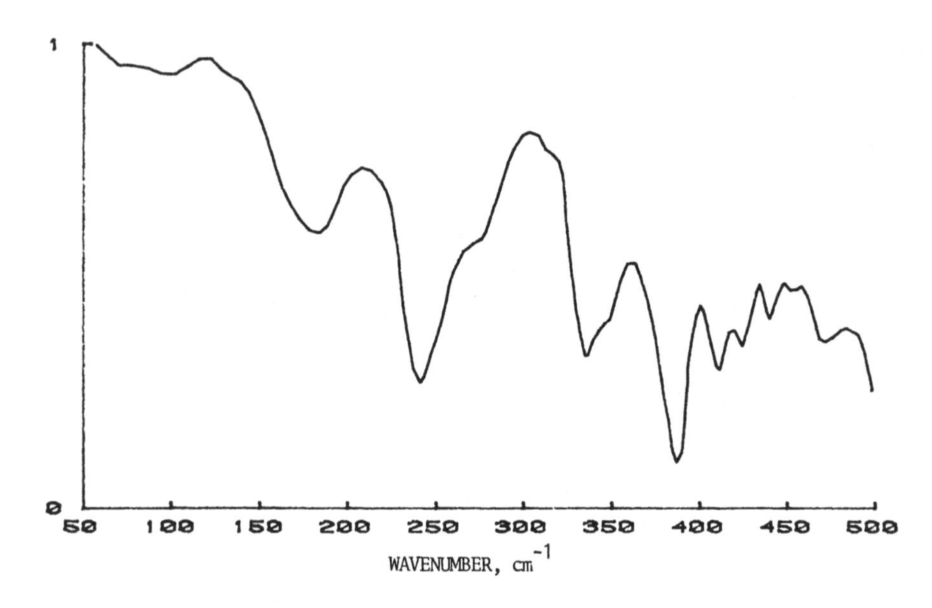

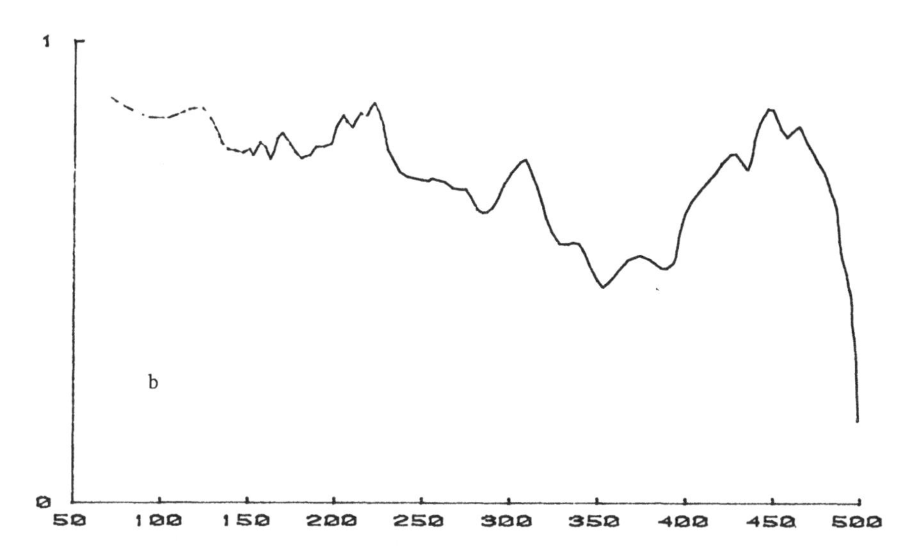

4,n-heptylphenylester of 4,n-butylbenzoic acid: a-nematic phase 70°C; b-isotropic liquid 80°C.

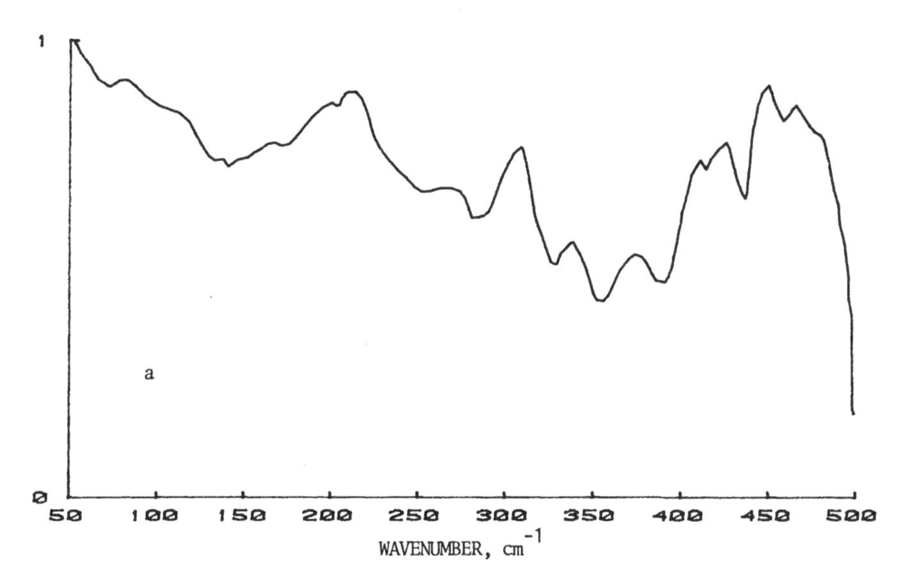

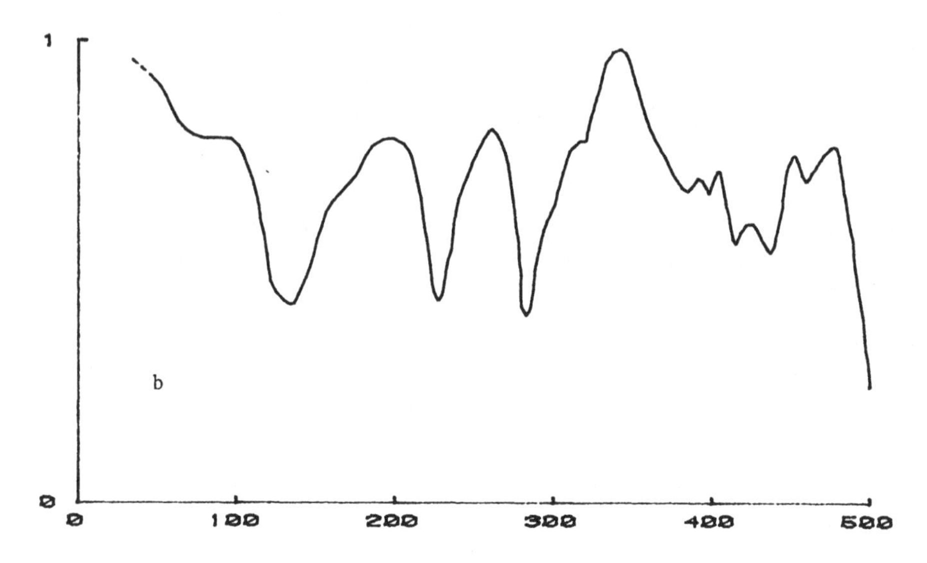

4-methylphenylester of 4,n-hexyloxybenzoic acid: a-crystal phase 20°C; b-nematic phase 60°C.

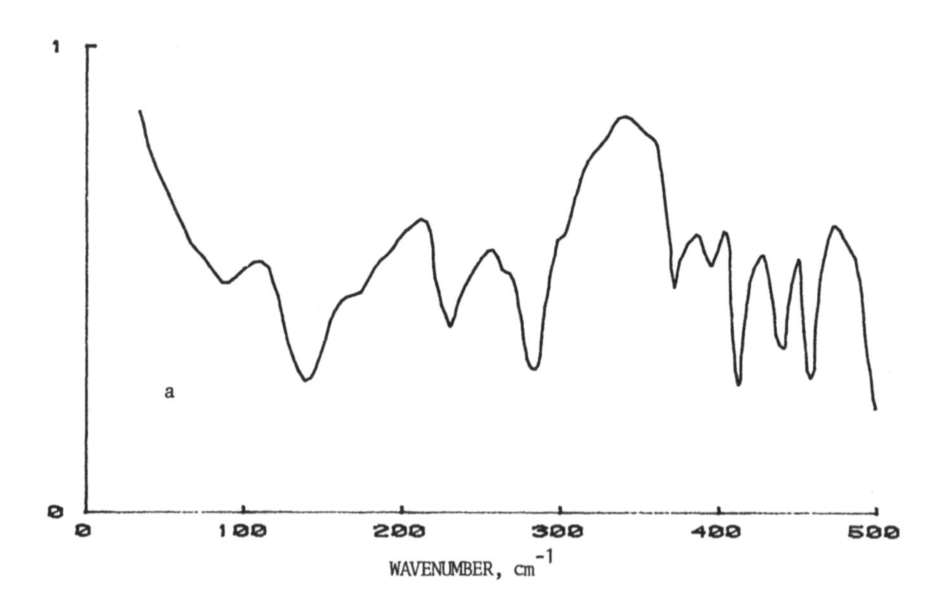

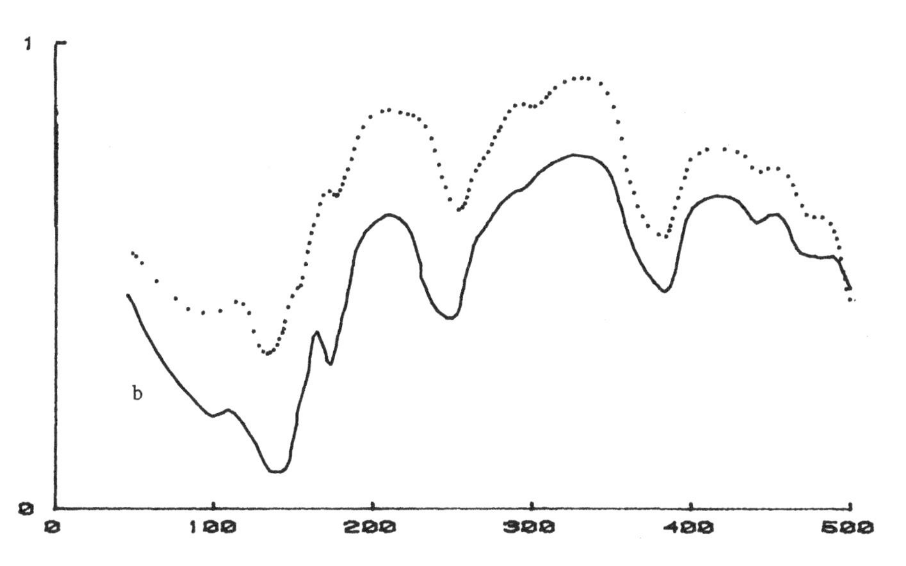

4,n-hexyloxyphenylester of 4,n-hexylbenzoic acid: a-crystal phase 20°C; b-full line-nematic phase 60°C, dotted line-isotropic liquid 100°C.

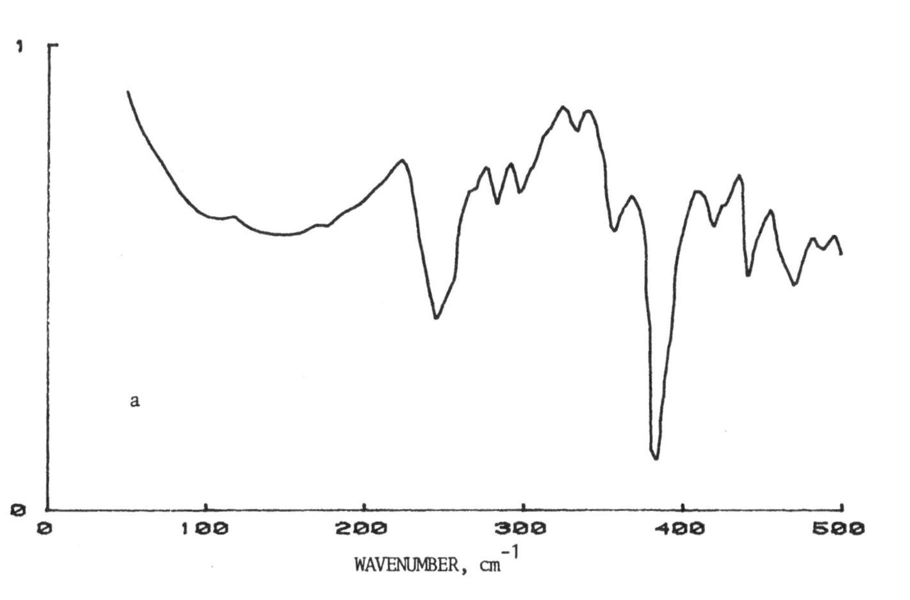

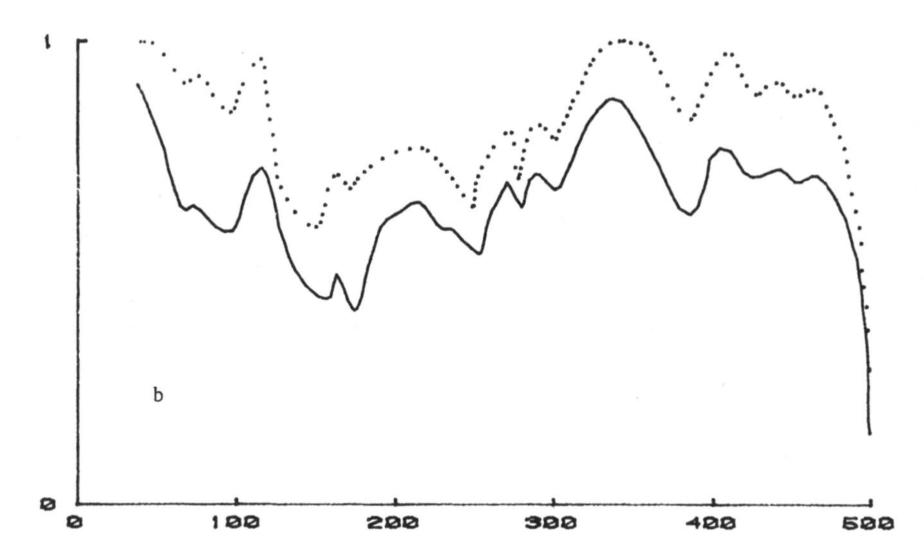

4,n-heptyloxyphenylester of 4,n-hexyloxybenzoic acid: a-crystal phase 20°C; b-full line-nematic phase 70°C, dotted line-isotropic liquid 95°C.

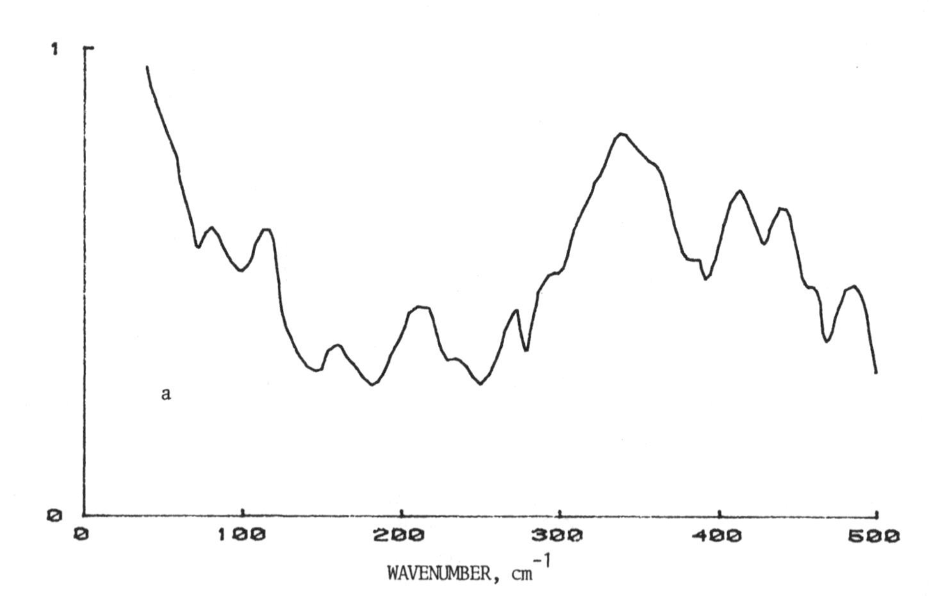

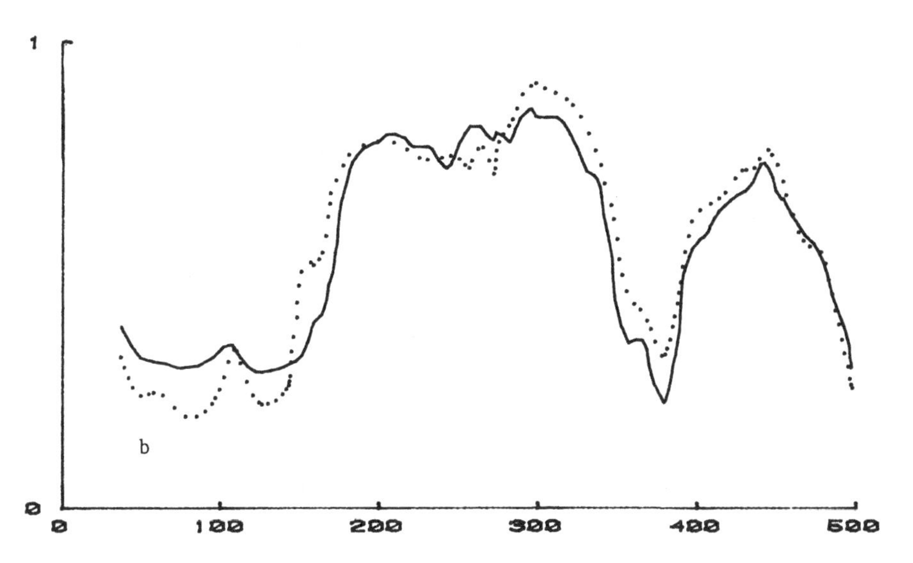

4,n-octyloxyphenylester of 4,n-hexylbenzoic acid: a-crystal phase 20°C; b-full line-nematic phase 60°C, dotted line-isotropic liquid 100°C.

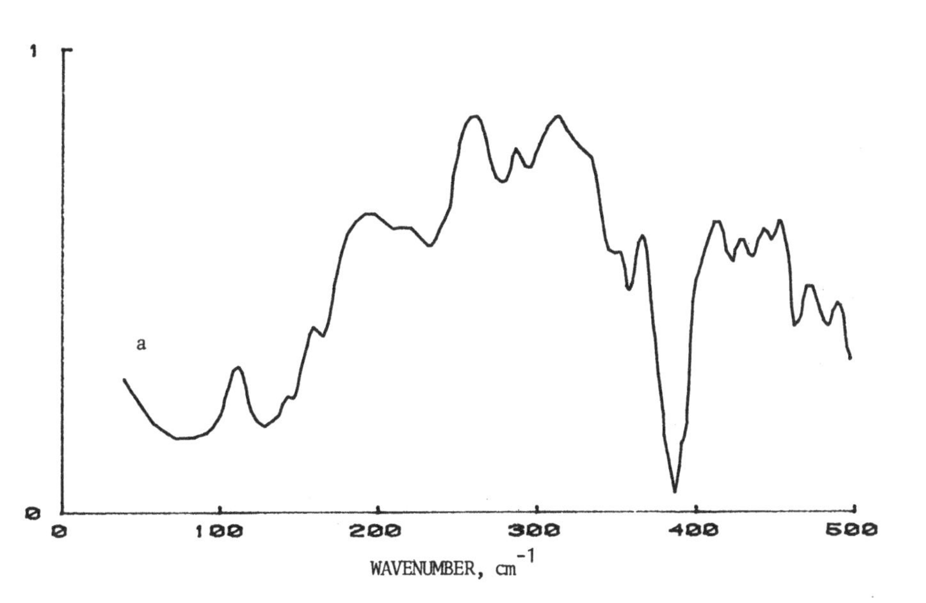

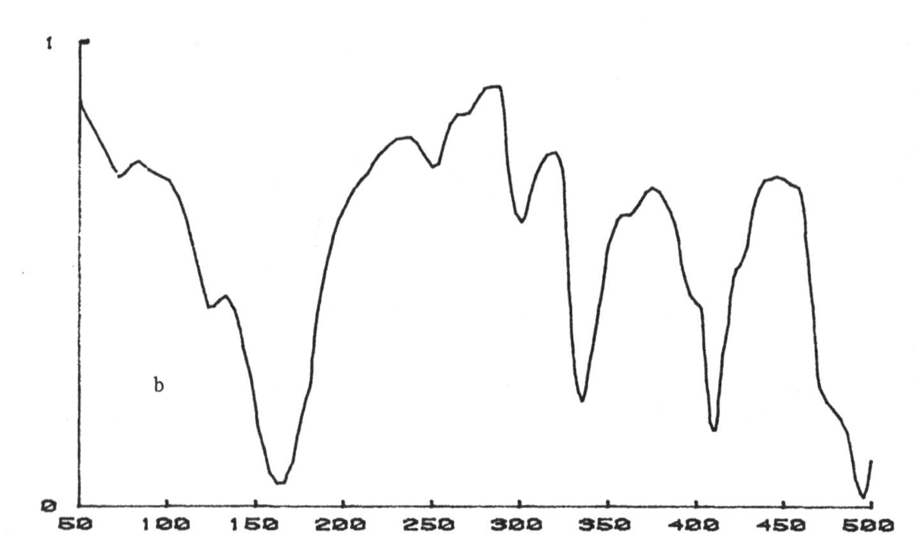

4-cyanophenylester of 4,n-heptylbenzoic acid: a-crystal phase 20°C; b-nematic phase 47°C.

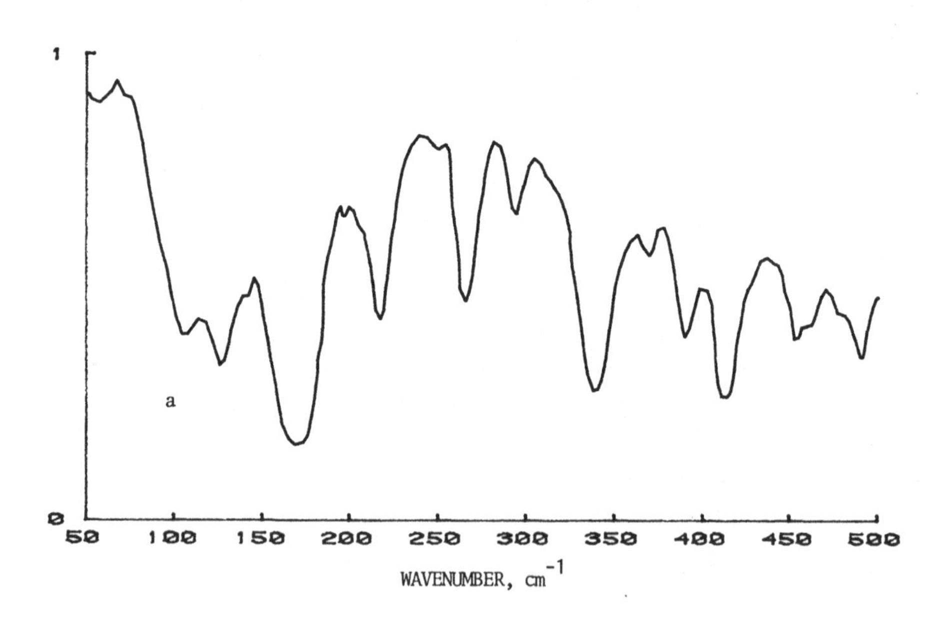

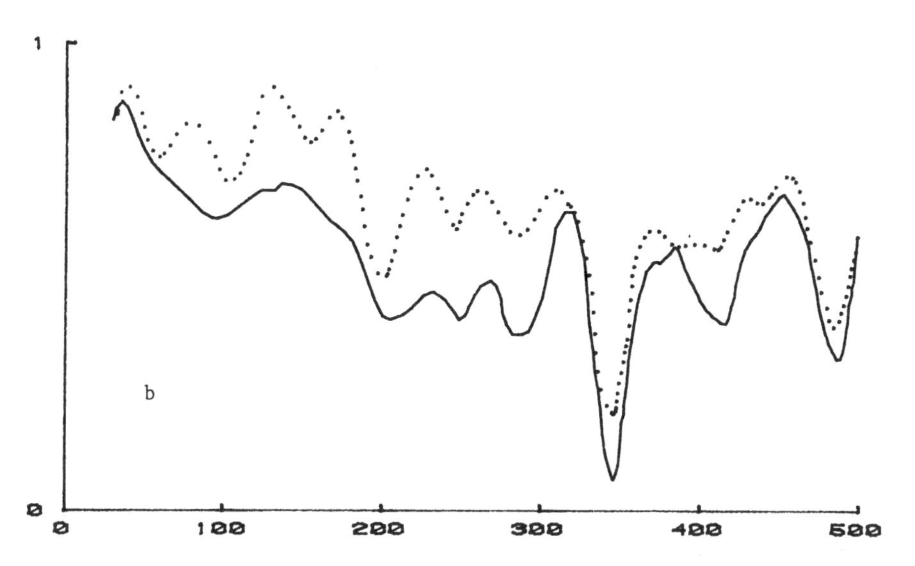

N-(4-ethoxybenzylidene) 4'-aminophenylester of capric acid: a-crystal phase 20°C; b-full line-nematic phase 90°C, dotted line-isotropic liquid 130°C.

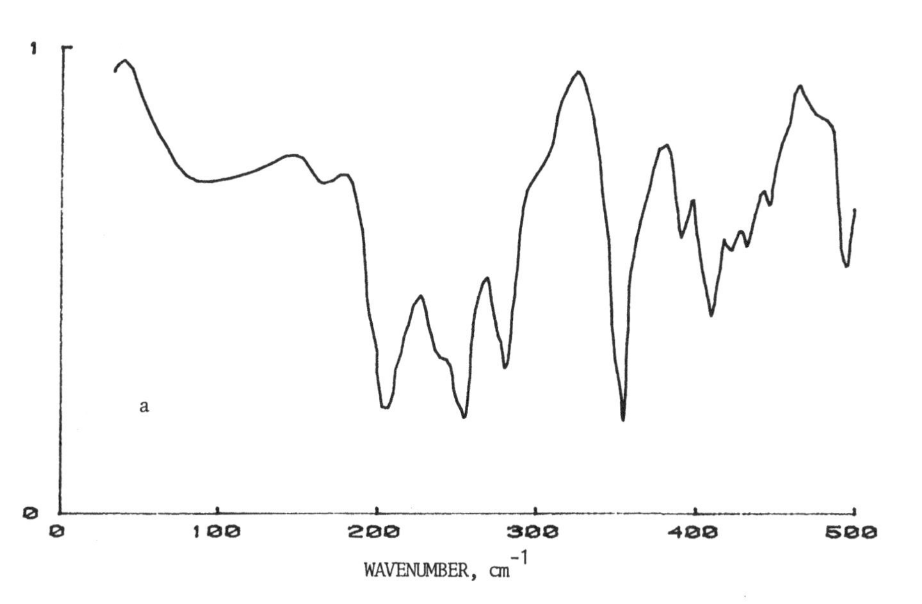

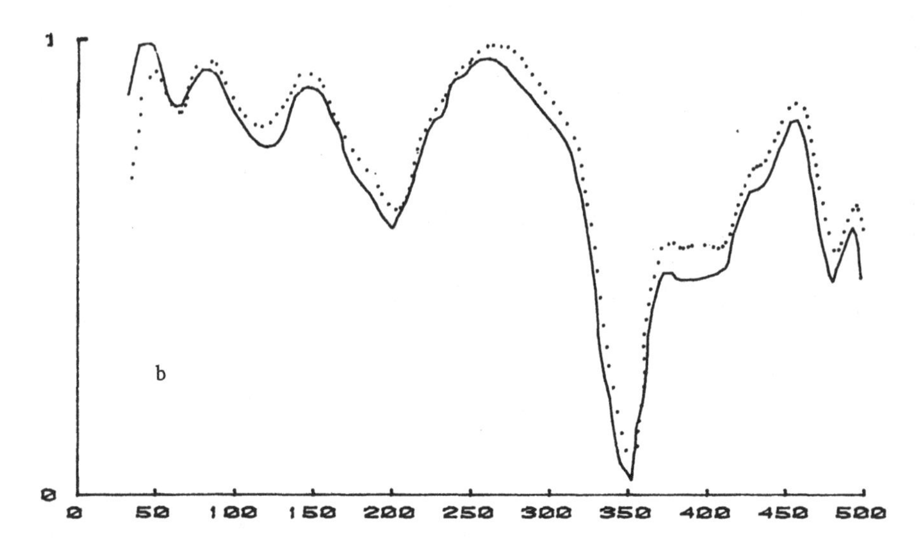

N-(4-ethoxybenzylidene) aminoester of cynnamic acid: a-full line-smectic B phase 90°C, dotted line-smectic A phase 130°C; b-full line-nematic phase 155°C, dotted line-isotropic liquid 165°C.

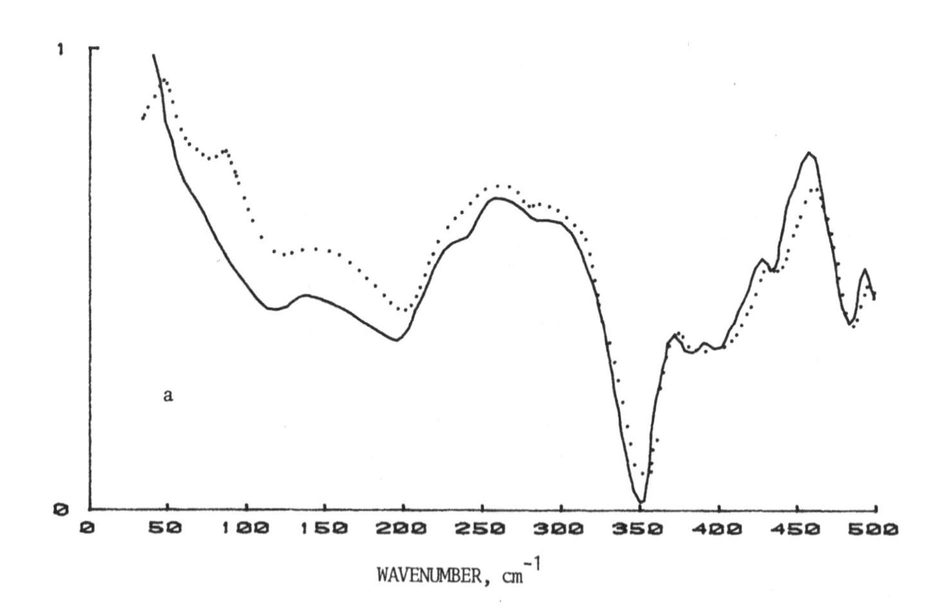

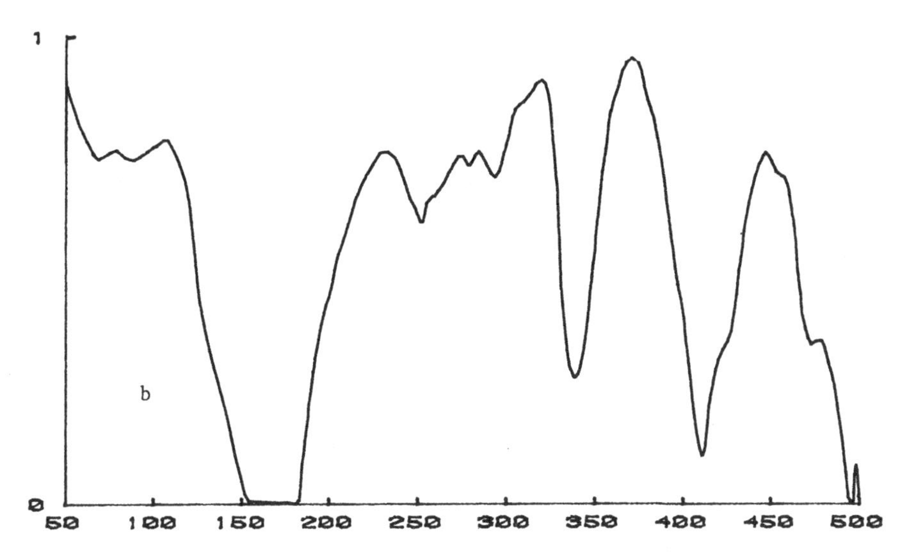

4-cyanobiphenylester of 4',n-heptylcynnamic acid: a-crystal phase; b-nematic phase 60°C.

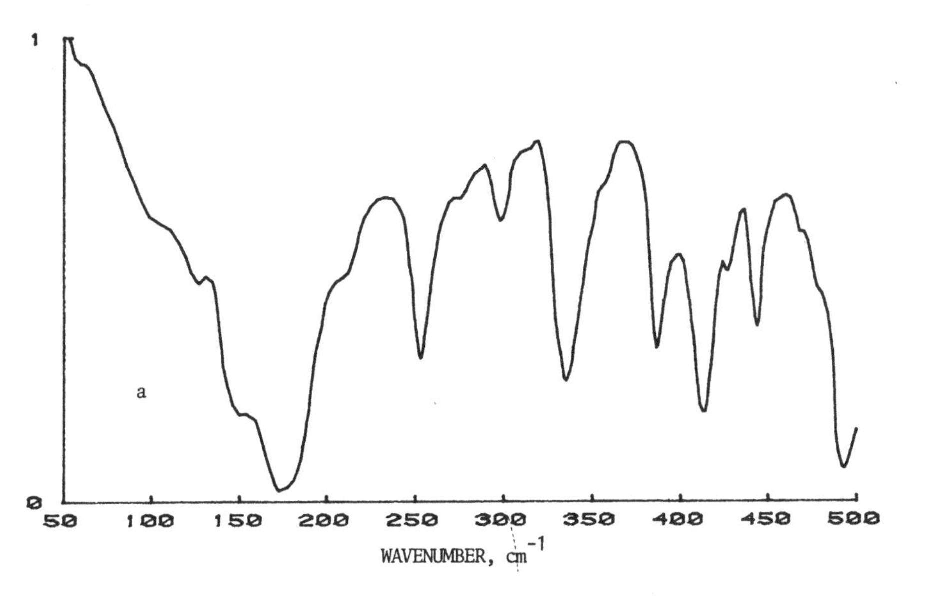

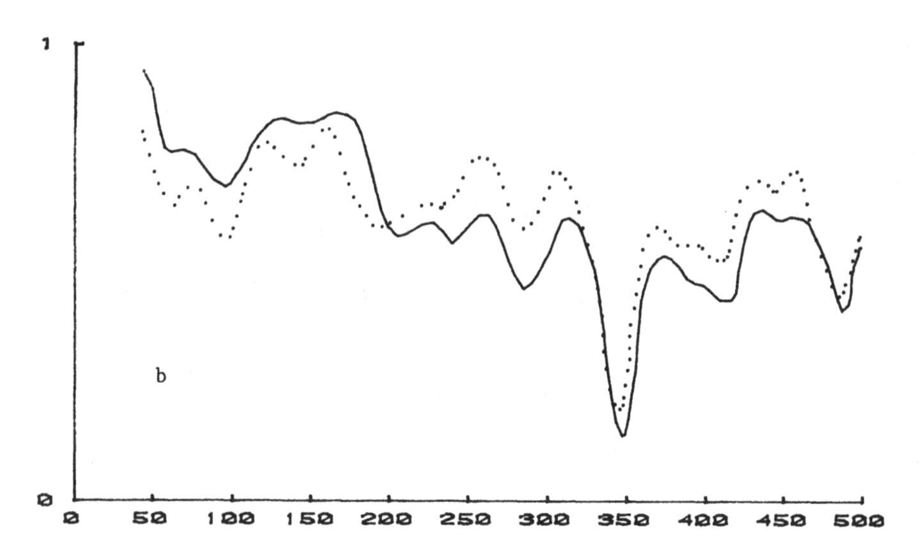

N-(4-ethoxybenzylidene) 4'-aminophenylester of enantic acid: a-crystal phase 20°C, b-full line nematic phase 90°C, dotted line-isotropic liquid 130°C.

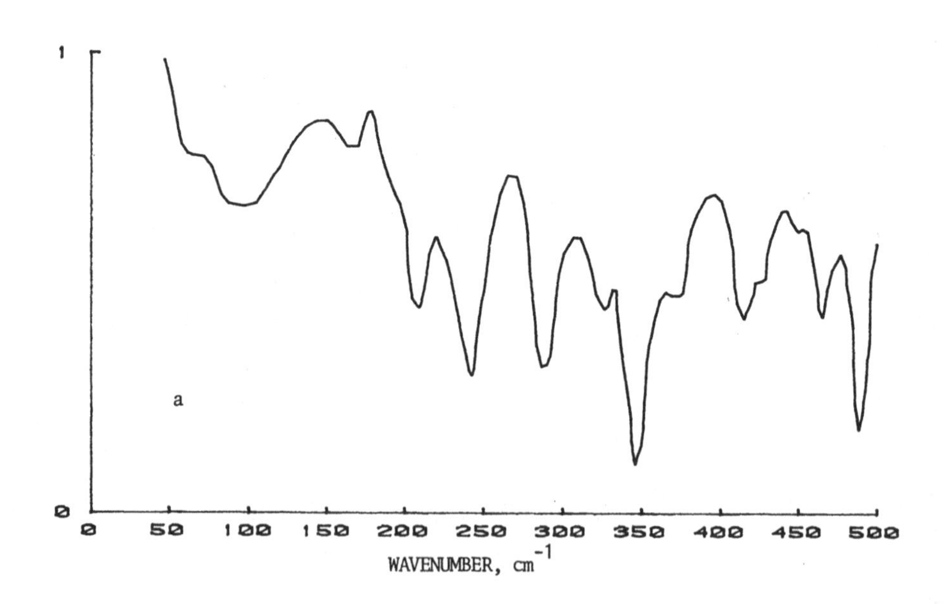

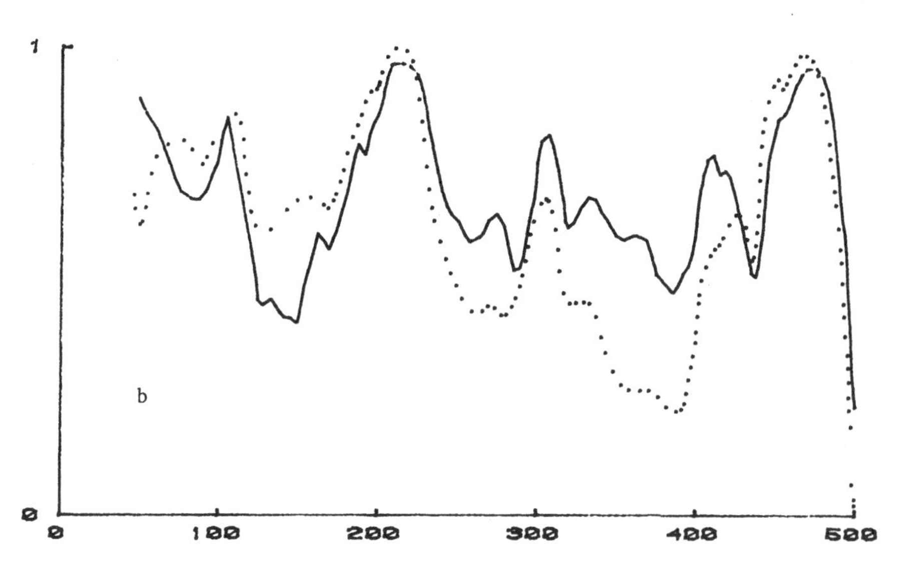

n-butyl-4-(4-ethoxyphenyloxy) carbonylphenylcarbonate: a-crystal phase 20°C; b-full line-nematic 70°C, dotted line-isotropic liquid 80°C.

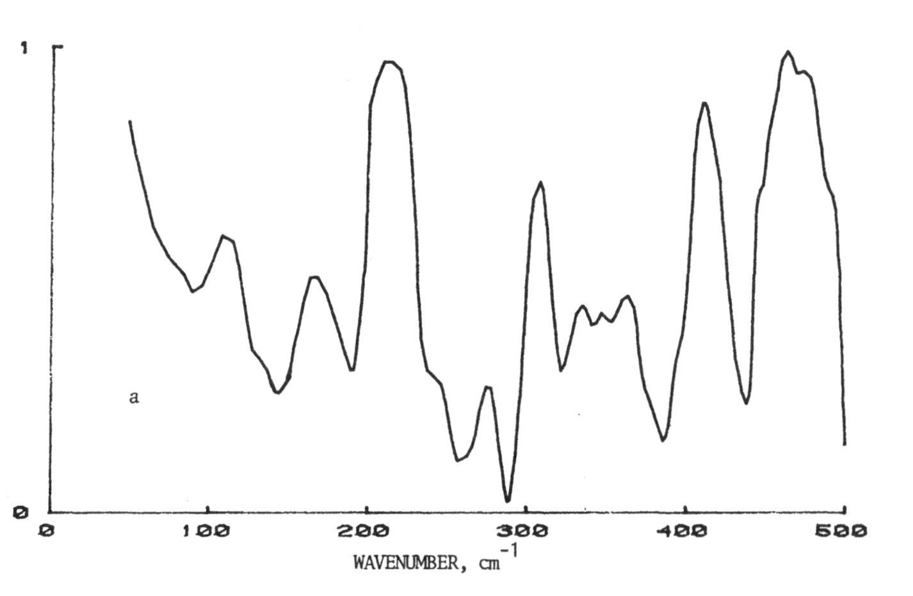

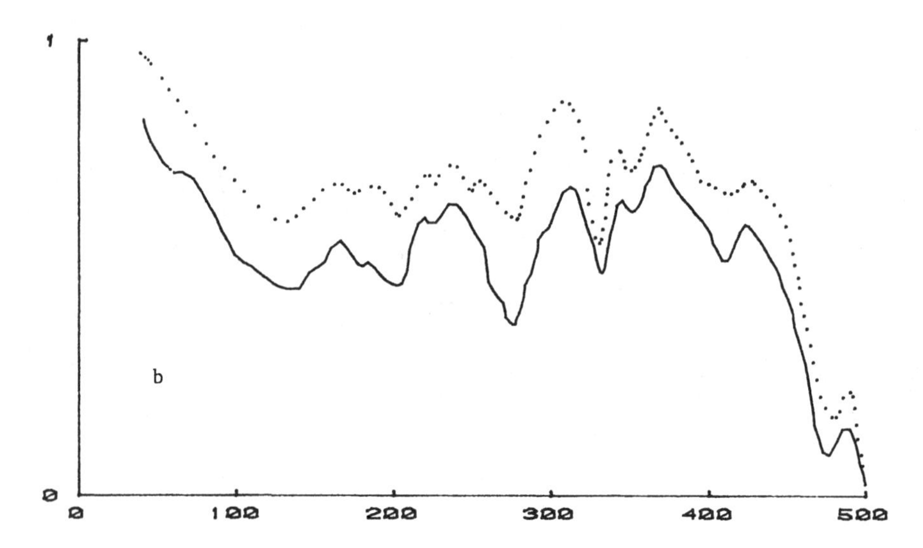

4,n-heptyl 4'-methyloxytolane: a-crystal phase 20°C; b-full line-nematic phase 45°C, dotted line-isotropic liquid 60°C.

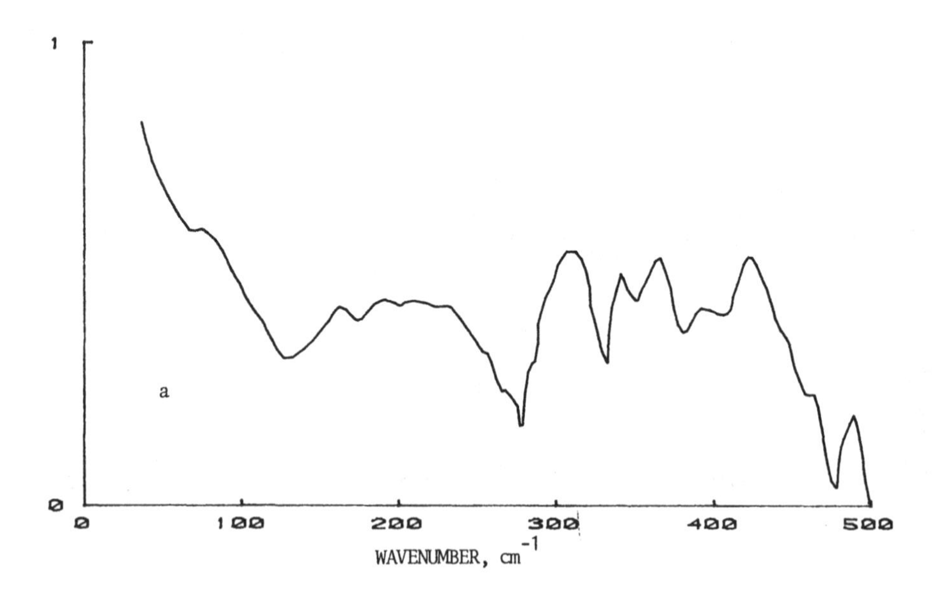

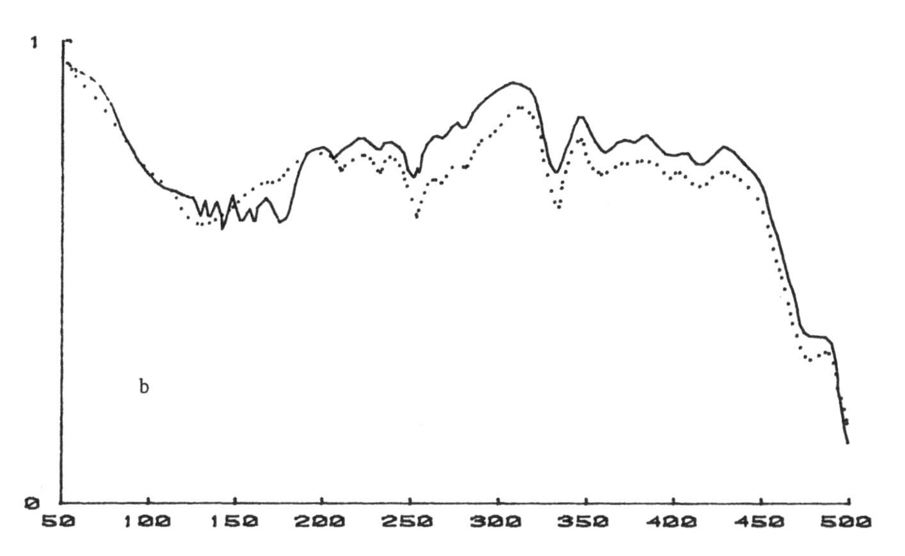

4,n-butyl 4'-ethoxytolane (BET): a-crystal phase 20°C; b-full line nematic phase 60°C,dotted line-isotropic liquid 90°C.

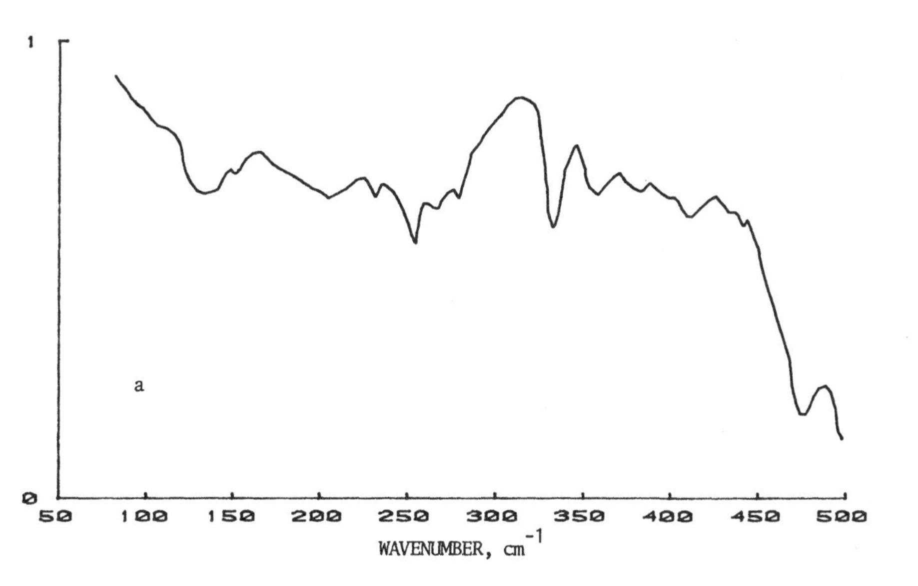

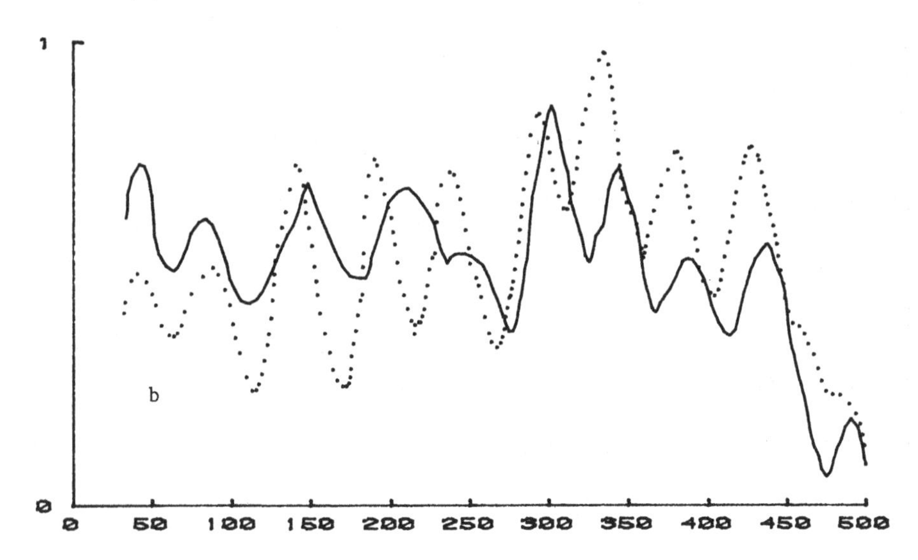

4,n-pentyl 4'-ethoxytolane: a-crystal phase 20°C; b-full line-nematic phase 70°C, dotted line-isotropic liquid 95°C.

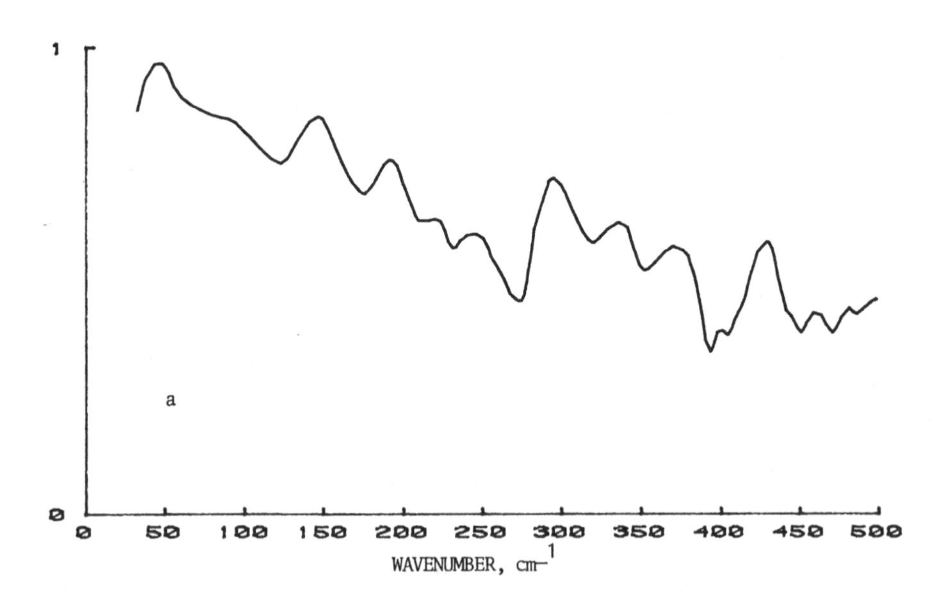

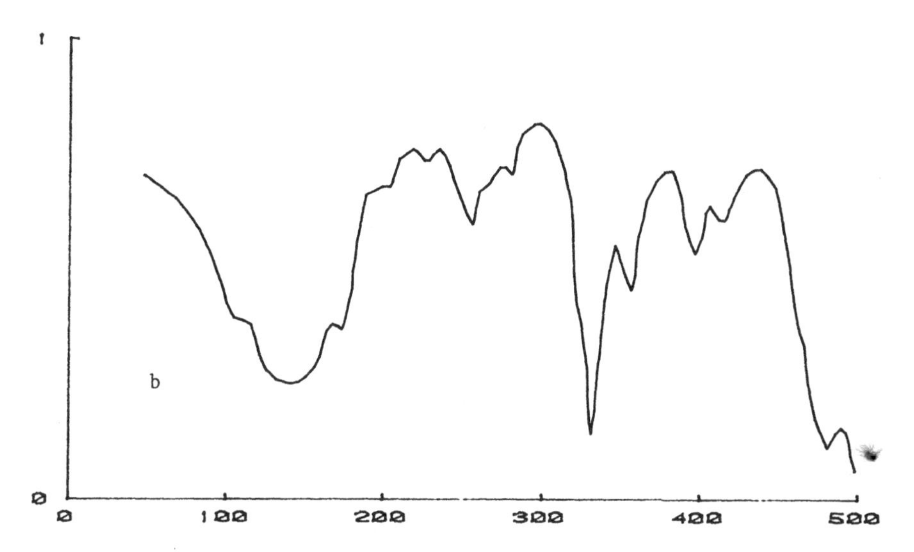

4,n-octyl 4'-ethoxytolane (OET): a-crystal phase 20°C; b-isotropic liquid 86°C.

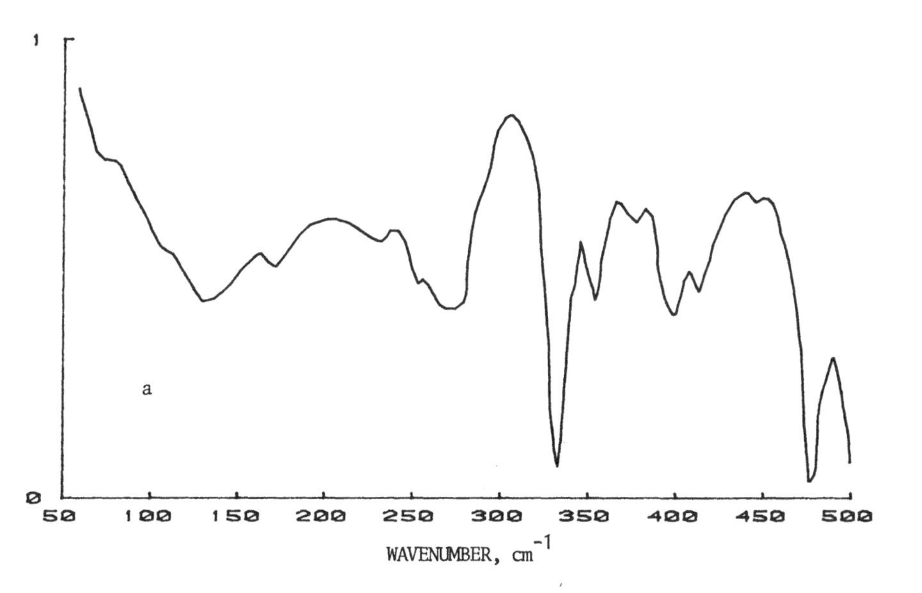

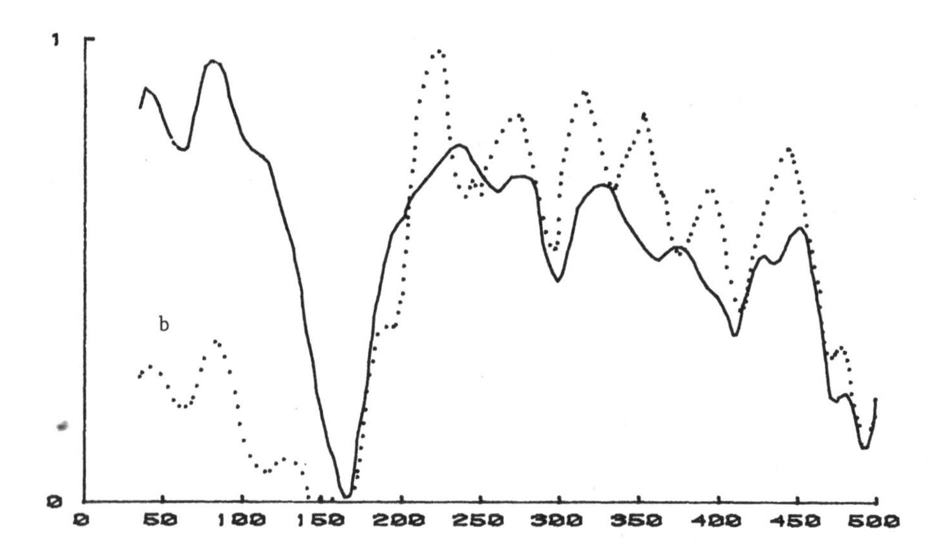

4,n-heptylbenzoate 4'-cyanobiphenyle: a-crystal phase 20°C; b-full line-nematic phase 50°C, dotted line-isotropic liquid 60°C.

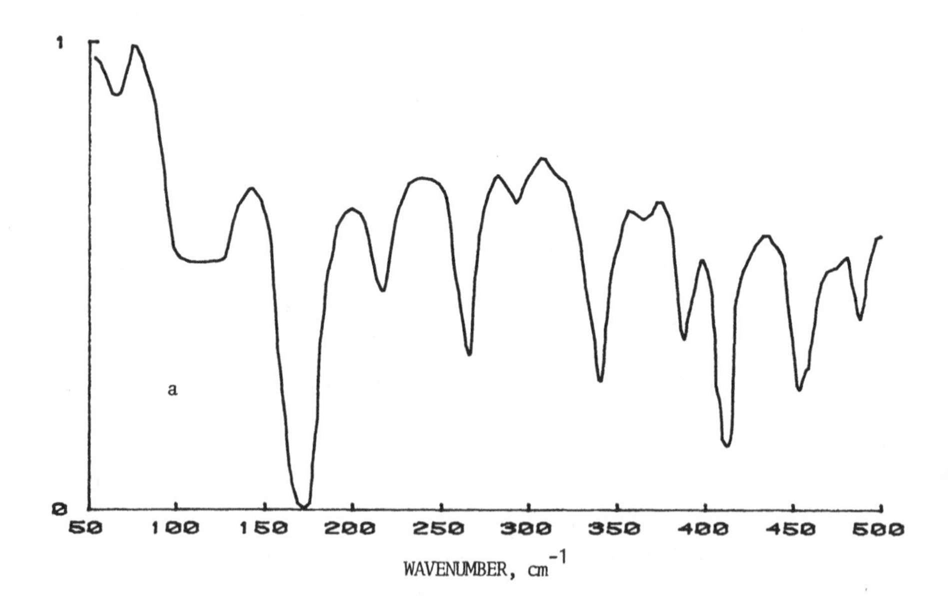

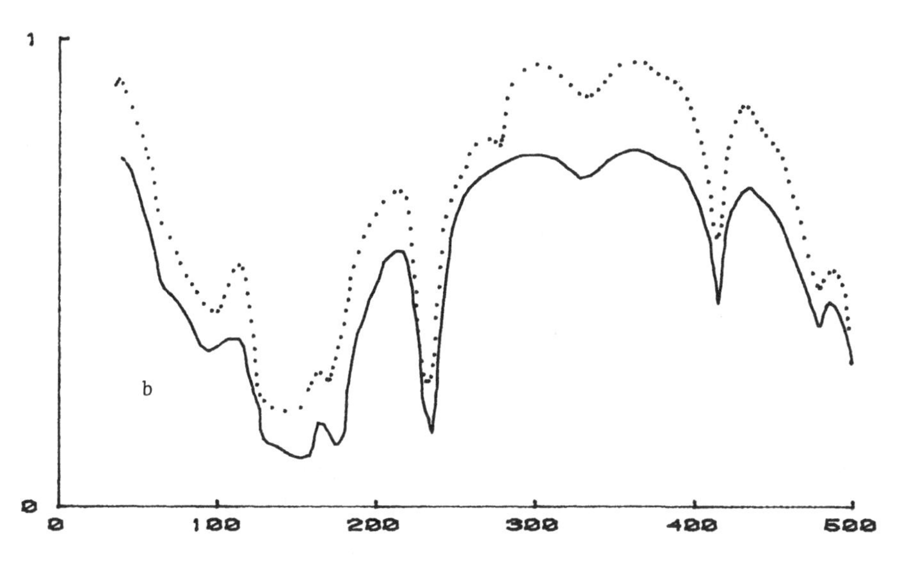

4,n-octyl 4'-cyanobiphenyle: a-crystal phase 20°C; b-full line-nematic phase 70°C, dotted line-isotropic liquid 85°C.

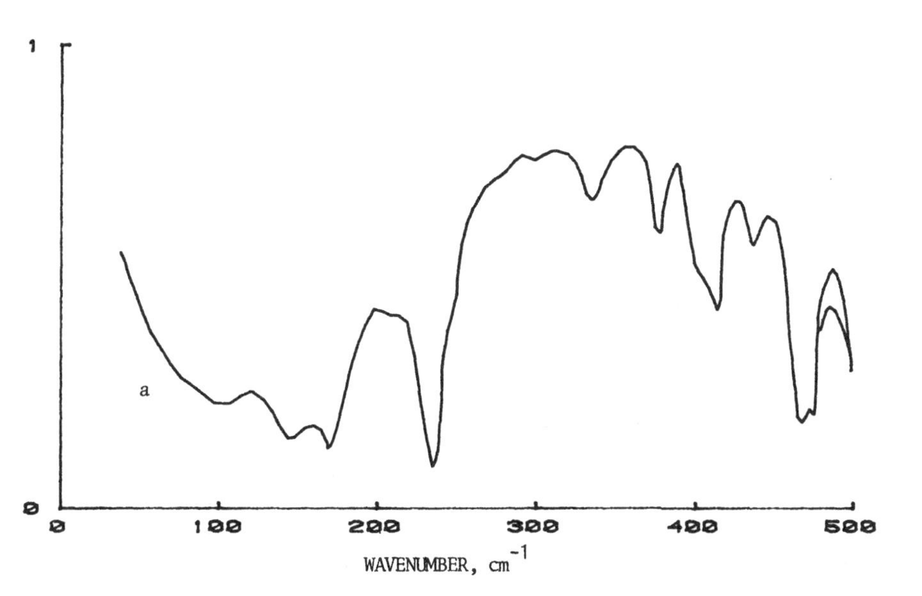

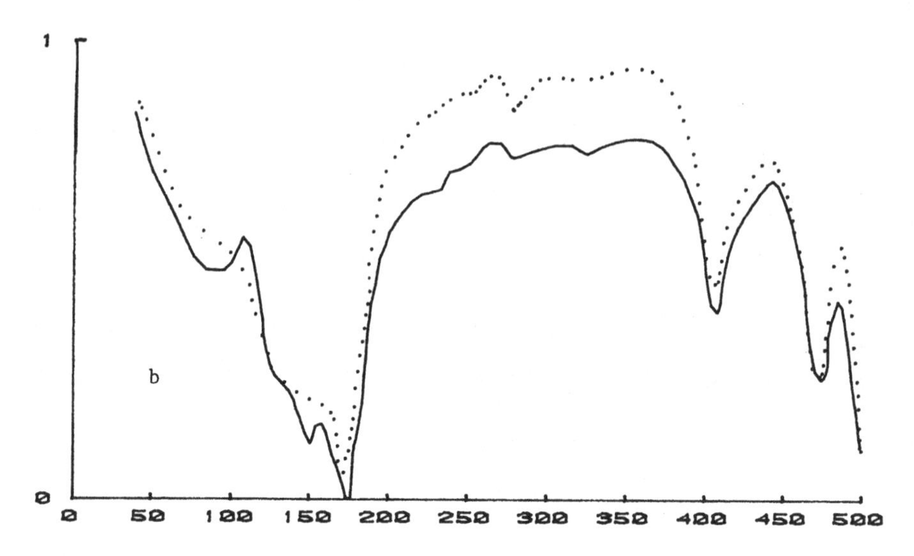

4,n-decyl 4'-cyanobiphenyle: a-crystal phase 20°C; b-full line-smecticA phase 45°C, dotted line-isotropic liquid 60°C.

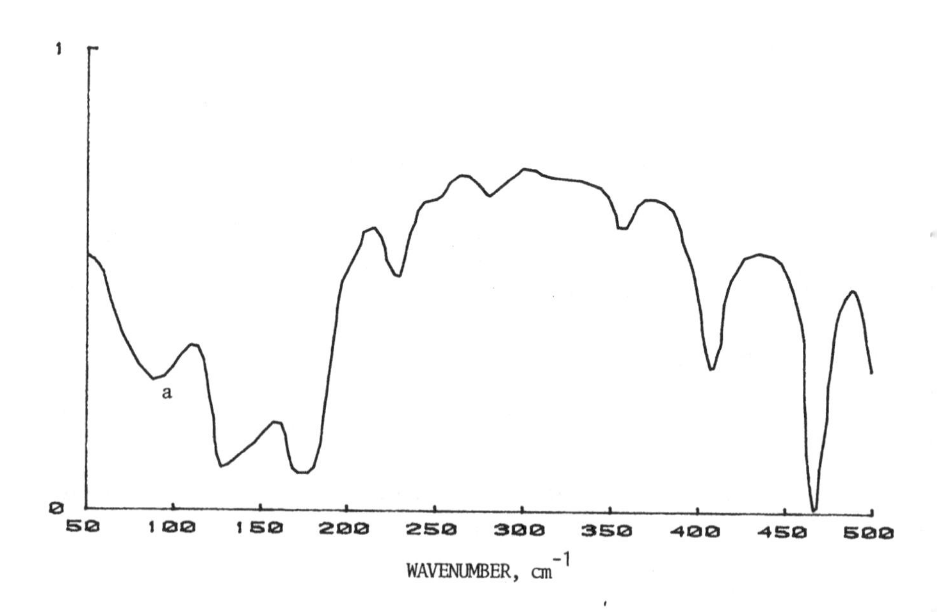

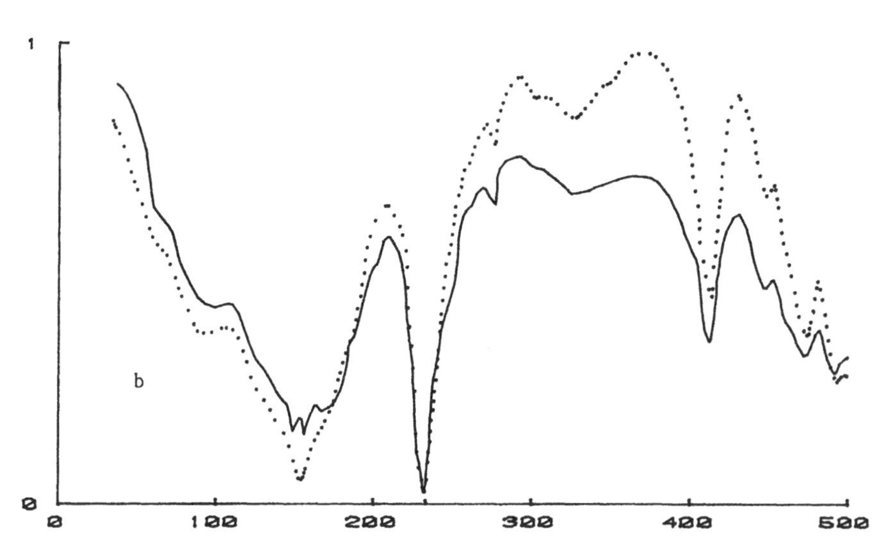

4,n-pentyloxy 4'-cyanobiphenyle (5OCB): a-crystal phase 20°C, b-full line - nematic phase 60°C, dotted line-isotropic liquid 75°C.

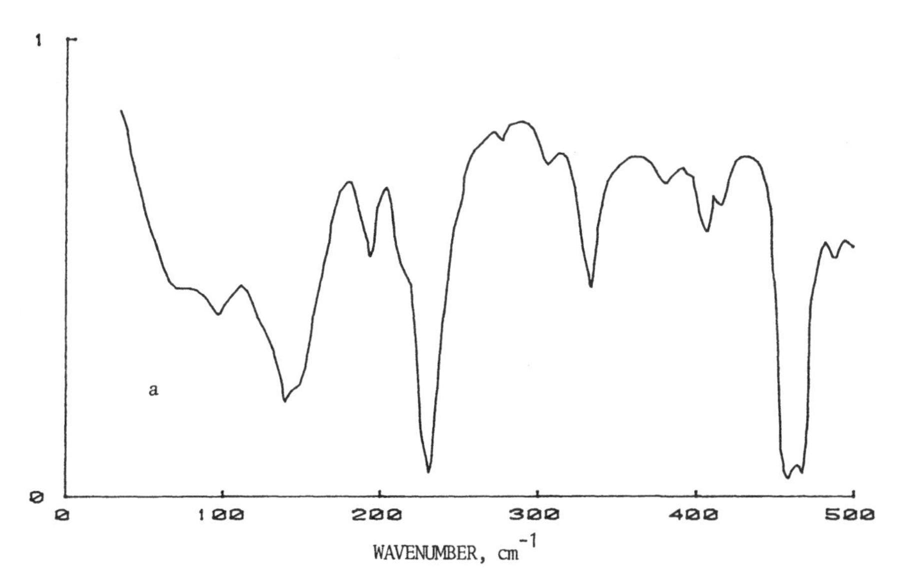

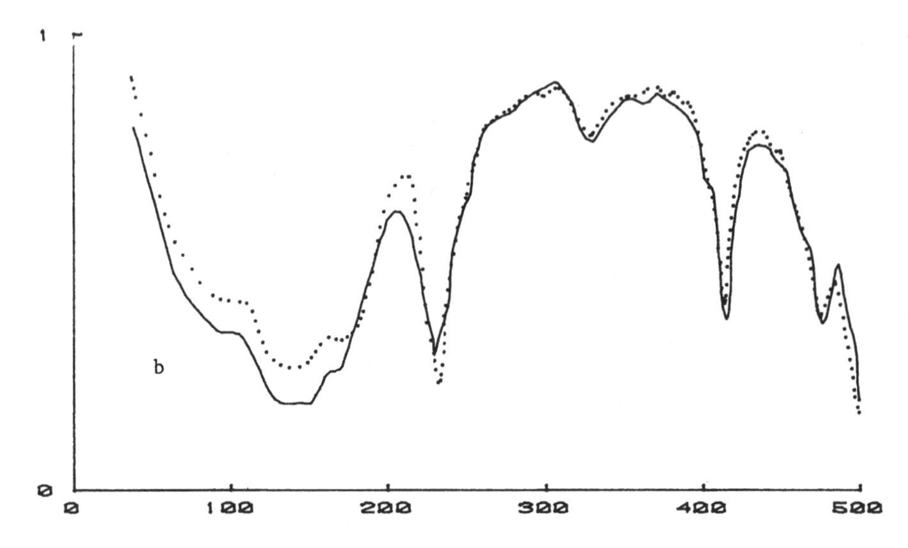

4,n-heptyloxy 4'-cyanobiphenyle (7OCB): a-crystal phase 20°C, b-full line-smectic A phase, dotted line-nematic phase 60°C.

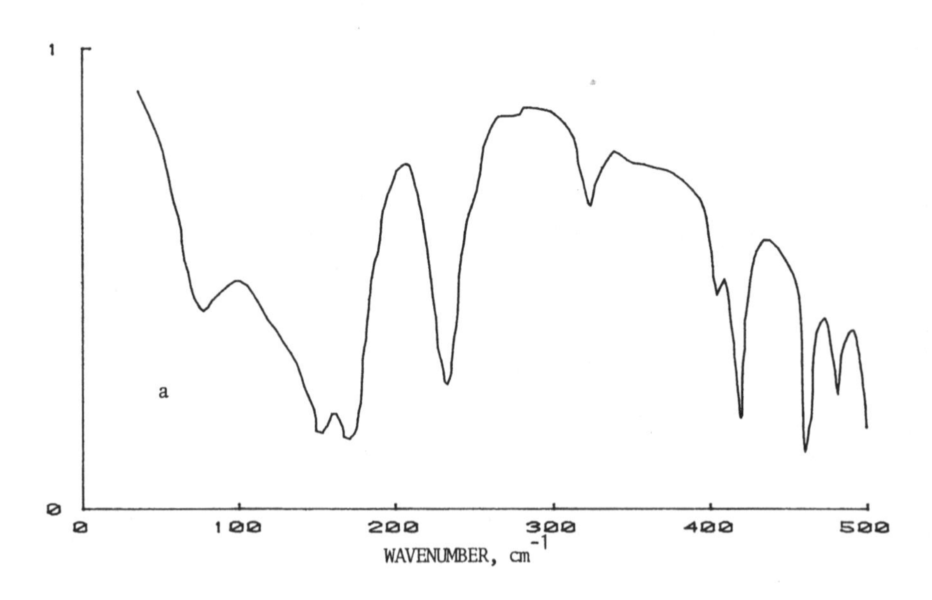

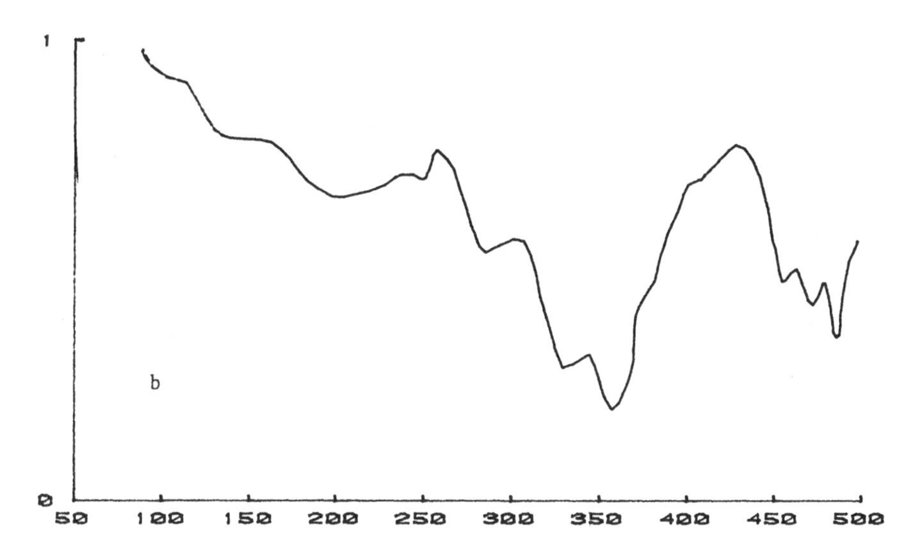

Cholesteryl oleylcarbonate: a-crystal phase 20°C; b-isotropic liquid 70°C.

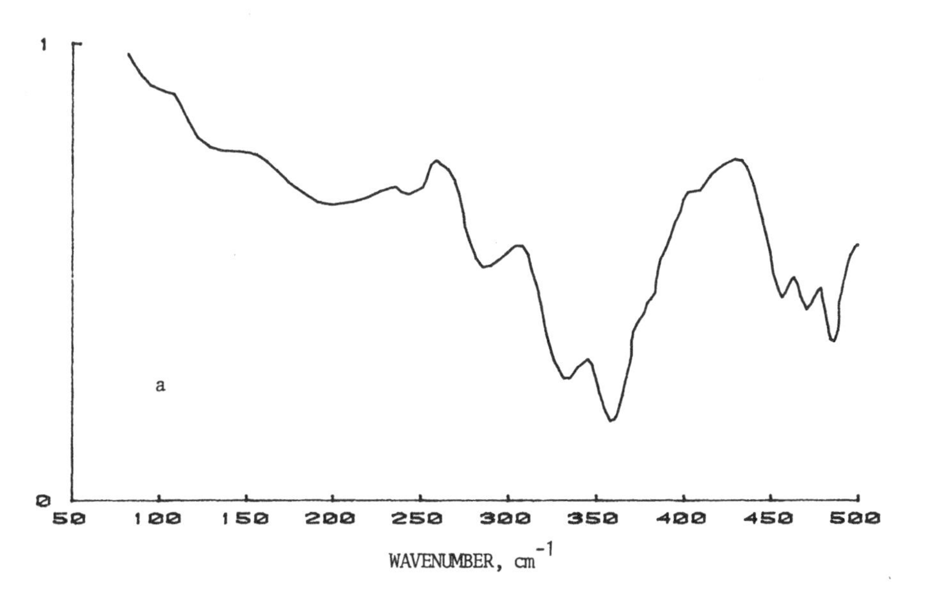

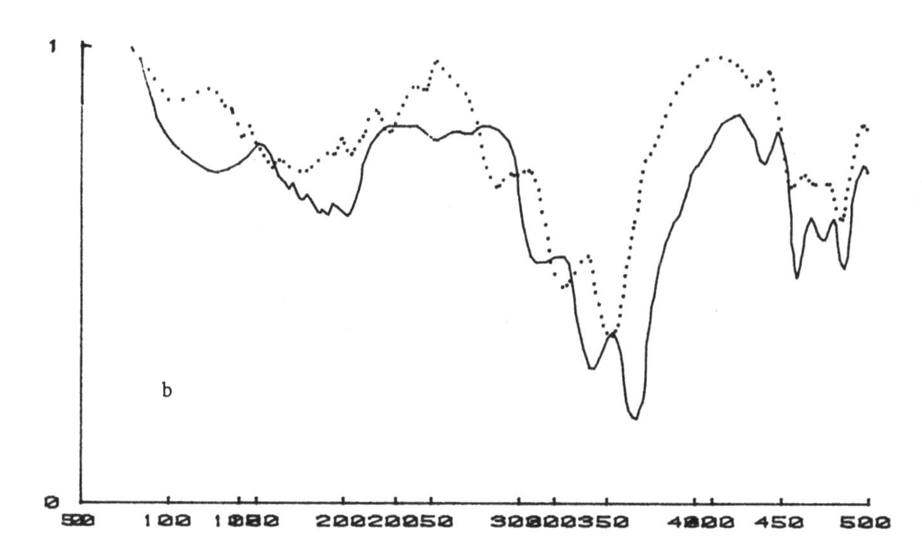

Cholesteryl mirist ate: a-full line-crystal phase 20°C, dotted line-smectic A phase 75°C;b-full line-cholesteric phase 80°C, dotted line-isotropic liquid 90°C.

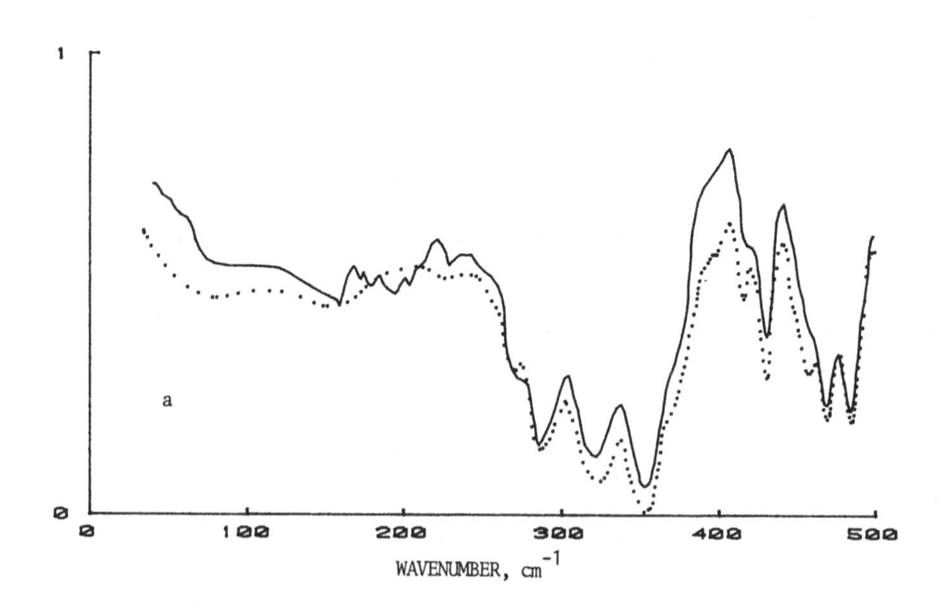

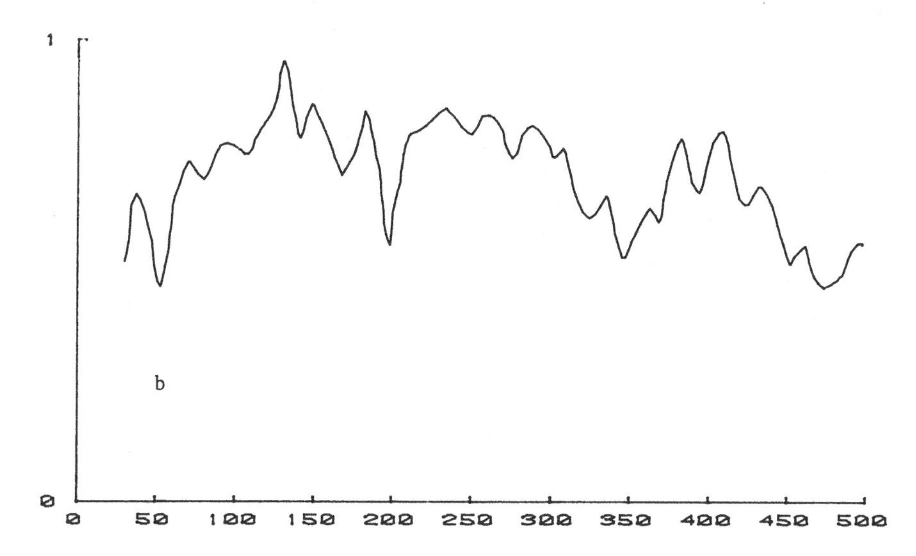

Cholesteryl propionate: a- cholesteric phase 100°C, b- isotropic liquid 120°C.

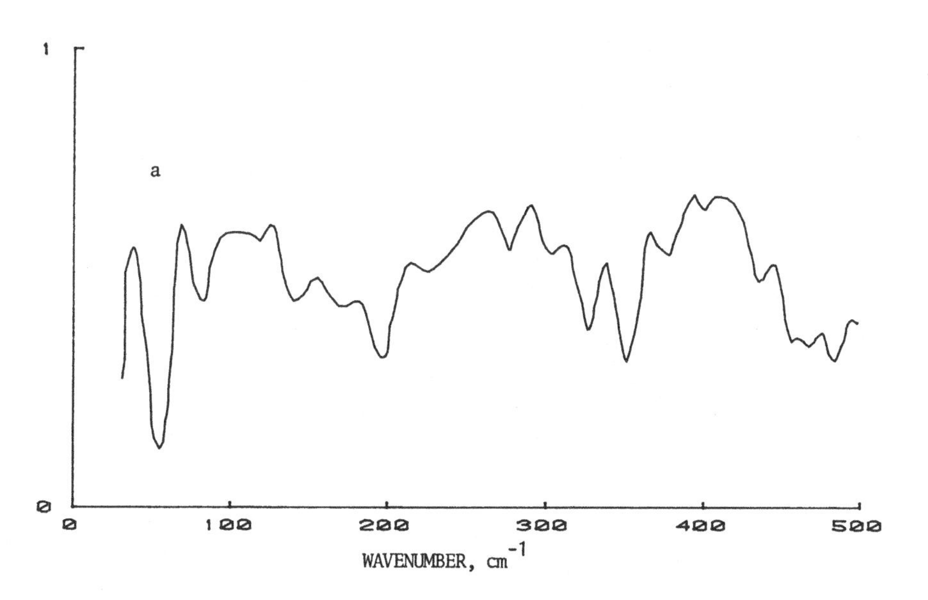

NEAR INFRARED SPECTRA

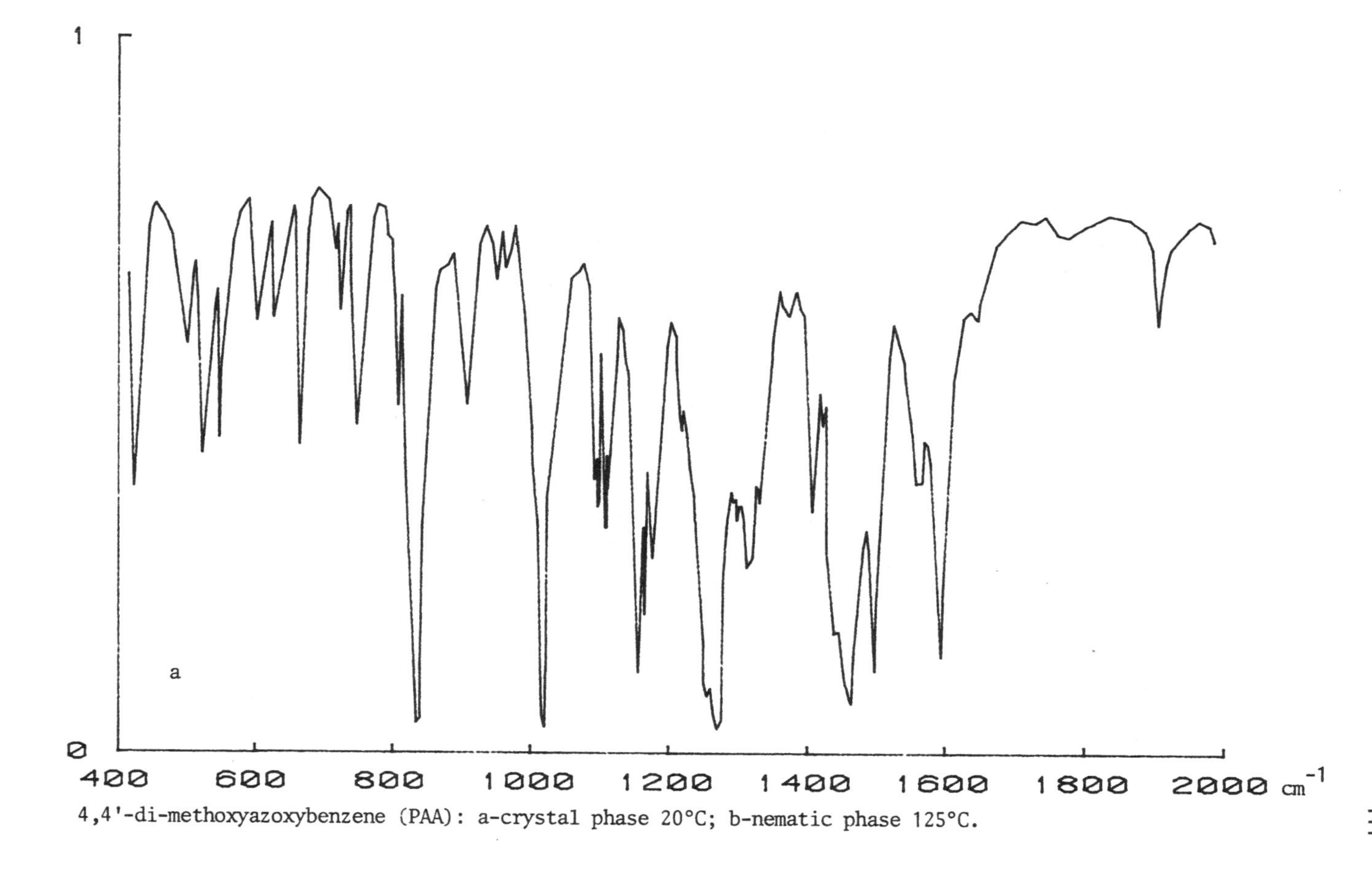

4,4'-di-methoxyazoxybenzene (PAA): a-crystal phase 20°C; b-nematic phase 125°C.

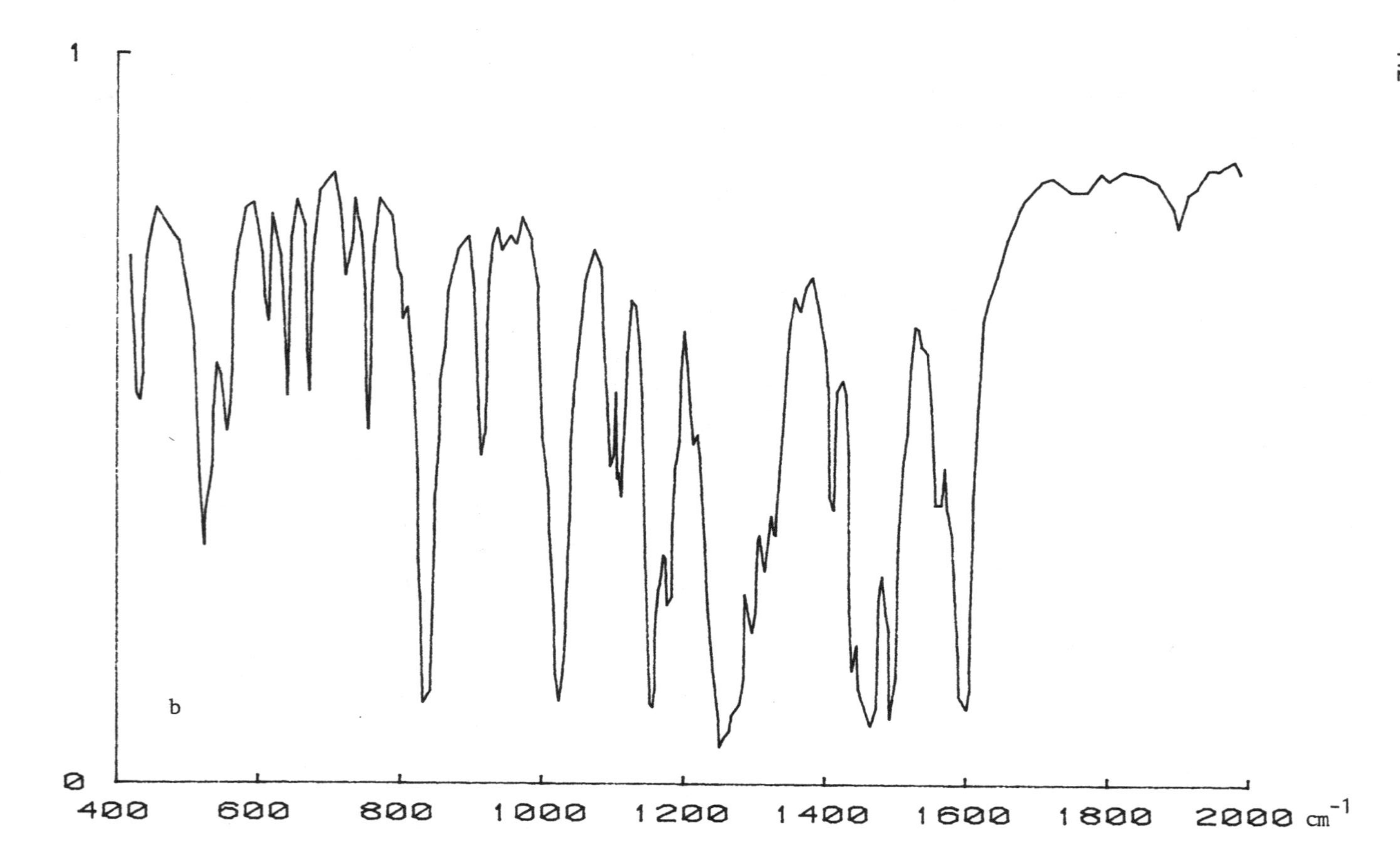

1
0
400
600
800
1000
1200
1400
1600
1800
2000 cm⁻¹
b

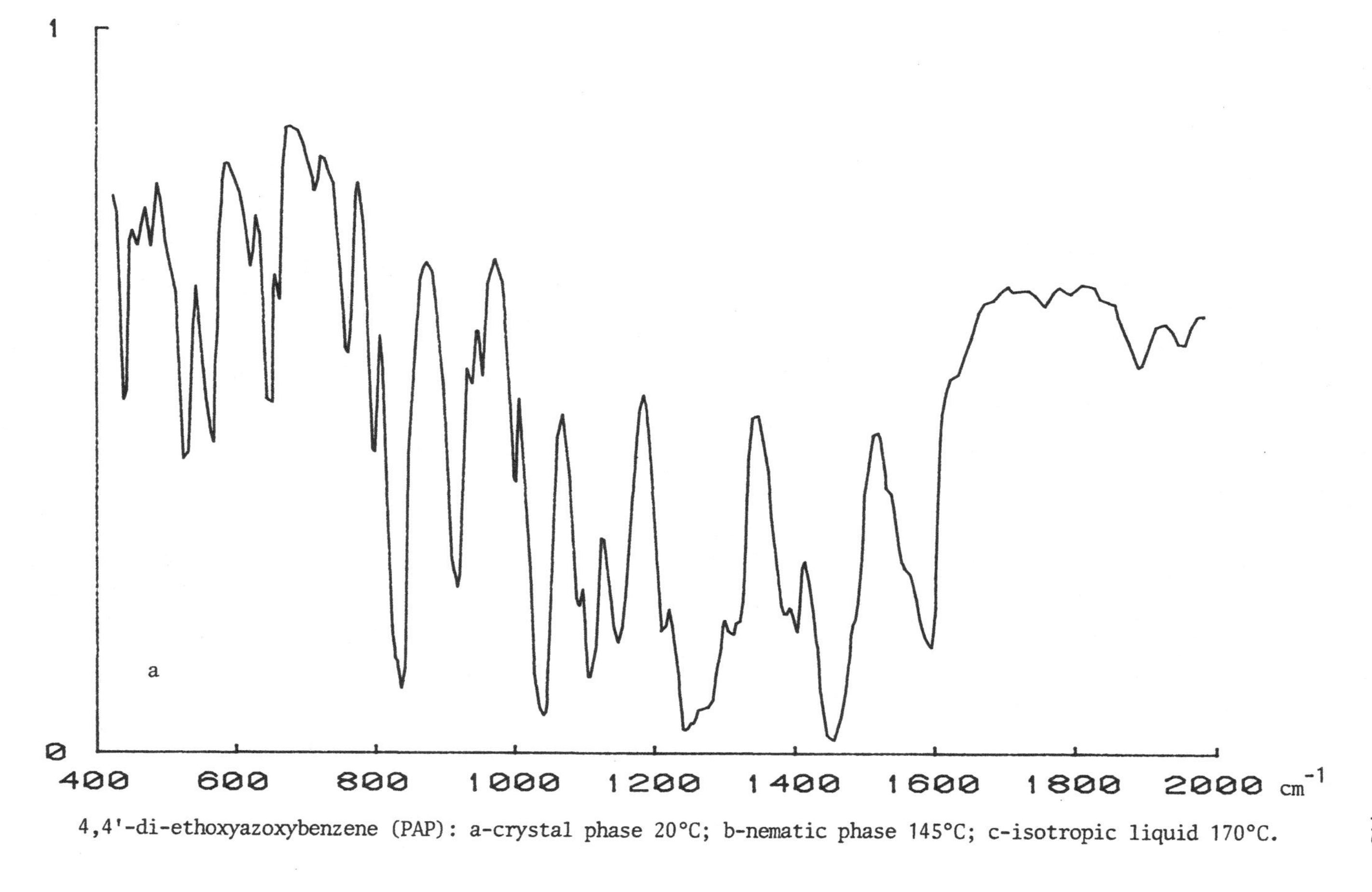

4,4'-di-ethoxyazoxybenzene (PAP): a-crystal phase 20°C; b-nematic phase 145°C; c-isotropic liquid 170°C.

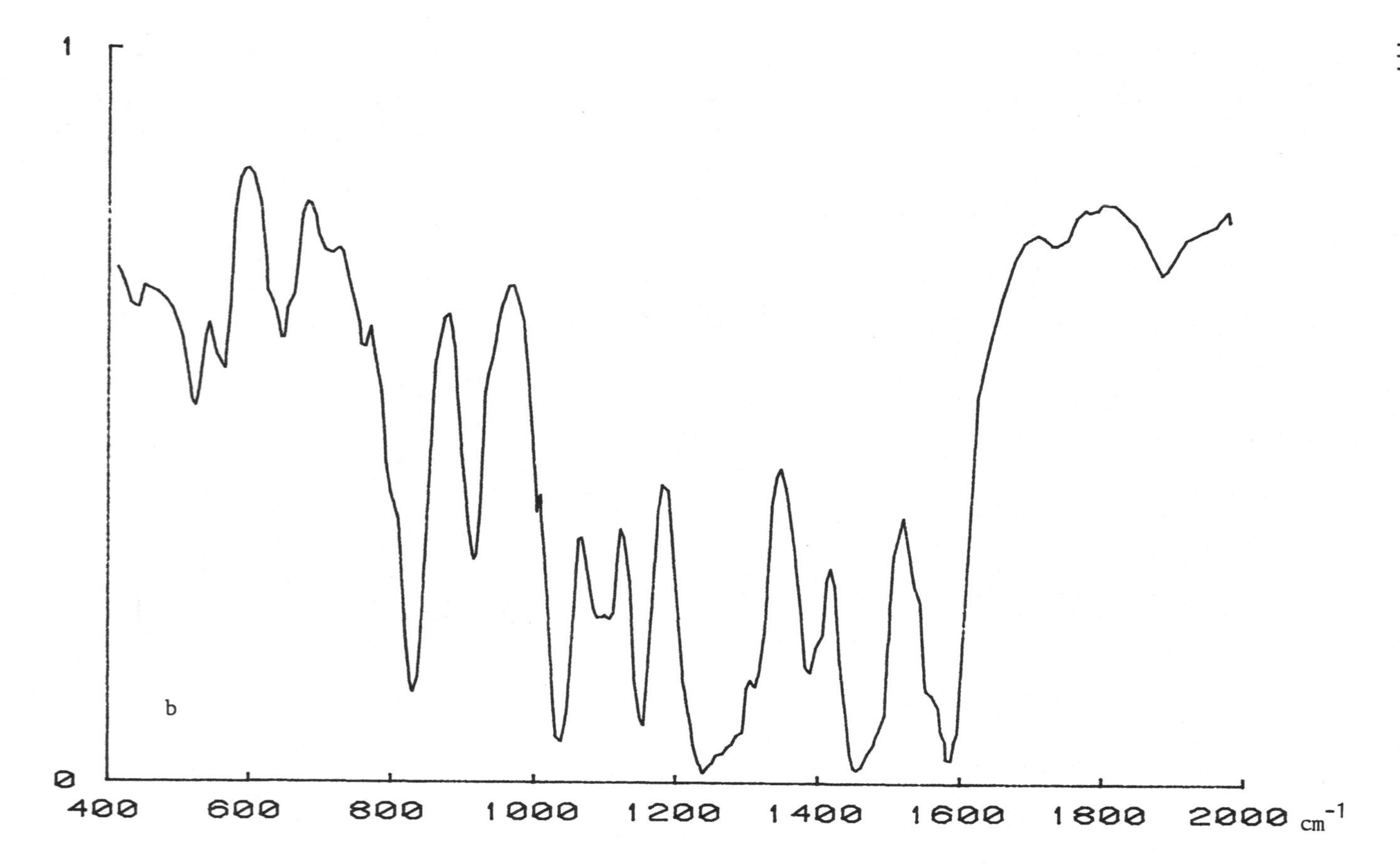

1
0
400
600
800
1000
1200
1400
1600
1800
2000
cm^{-1}
b

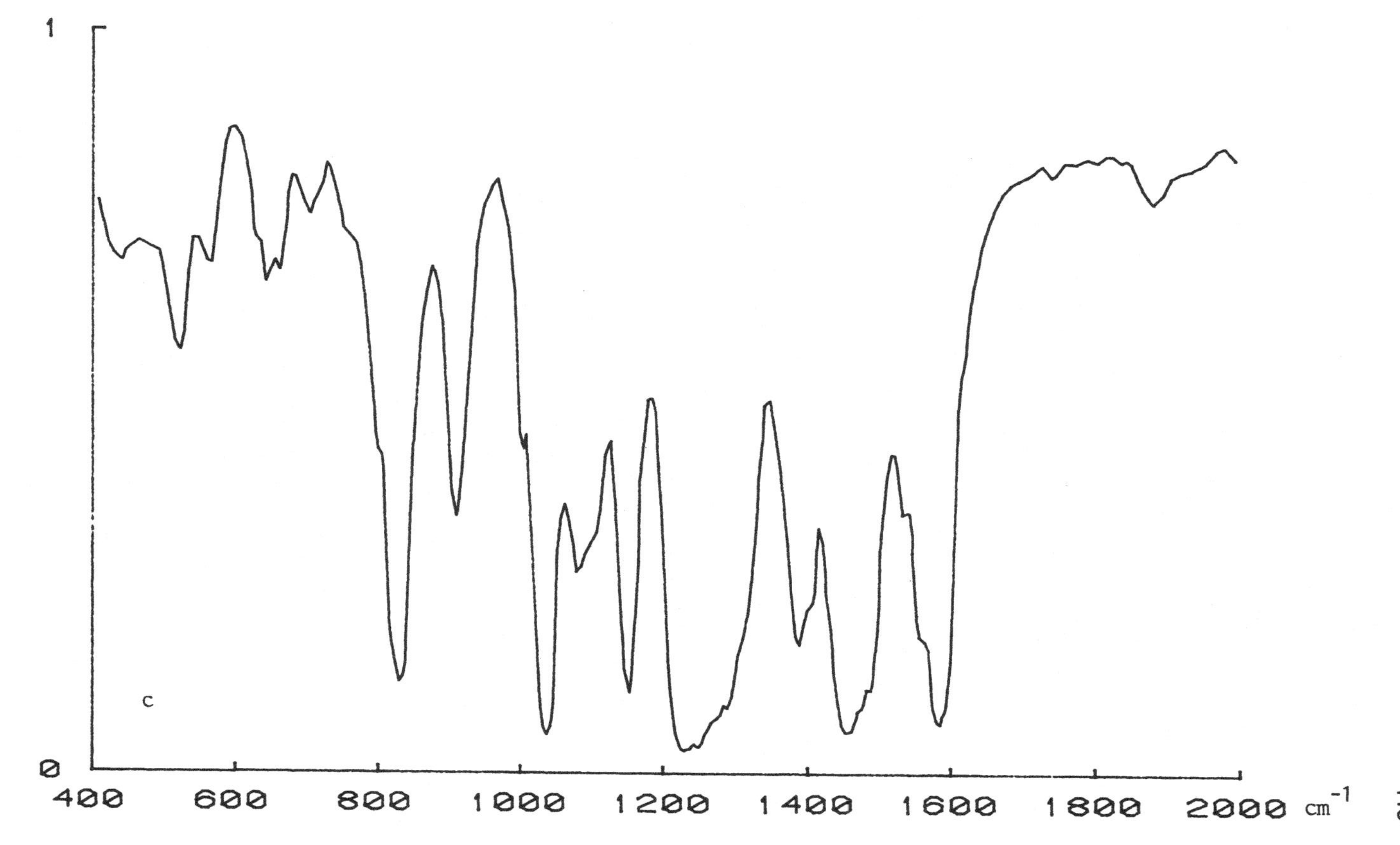
1
0
400
600
800
1000
1200
1400
1600
1800
2000
cm-1
c

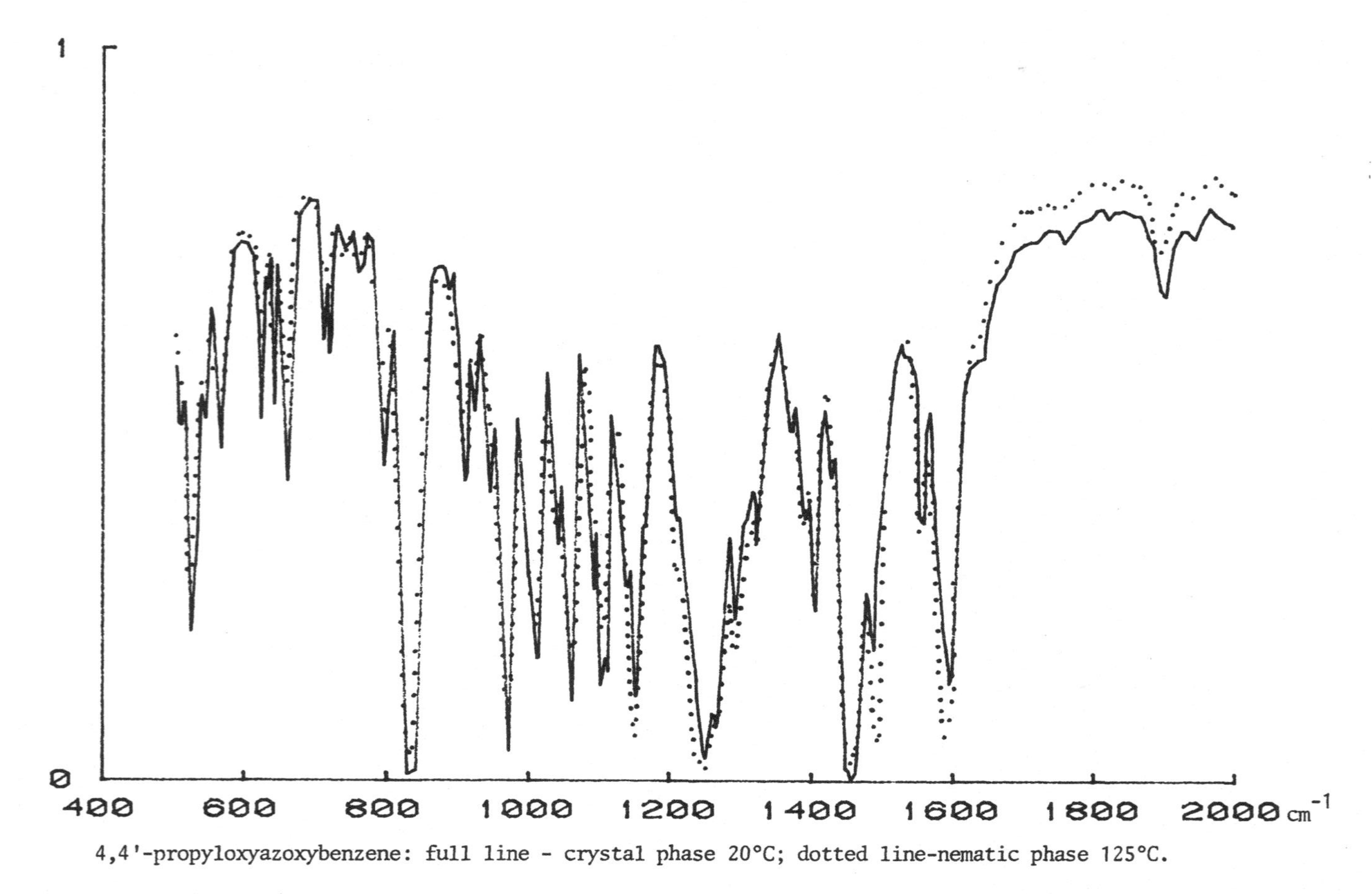

4,4'-propyloxyazoxybenzene: full line - crystal phase 20°C; dotted line-nematic phase 125°C.

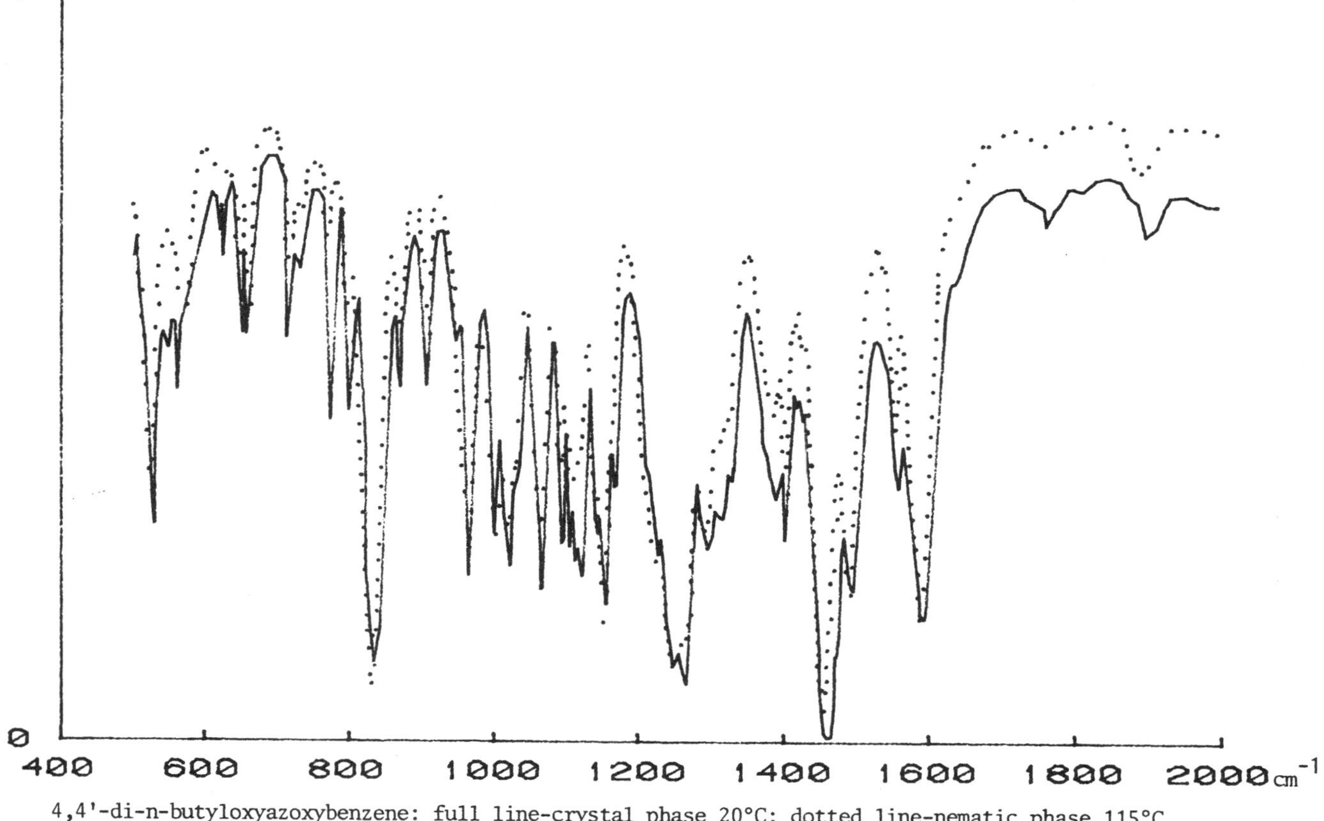

4,4'-di-n-butyloxyazoxybenzene: full line-crystal phase 20°C; dotted line-nematic phase 115°C.

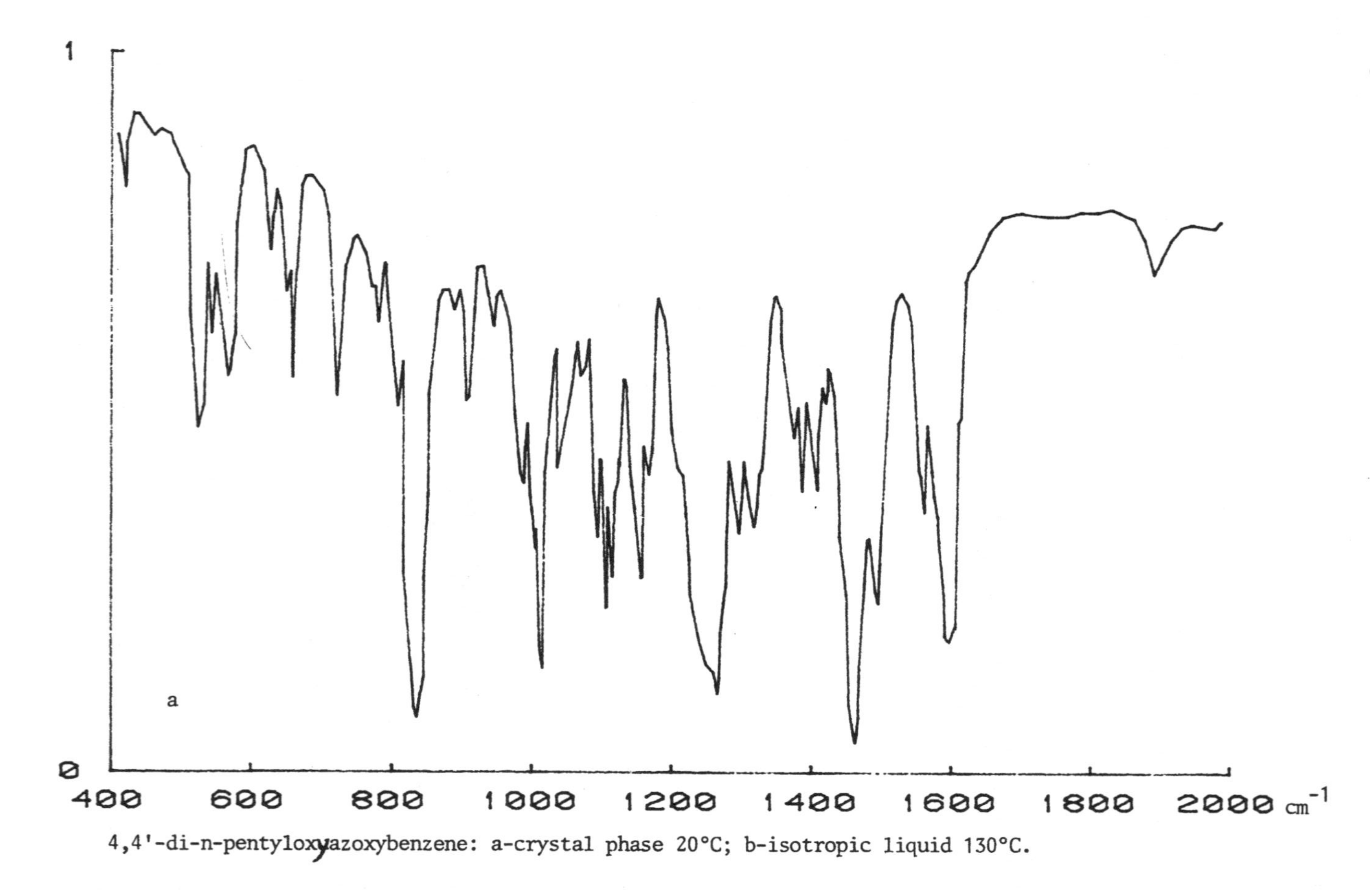

4,4'-di-n-pentyloxyazoxybenzene: a-crystal phase 20°C; b-isotropic liquid 130°C.

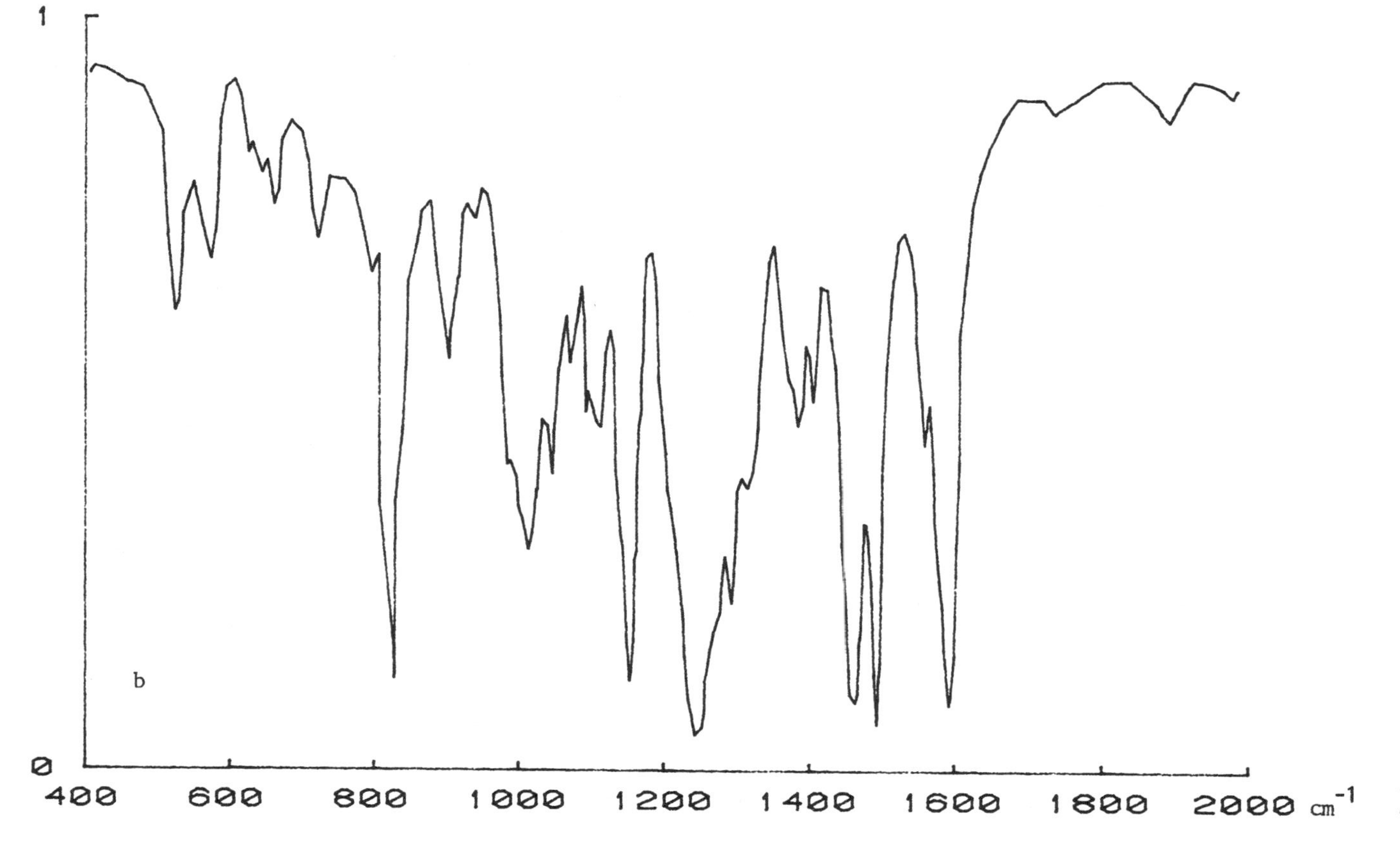

1
b
0
400
600
800
1000
1200
1400
1600
1800
2000
cm-1

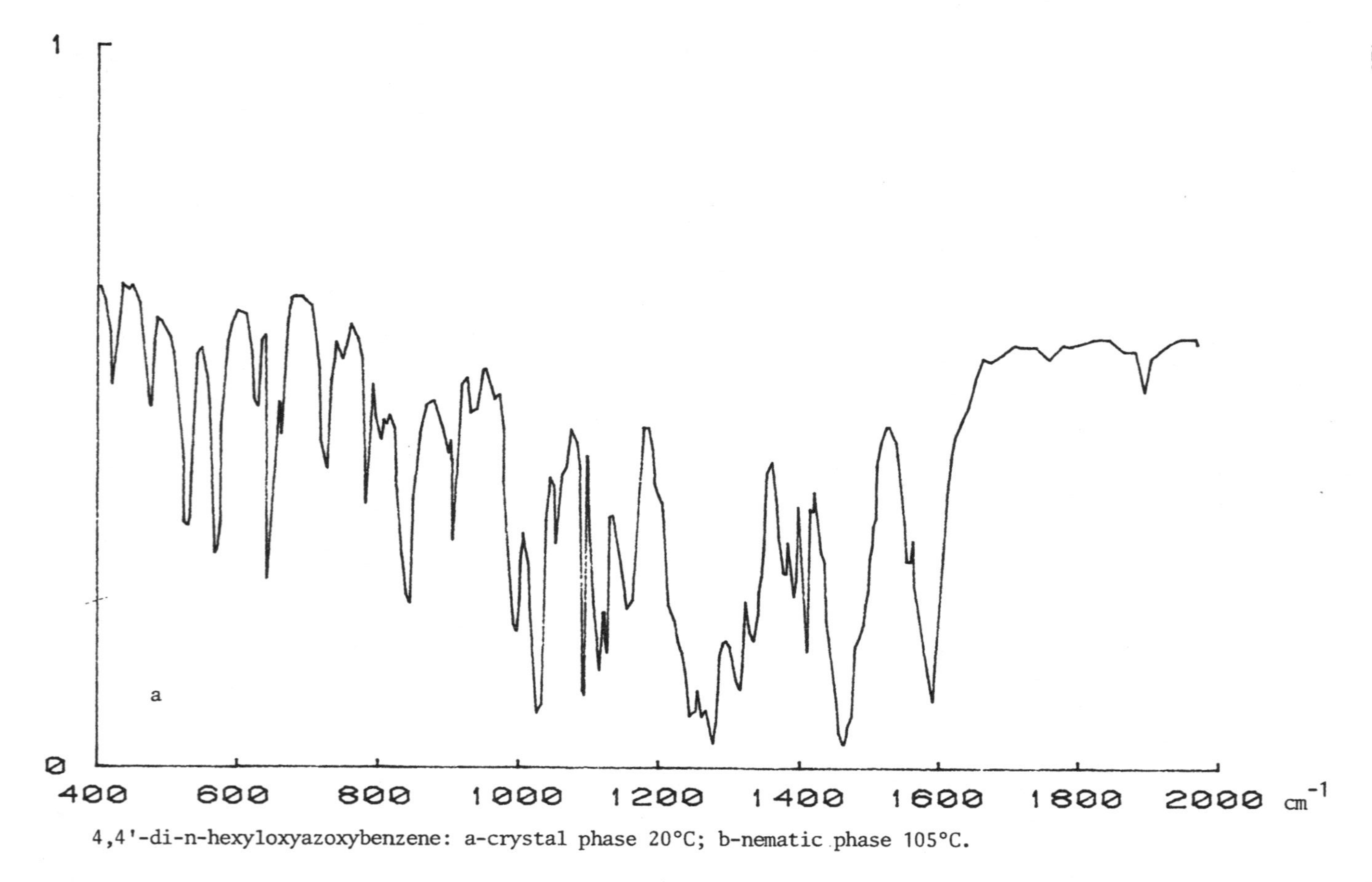

4,4'-di-n-hexyloxyazoxybenzene: a-crystal phase 20°C; b-nematic phase 105°C.

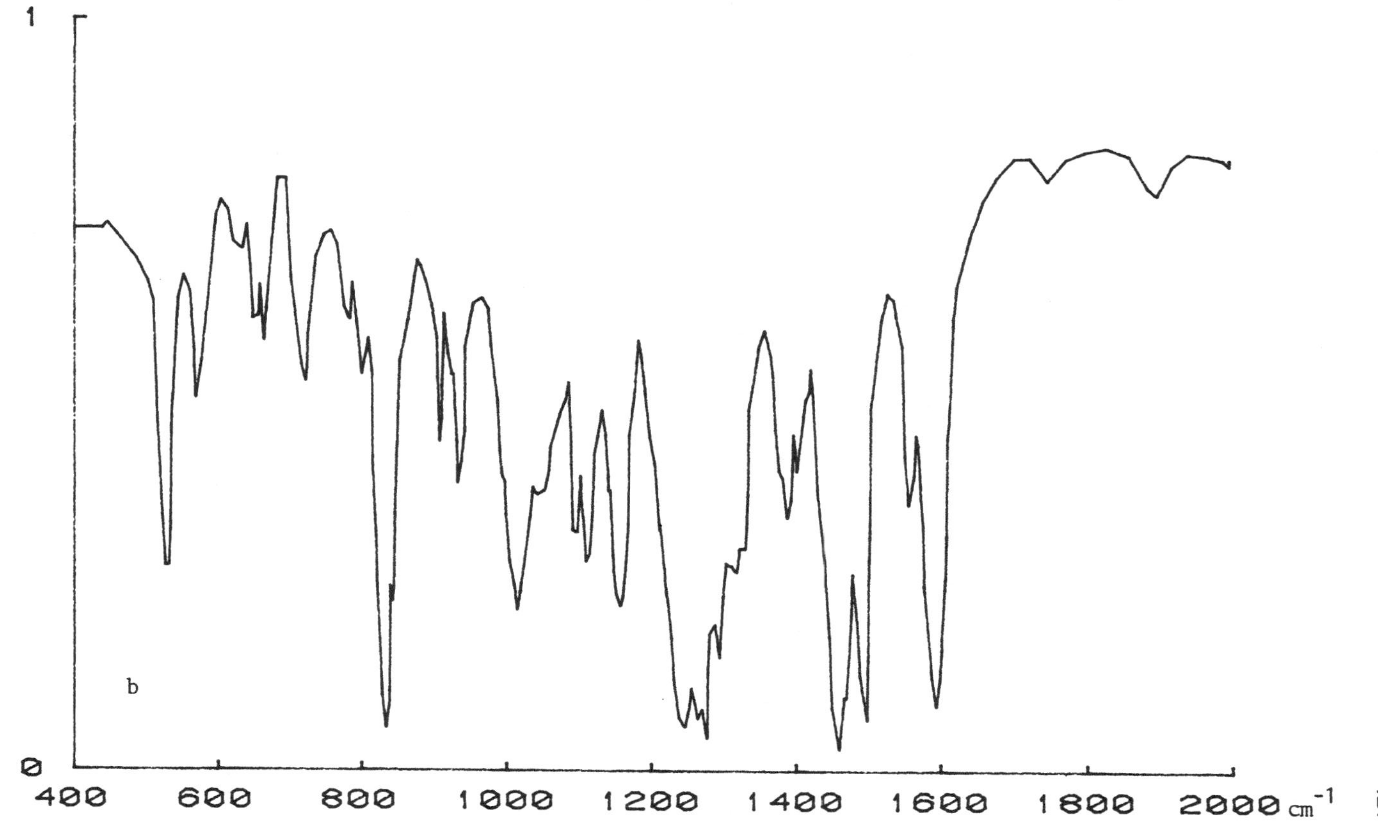

1
0
400
600
800
1000
1200
1400
1600
1800
2000 cm-1
b

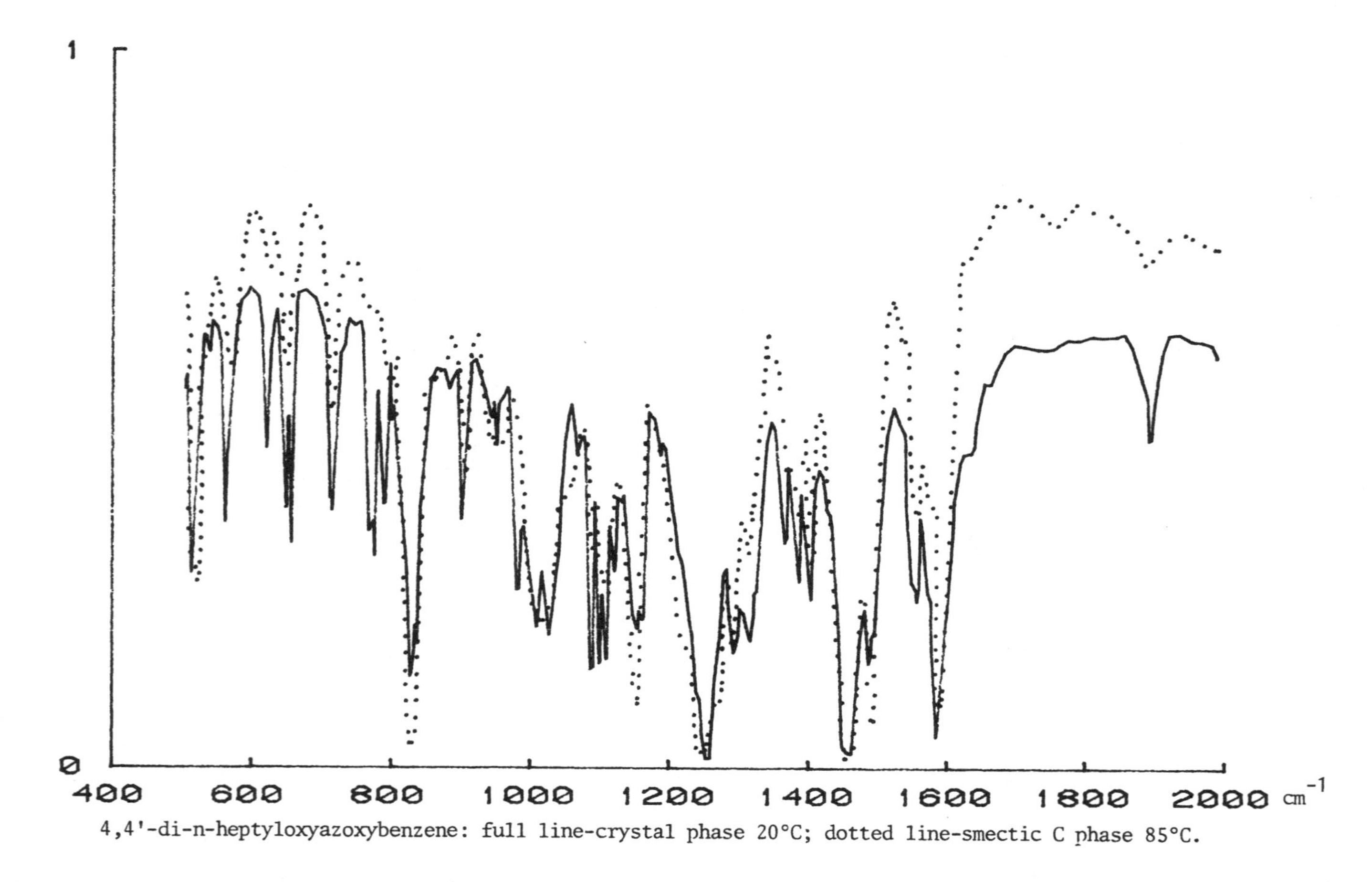

4,4'-di-n-heptyloxyazoxybenzene: full line-crystal phase 20°C; dotted line-smectic C phase 85°C.

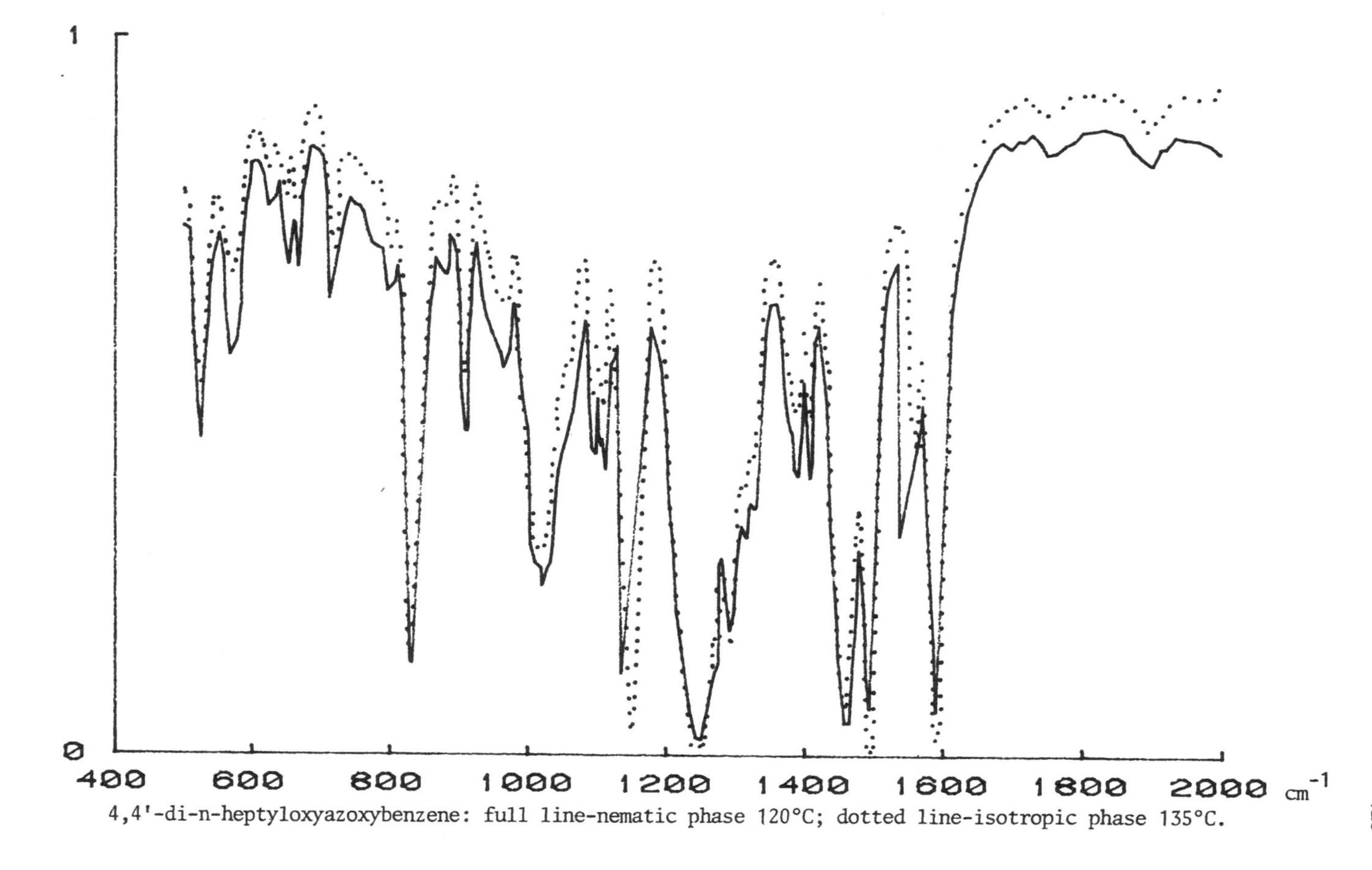

4,4'-di-n-heptyloxyazoxybenzene: full line-nematic phase 120°C; dotted line-isotropic phase 135°C.

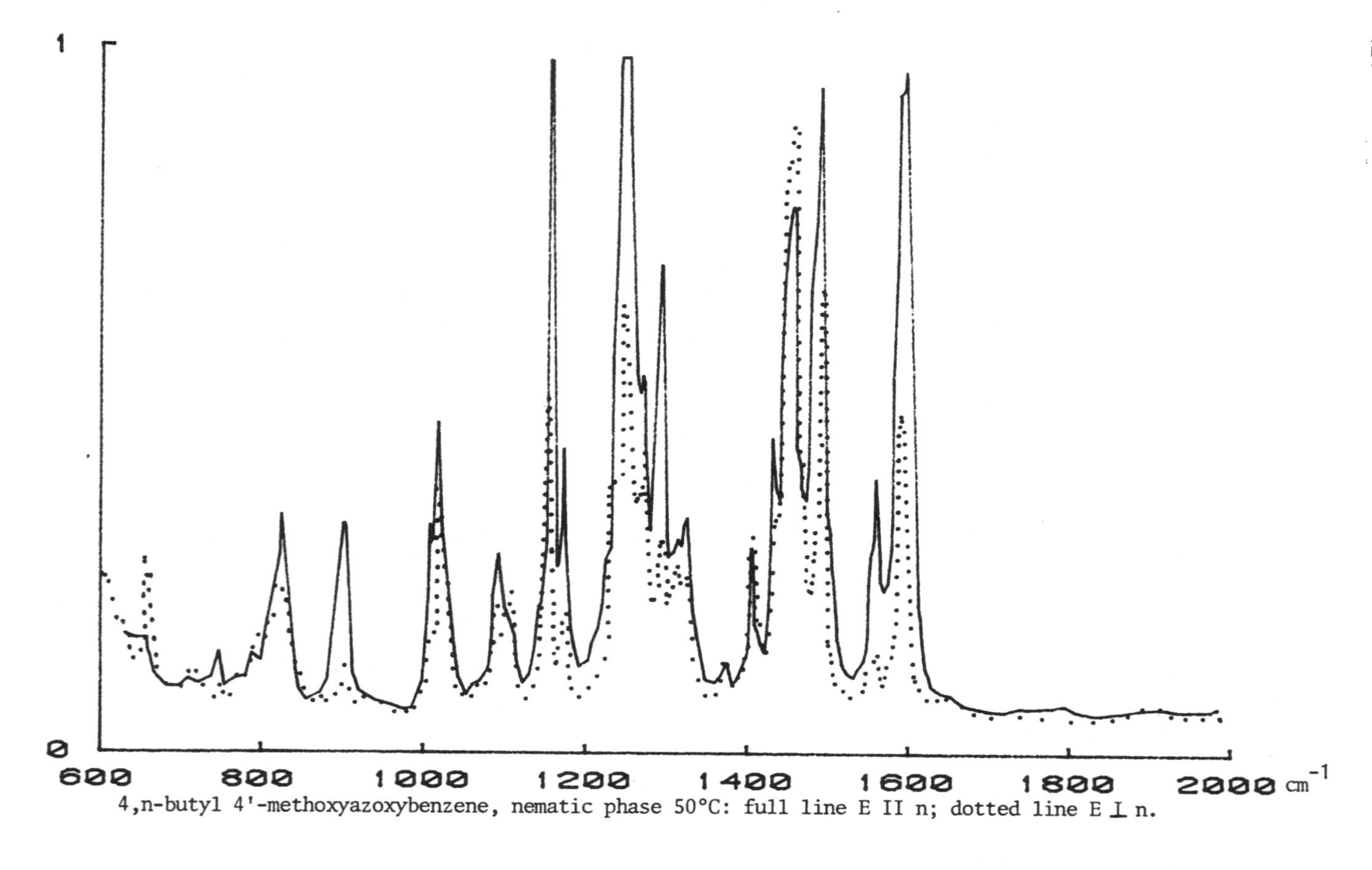

4,n-butyl 4'-methoxyazoxybenzene, nematic phase 50°C: full line E II n; dotted line E ⊥ n.

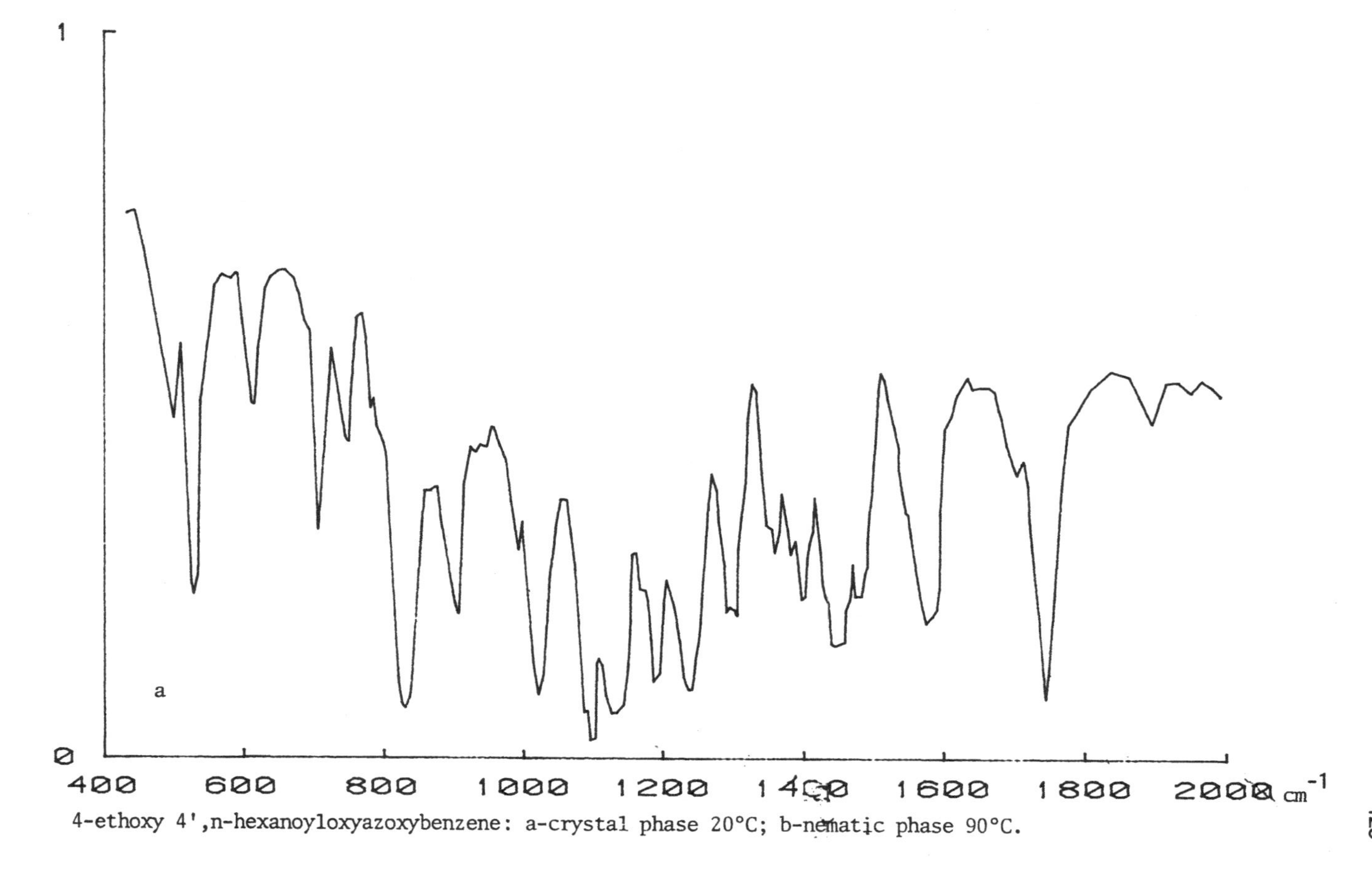

4-ethoxy 4',n-hexanoyloxyazoxybenzene: a-crystal phase 20°C; b-nematic phase 90°C.

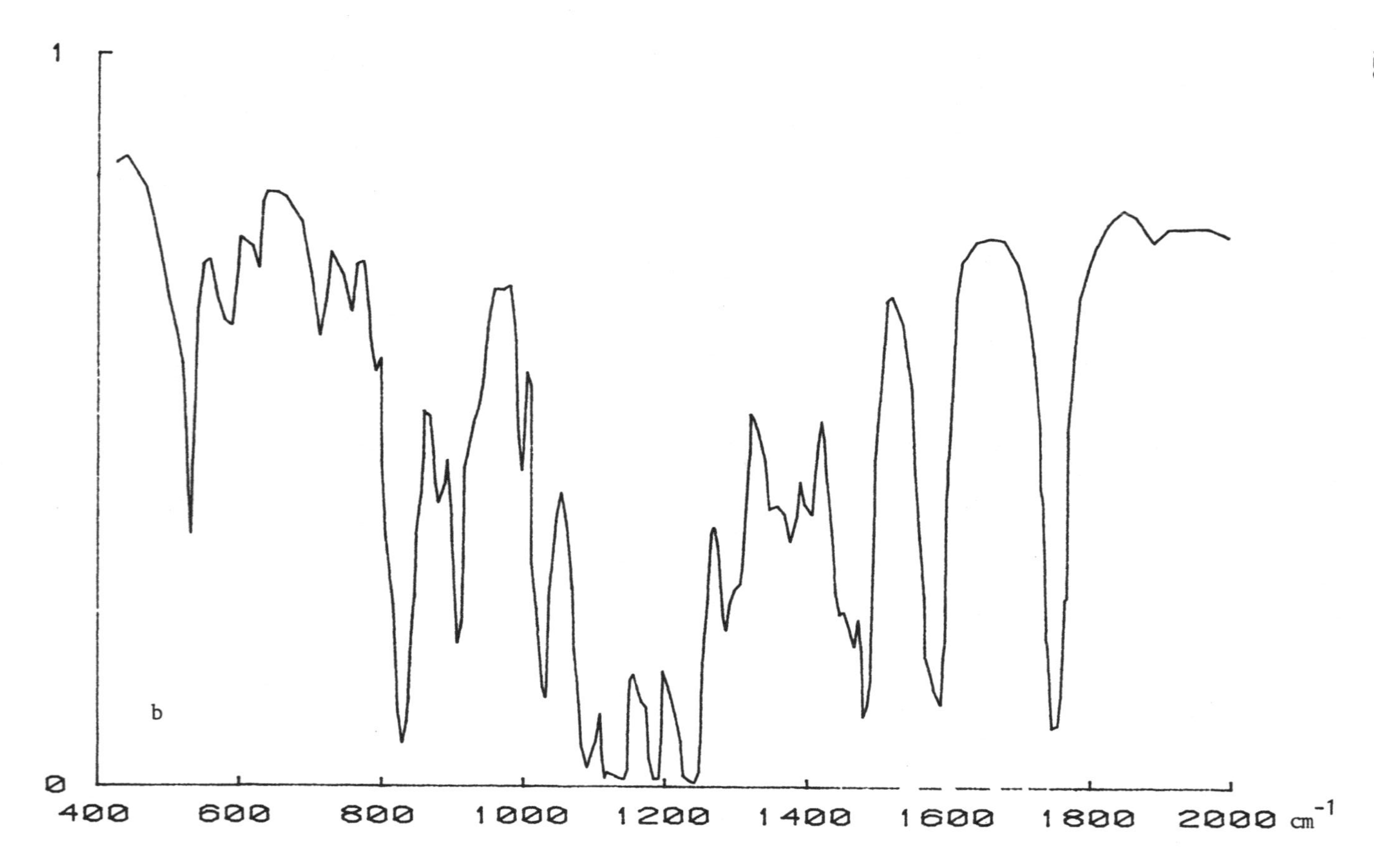

1
0
400
600
800
1000
1200
1400
1600
1800
2000
cm-1
b

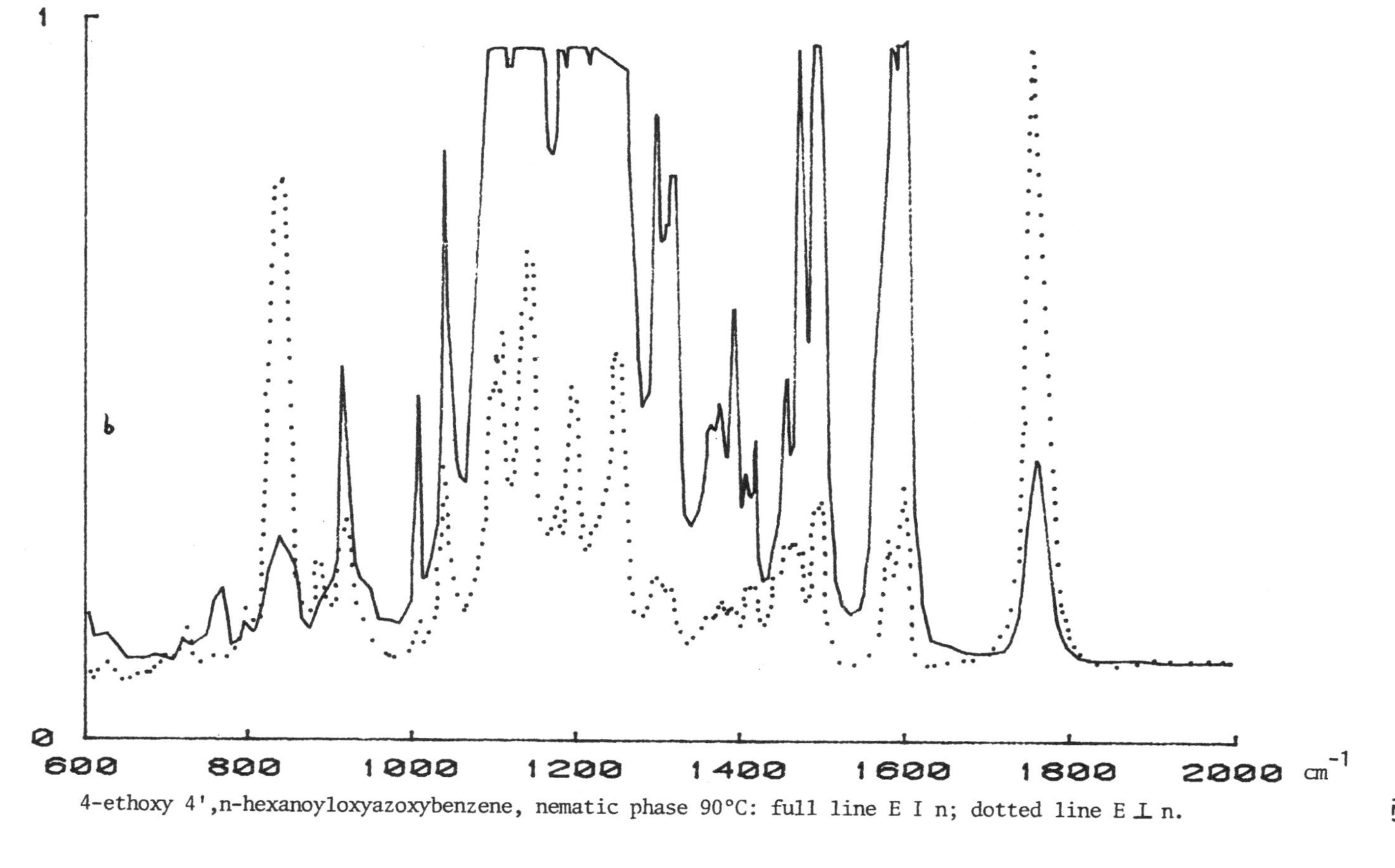

4-ethoxy 4',n-hexanoyloxyazoxybenzene, nematic phase 90°C: full line E I n; dotted line E ⊥ n.

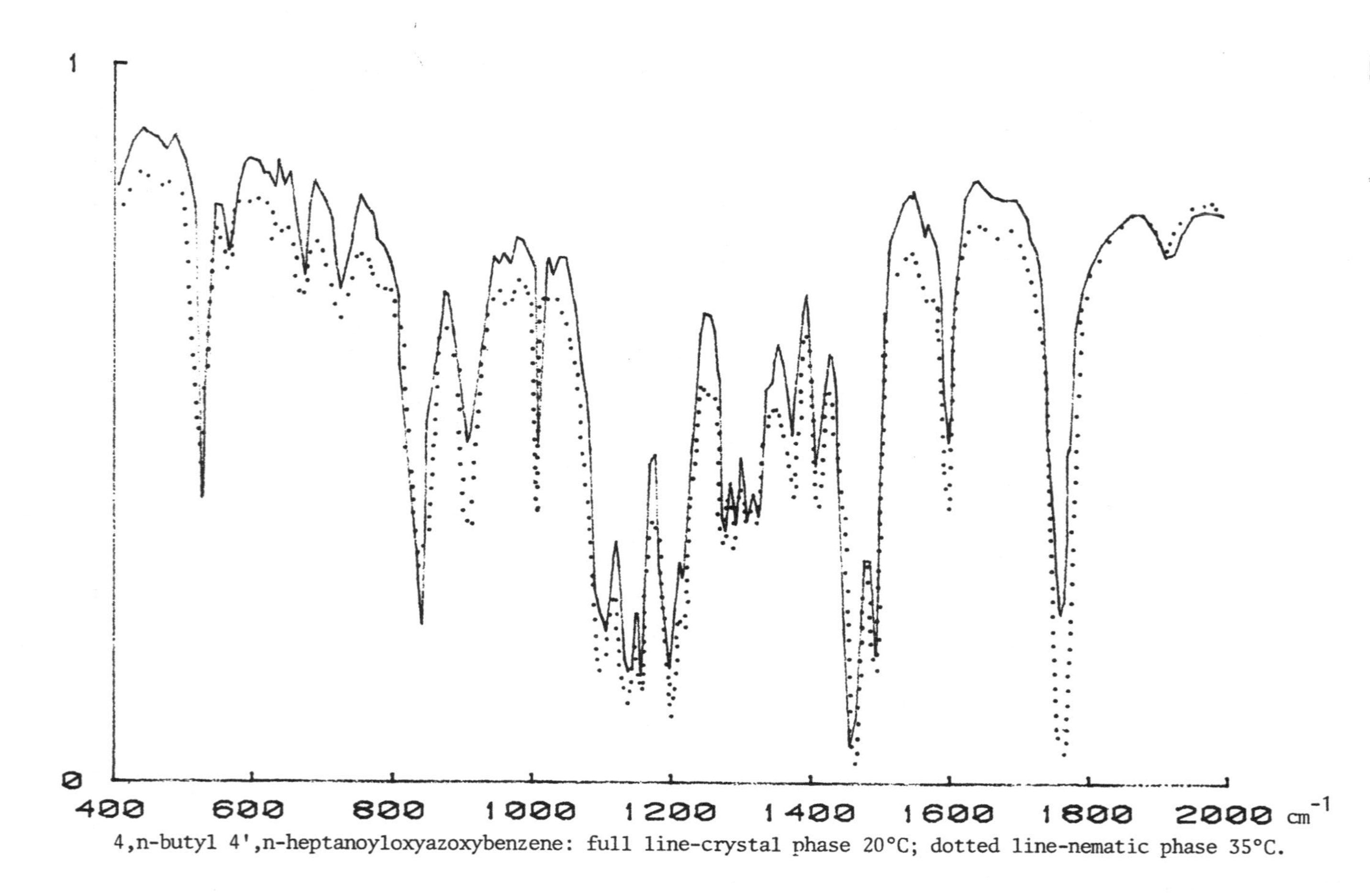

4,n-butyl 4',n-heptanoyloxyazoxybenzene: full line-crystal phase 20°C; dotted line-nematic phase 35°C.

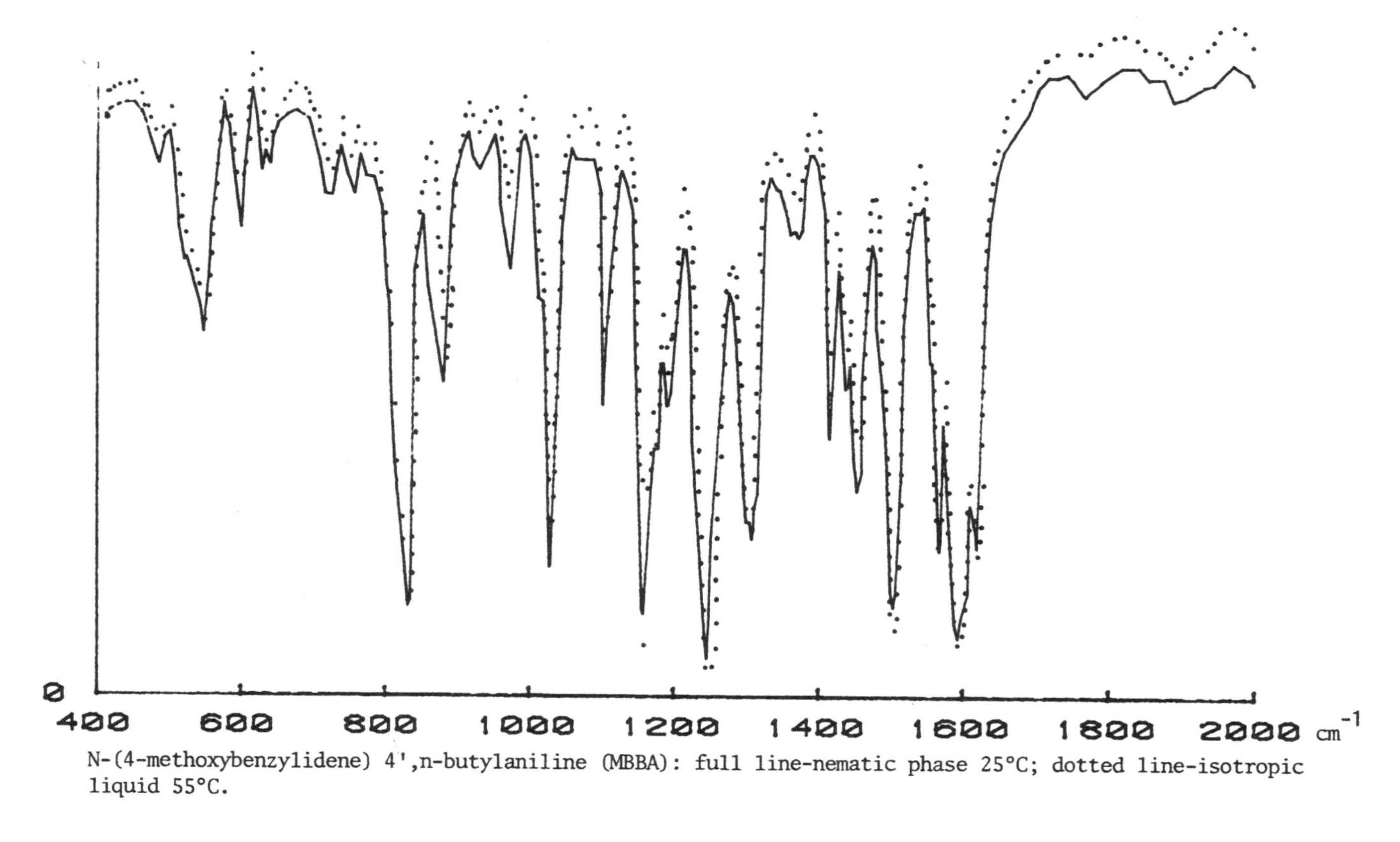

N-(4-methoxybenzylidene) 4',n-butylaniline (MBBA): full line-nematic phase 25°C; dotted line-isotropic liquid 55°C.

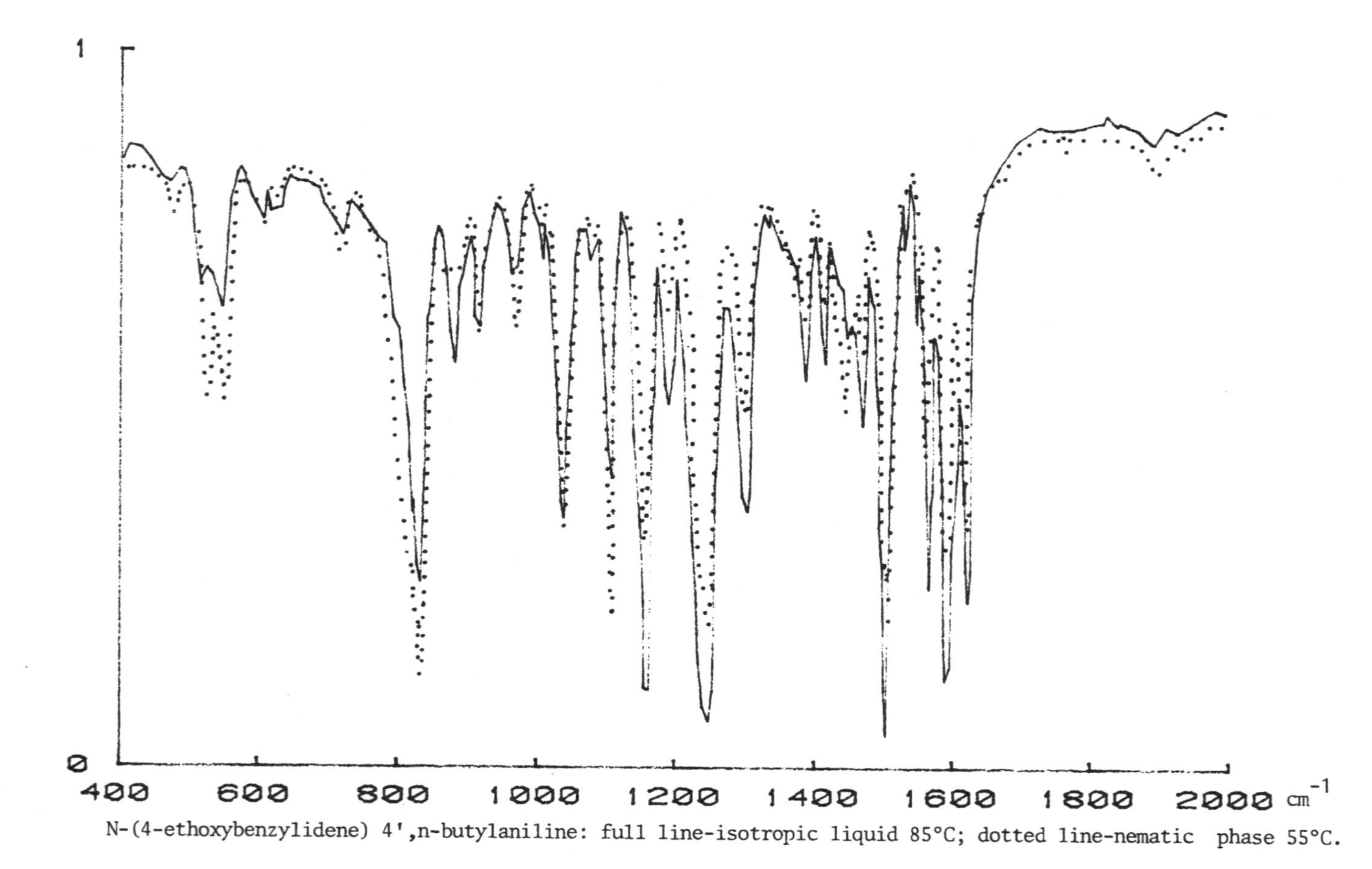

N-(4-ethoxybenzylidene) 4',n-butylaniline: full line-isotropic liquid 85°C; dotted line-nematic phase 55°C.

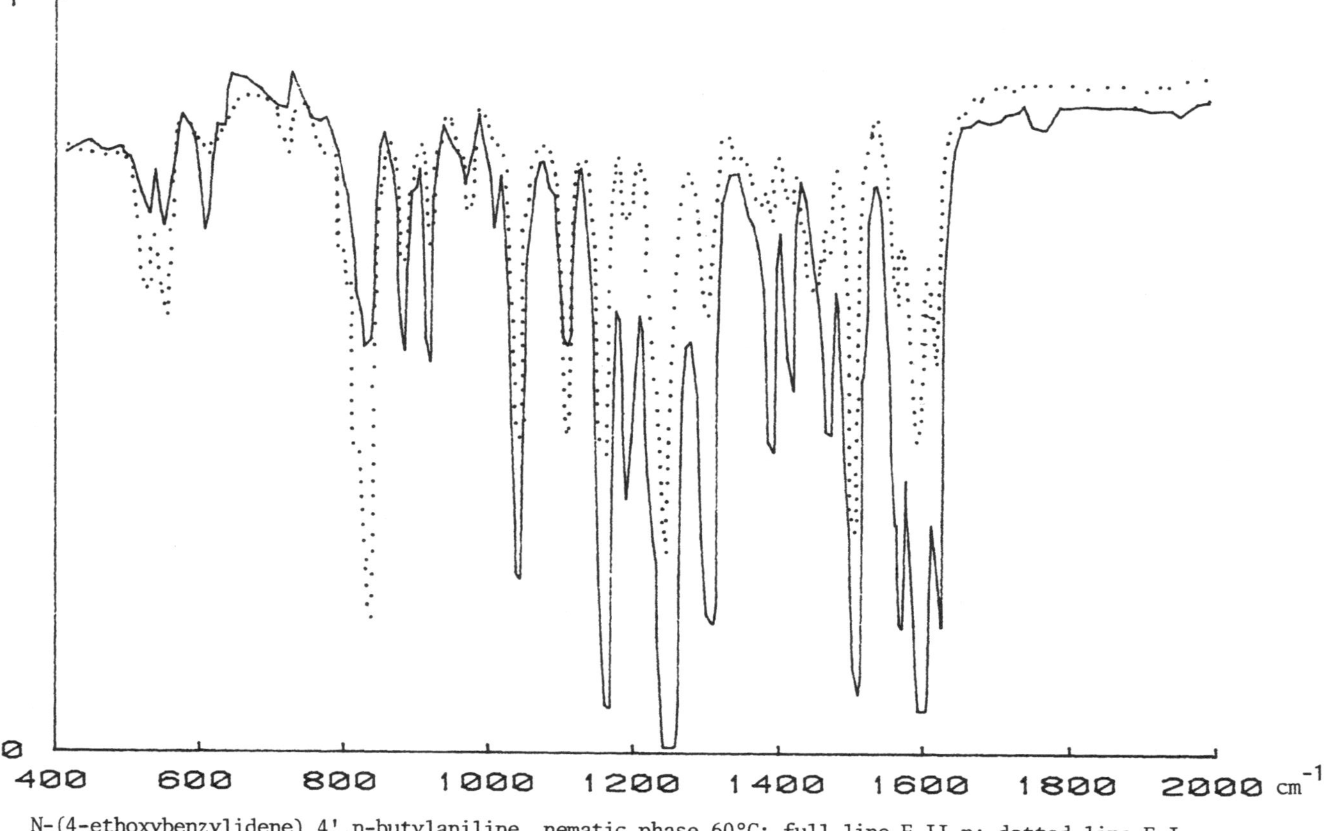

N-(4-ethoxybenzylidene) 4',n-butylaniline, nematic phase 60°C: full line E II n; dotted line E I n.

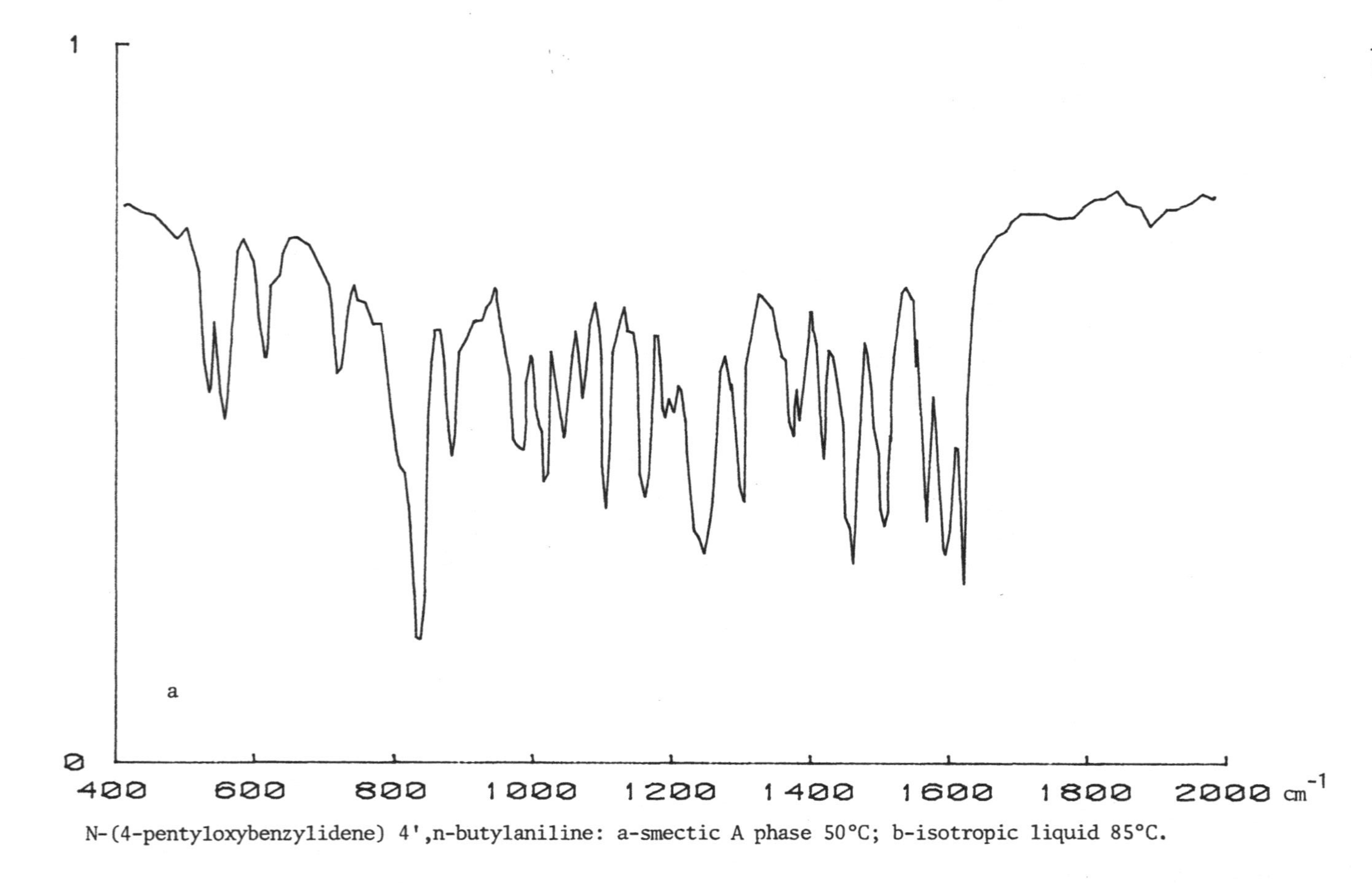

N-(4-pentyloxybenzylidene) 4',n-butylaniline: a-smectic A phase 50°C; b-isotropic liquid 85°C.

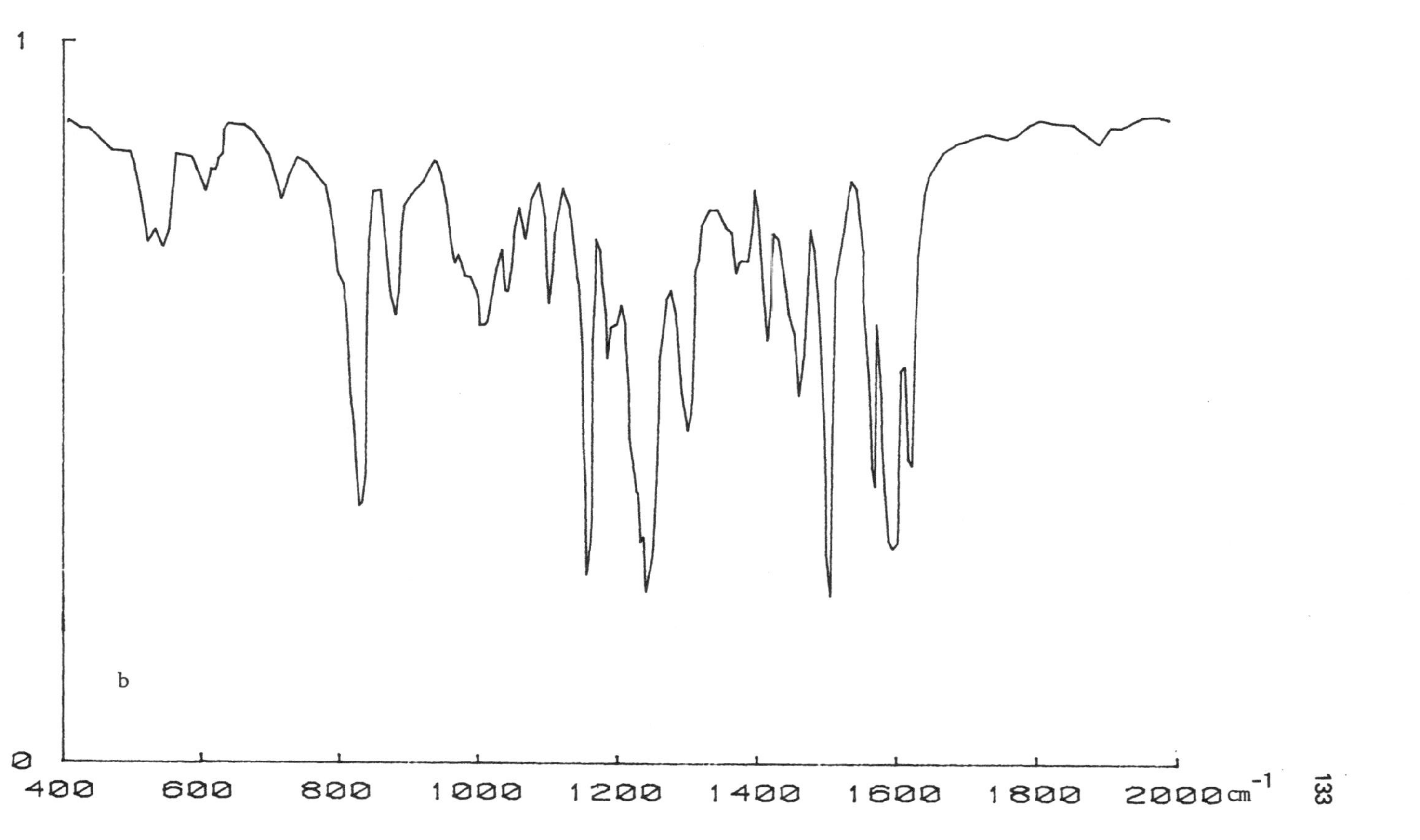
1
0
b
400
600
800
1000
1200
1400
1600
1800
2000cm-1

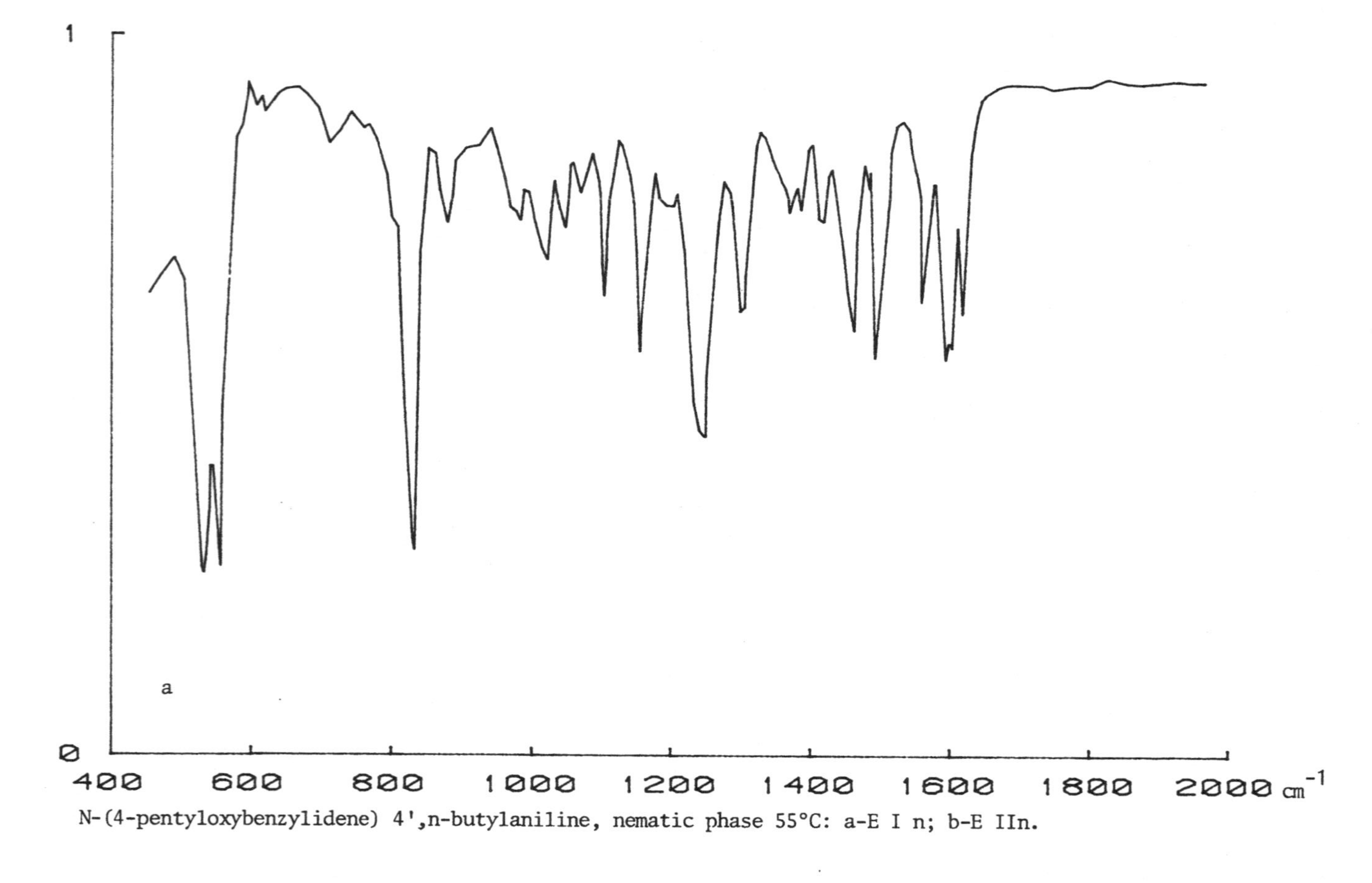

N-(4-pentyloxybenzylidene) 4',n-butylaniline, nematic phase 55°C: a-E I n; b-E IIn.

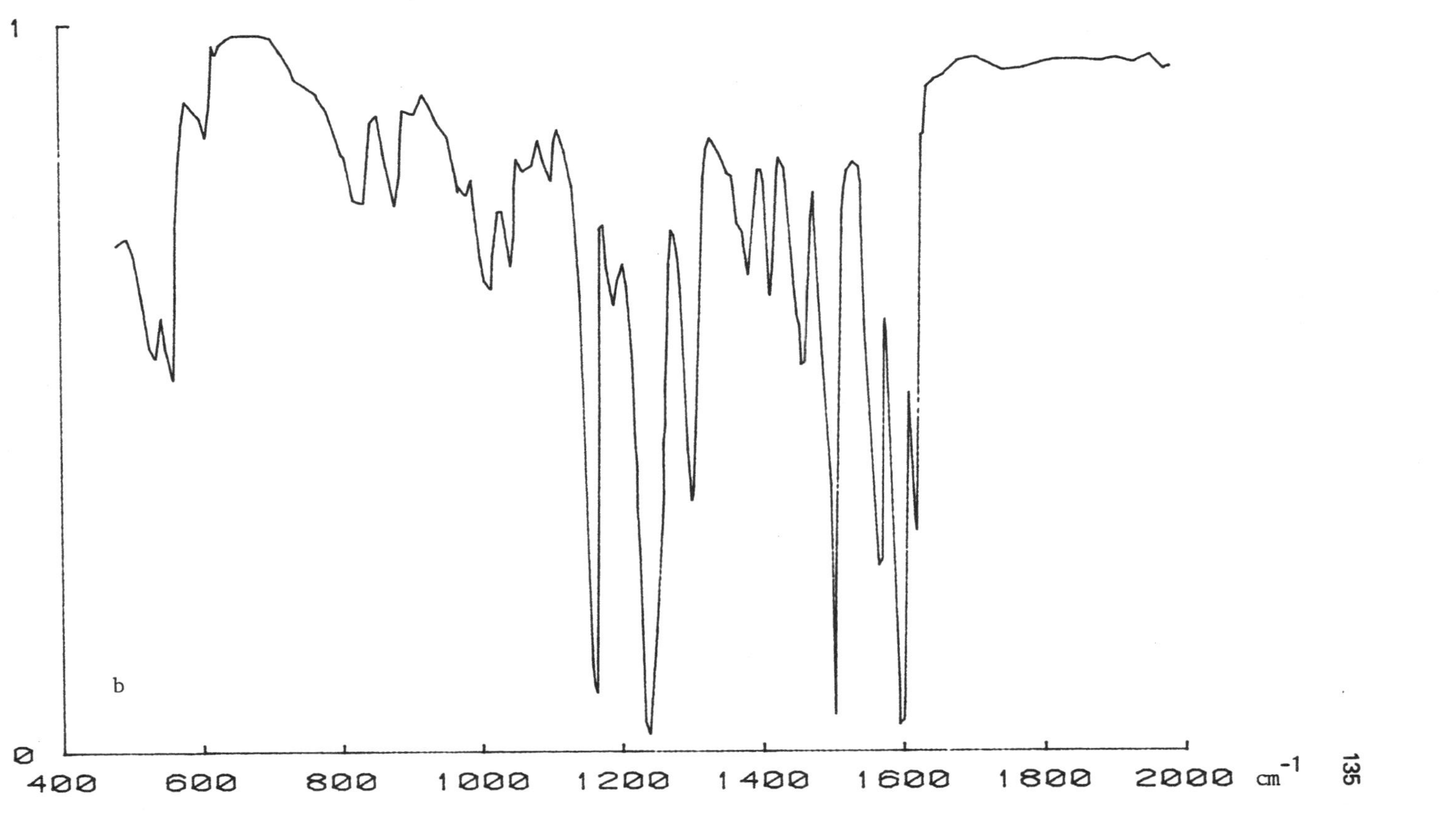

1
0
b
400
600
800
1000
1200
1400
1600
1800
2000
cm-1

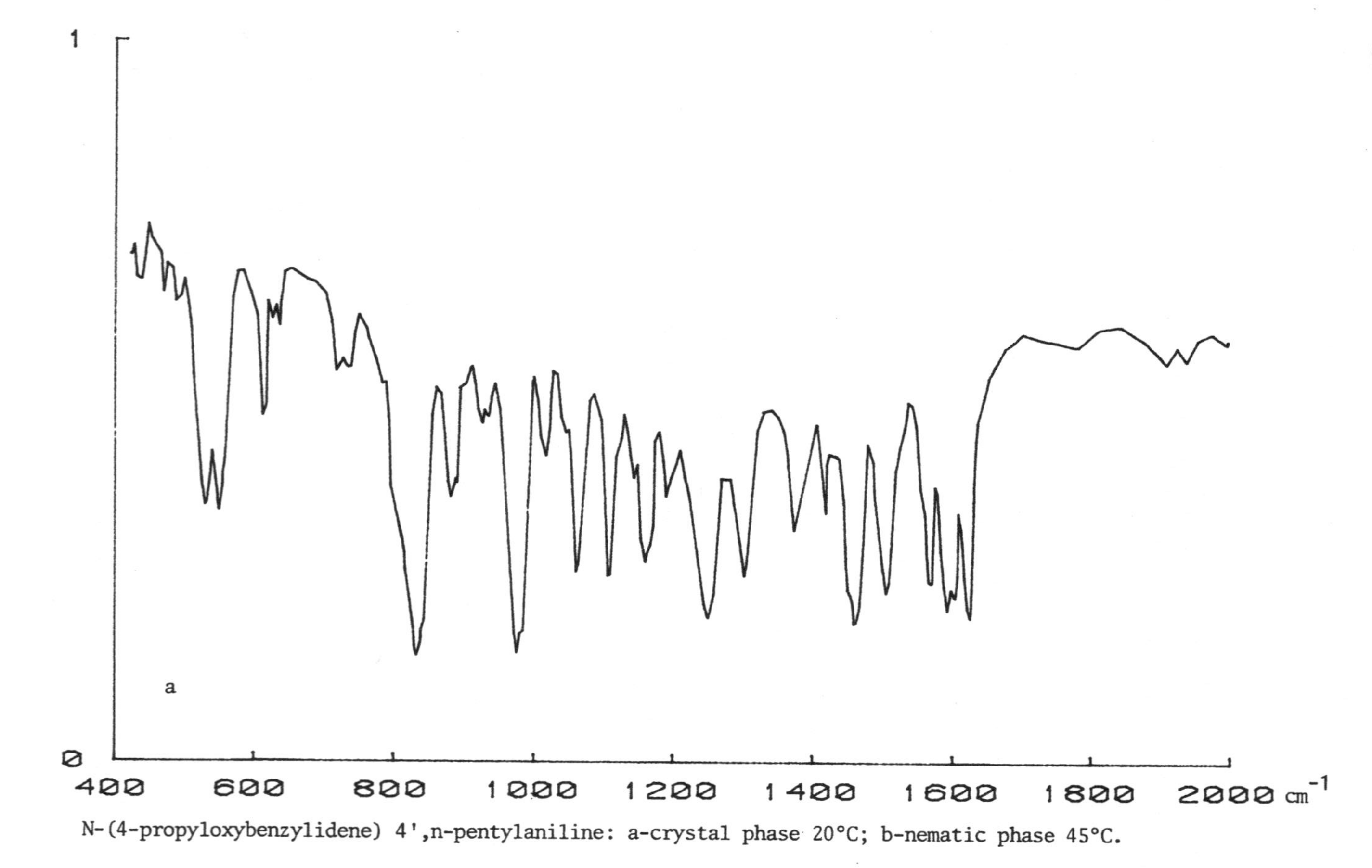

N-(4-propyloxybenzylidene) 4',n-pentylaniline: a-crystal phase 20°C; b-nematic phase 45°C.

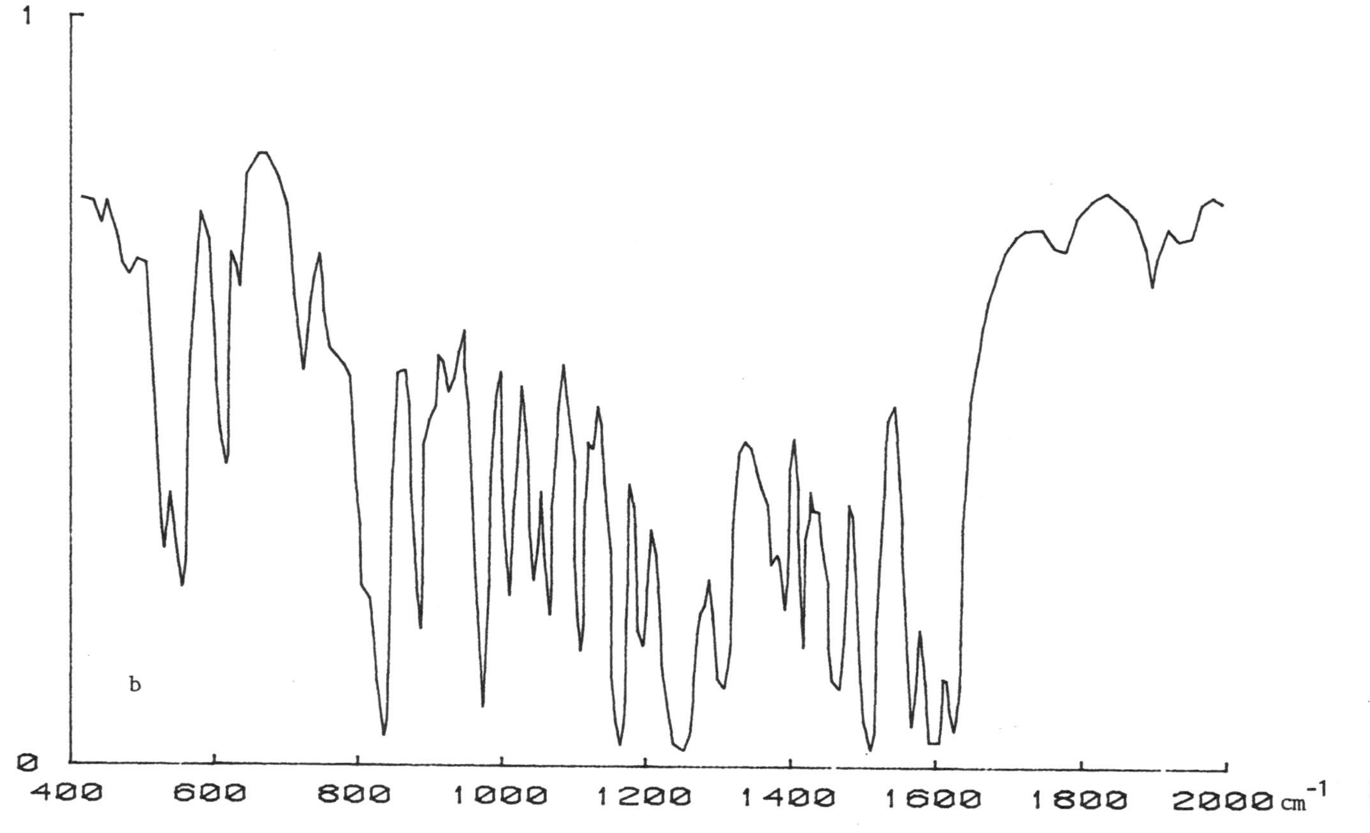
1
0
b
400
600
800
1000
1200
1400
1600
1800
2000 cm-1

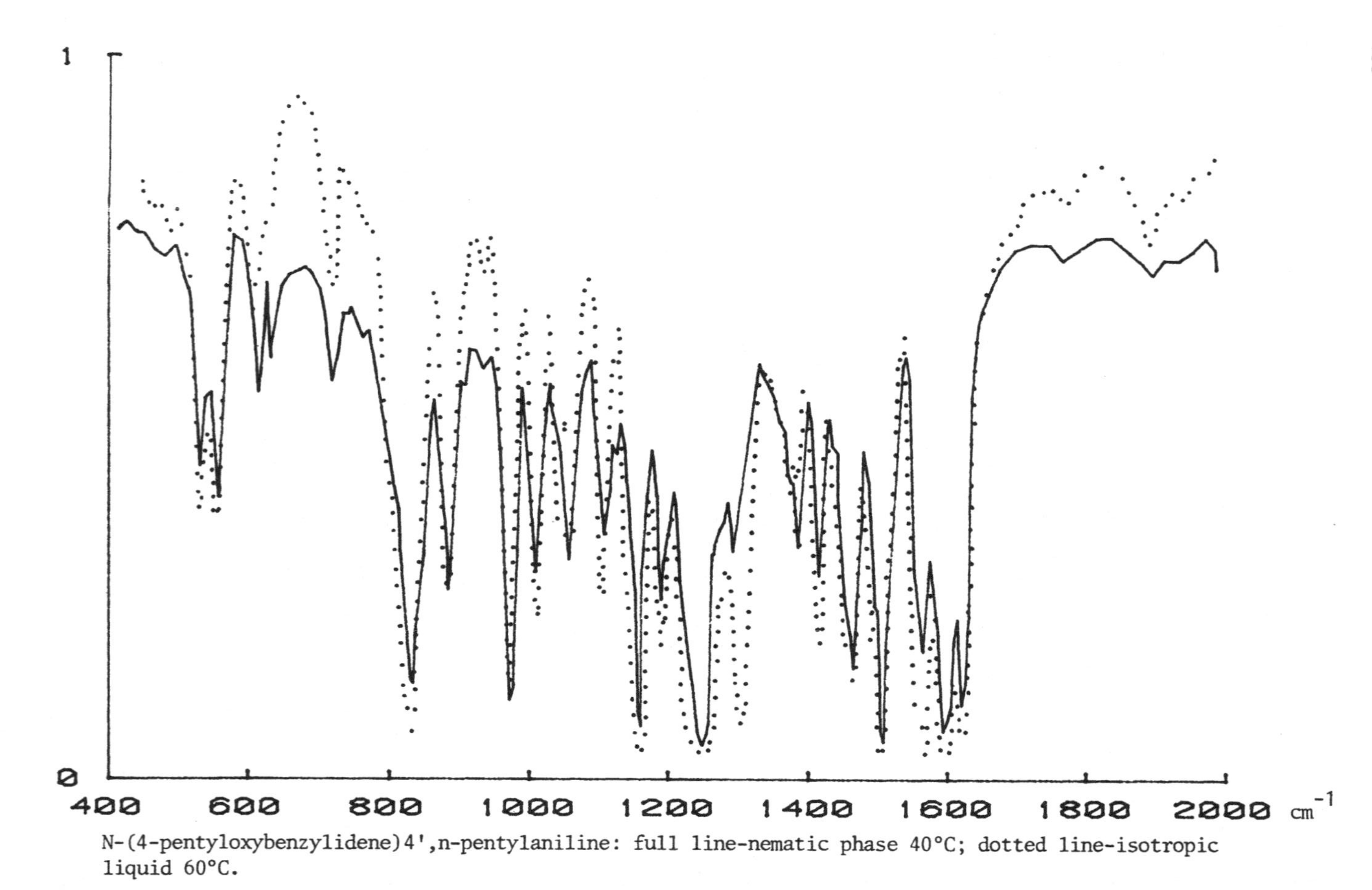

N-(4-pentyloxybenzylidene)4',n-pentylaniline: full line-nematic phase 40°C; dotted line-isotropic liquid 60°C.

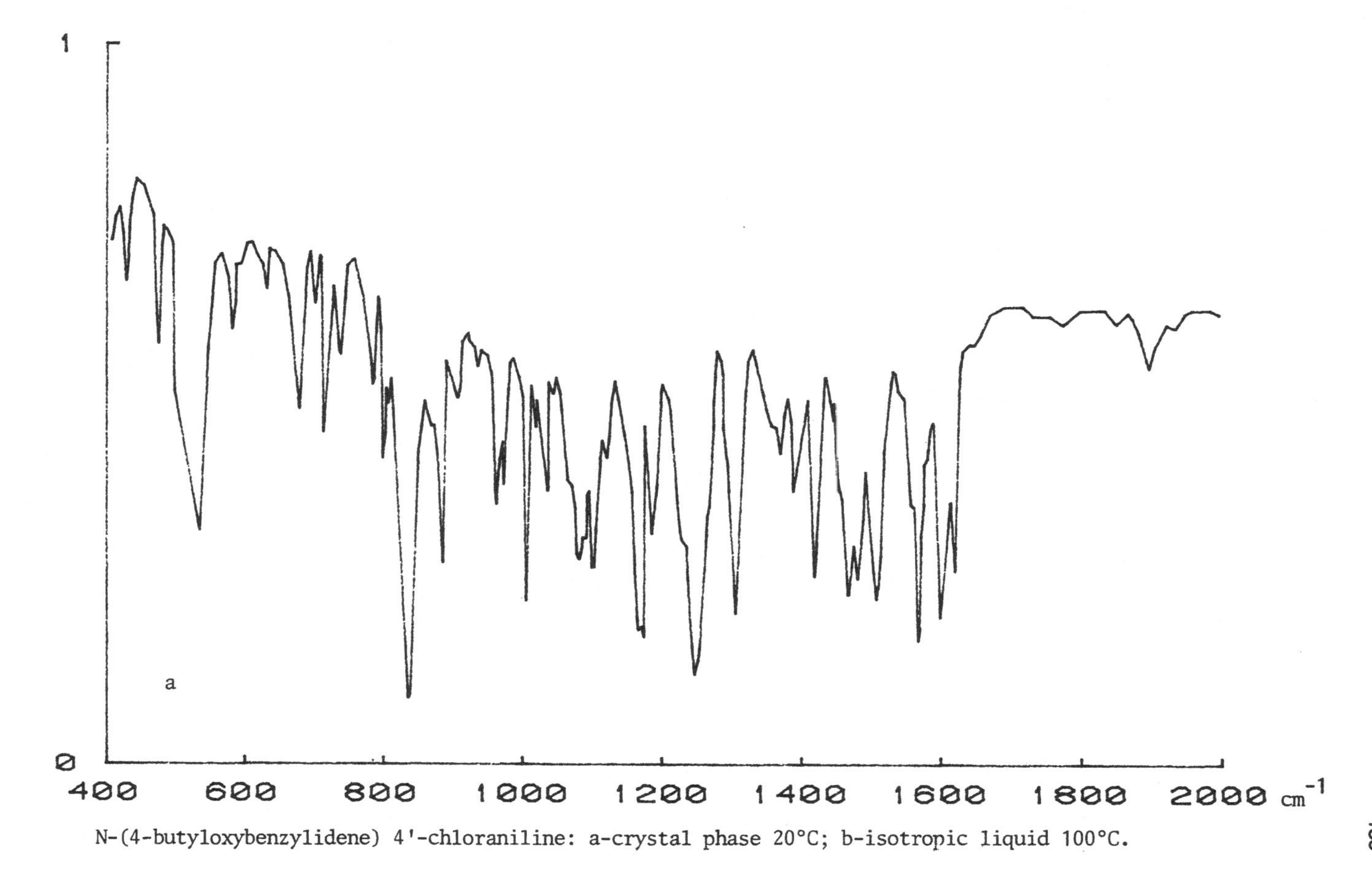

N-(4-butyloxybenzylidene) 4'-chloraniline: a-crystal phase 20°C; b-isotropic liquid 100°C.

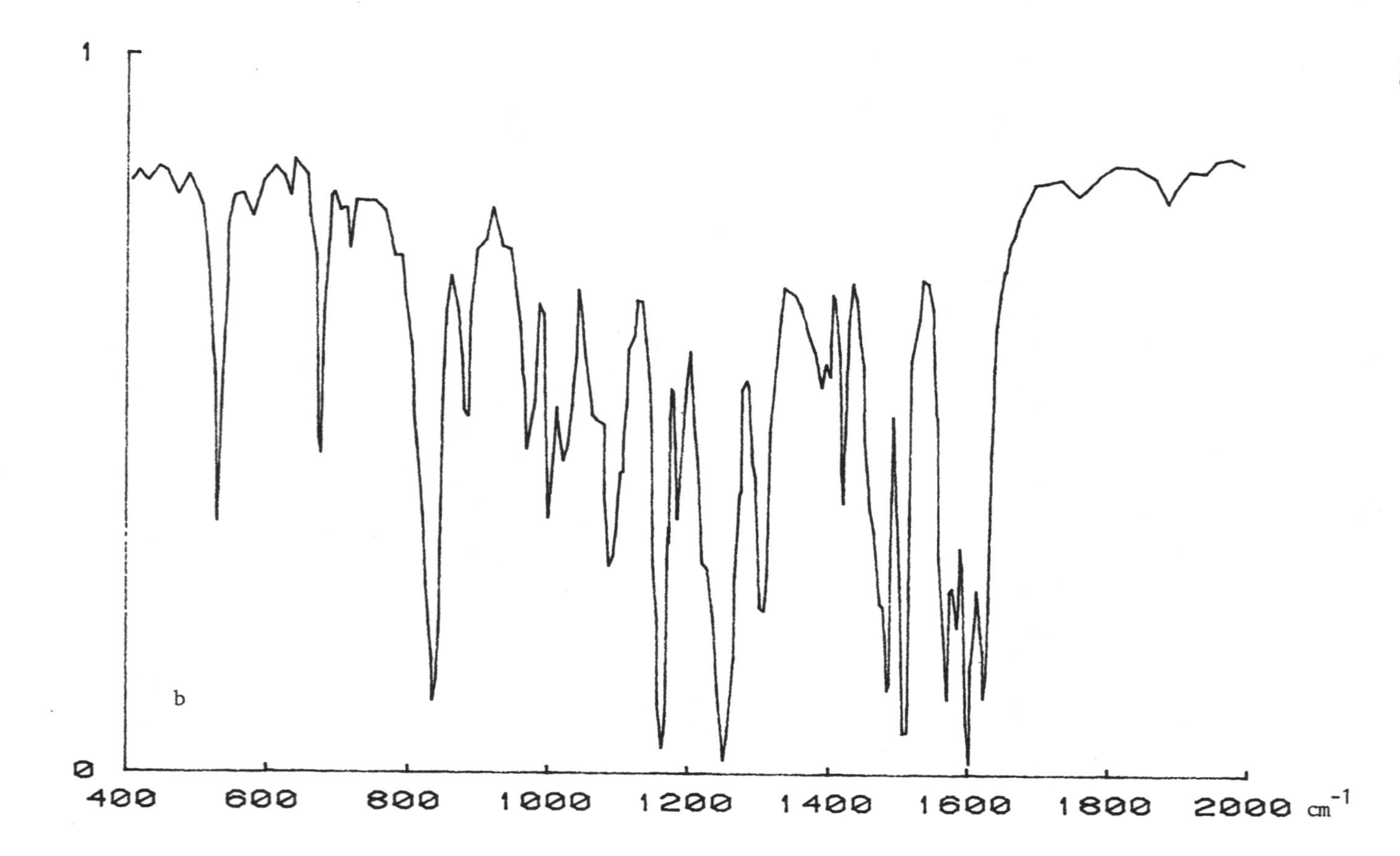

1
0
b
400
600
800
1000
1200
1400
1600
1800
2000
cm-1

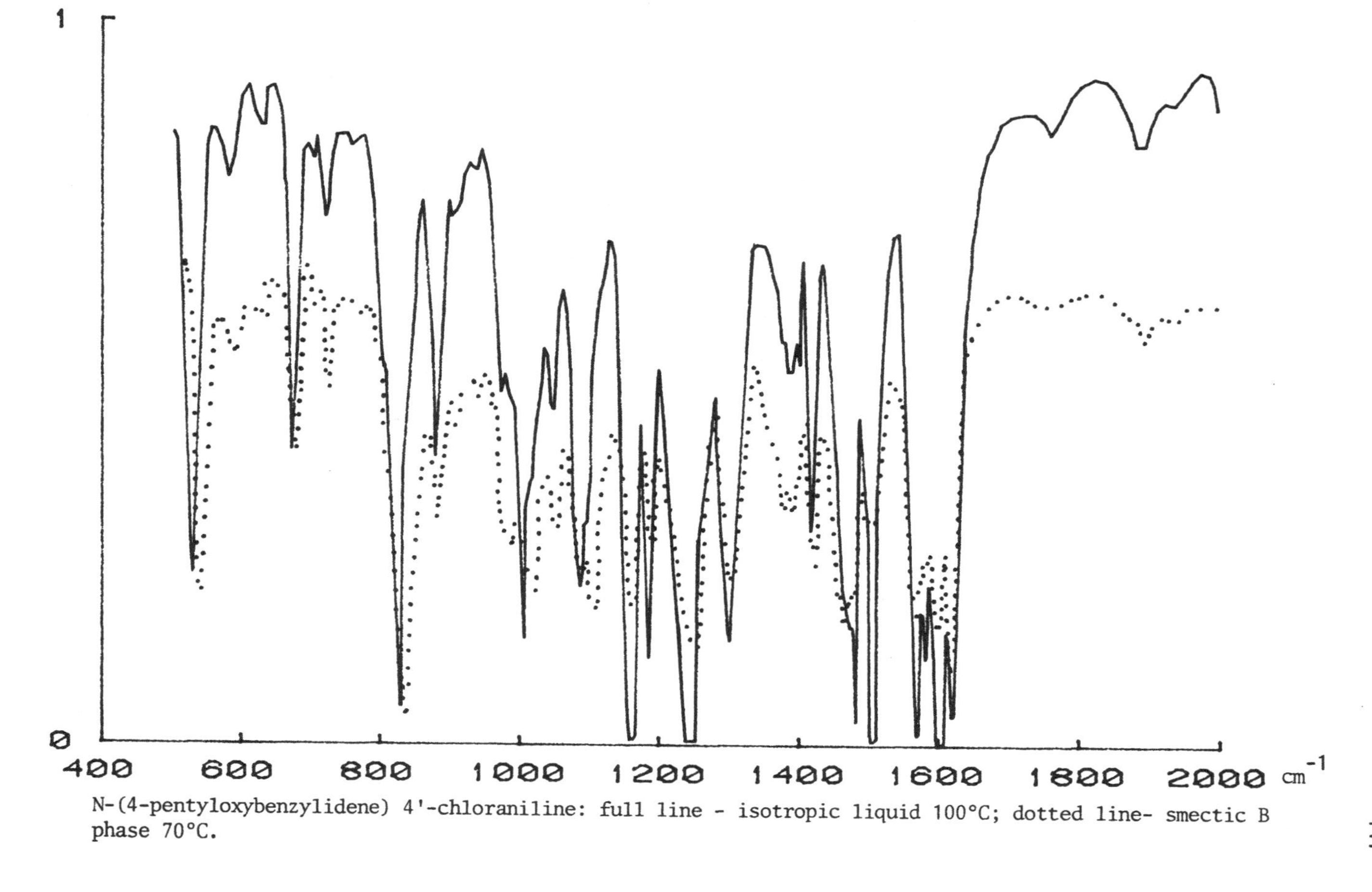

N-(4-pentyloxybenzylidene) 4'-chloraniline: full line - isotropic liquid 100°C; dotted line- smectic B phase 70°C.

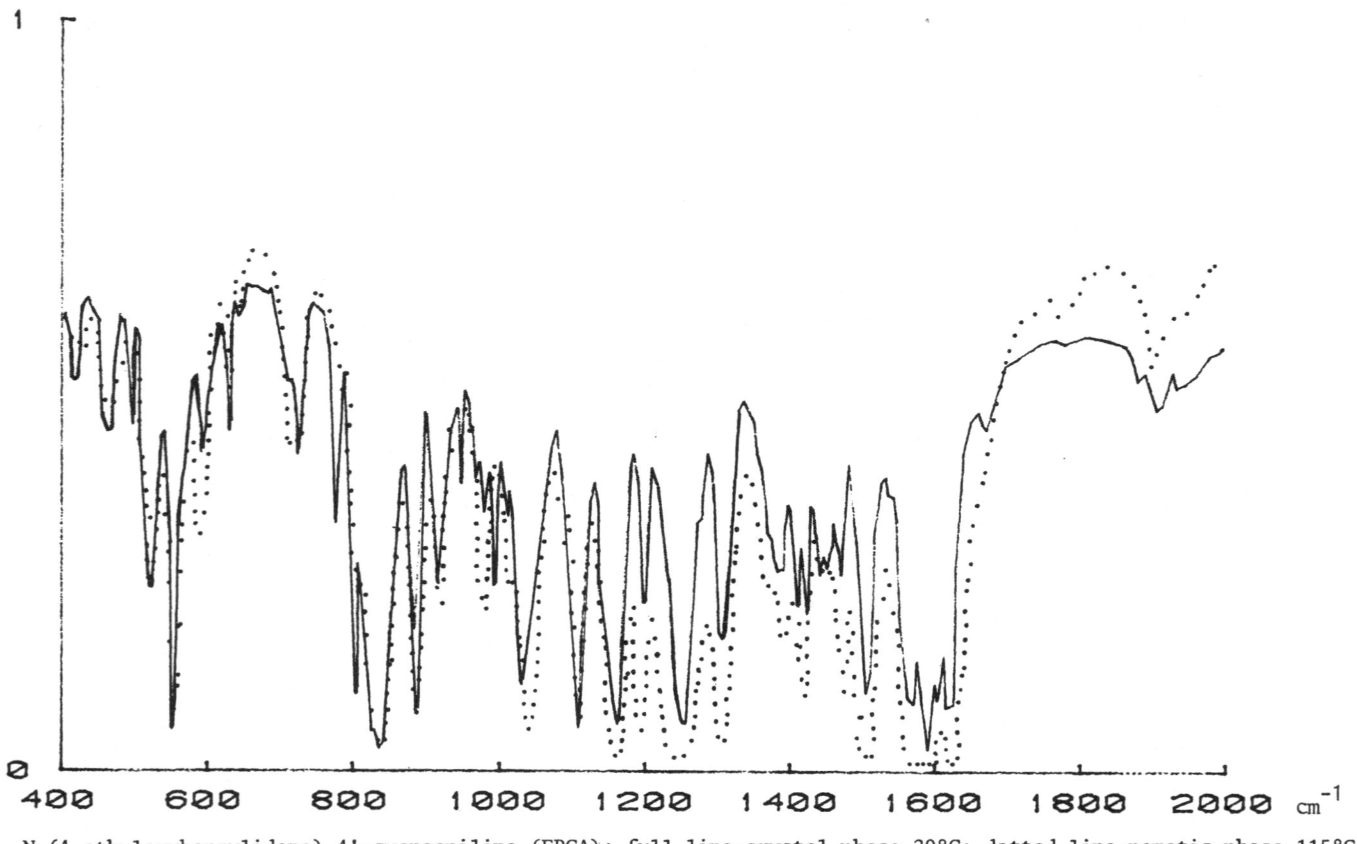

N-(4-ethyloxybenzylidene) 4'-cyanoaniline (EBCA): full line-crystal phase 20°C; dotted line-nematic phase 115°C.

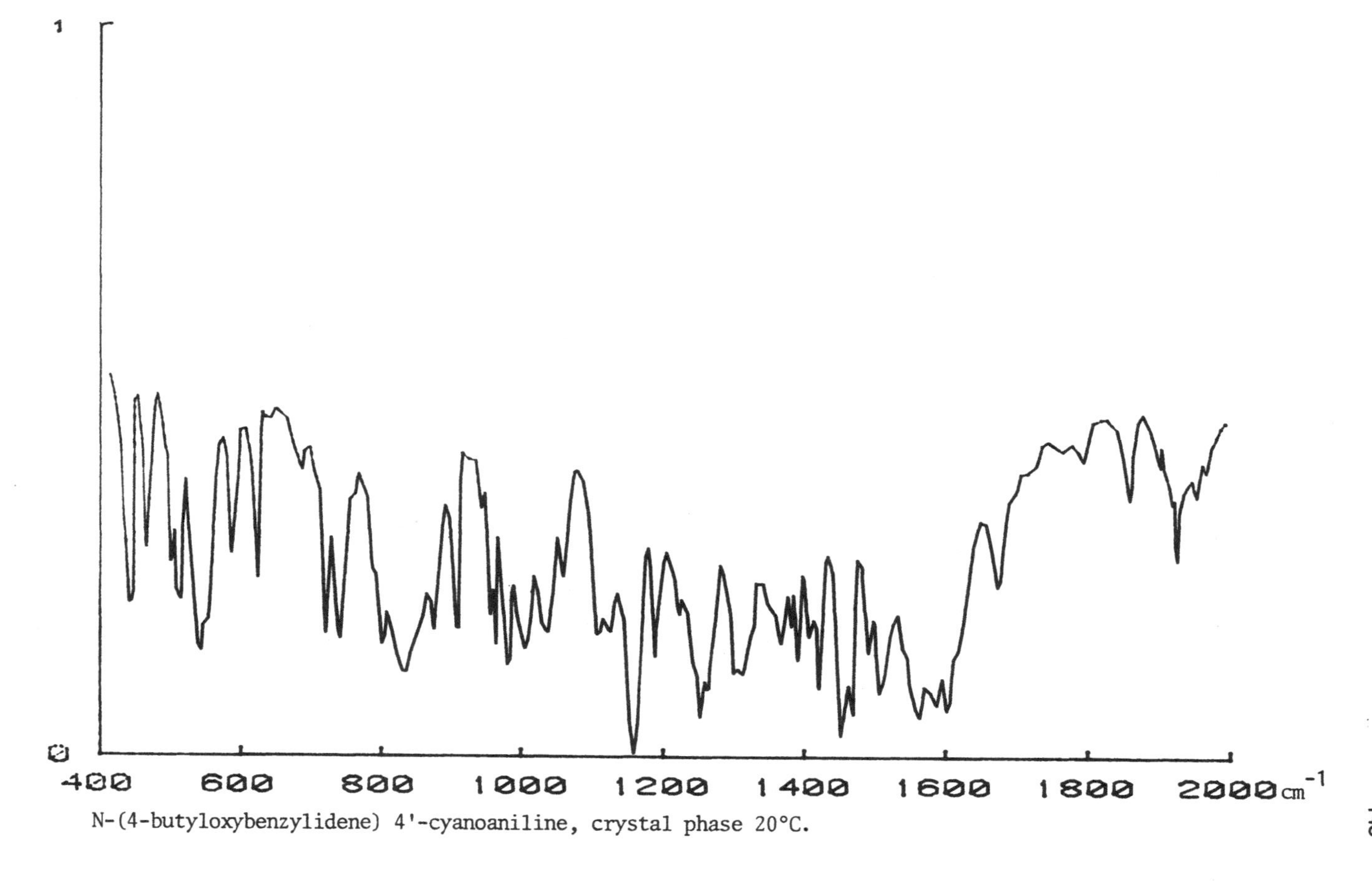

N-(4-butyloxybenzylidene) 4'-cyanoaniline, crystal phase 20°C.

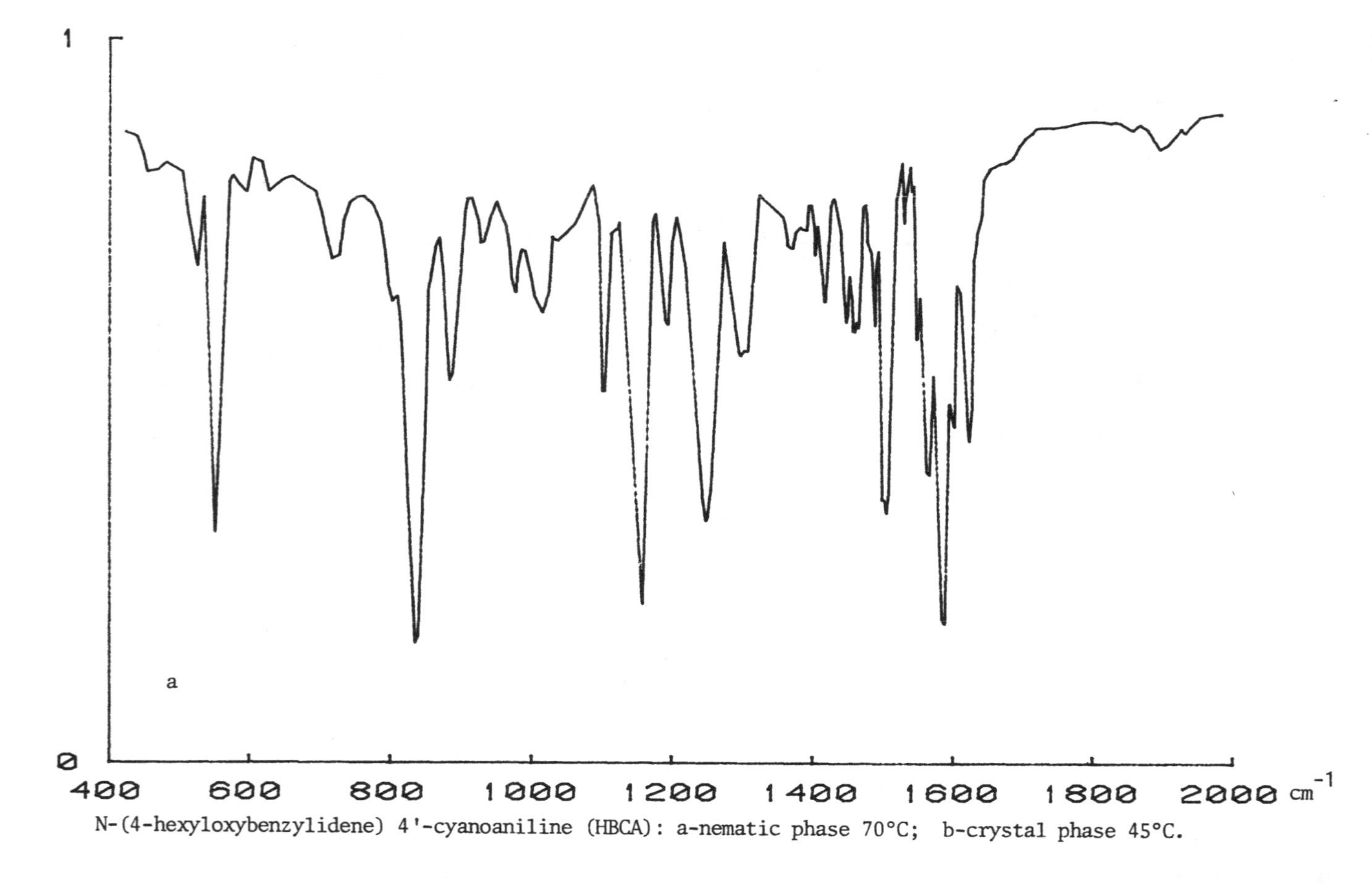

N-(4-hexyloxybenzylidene) 4'-cyanoaniline (HBCA): a-nematic phase 70°C; b-crystal phase 45°C.

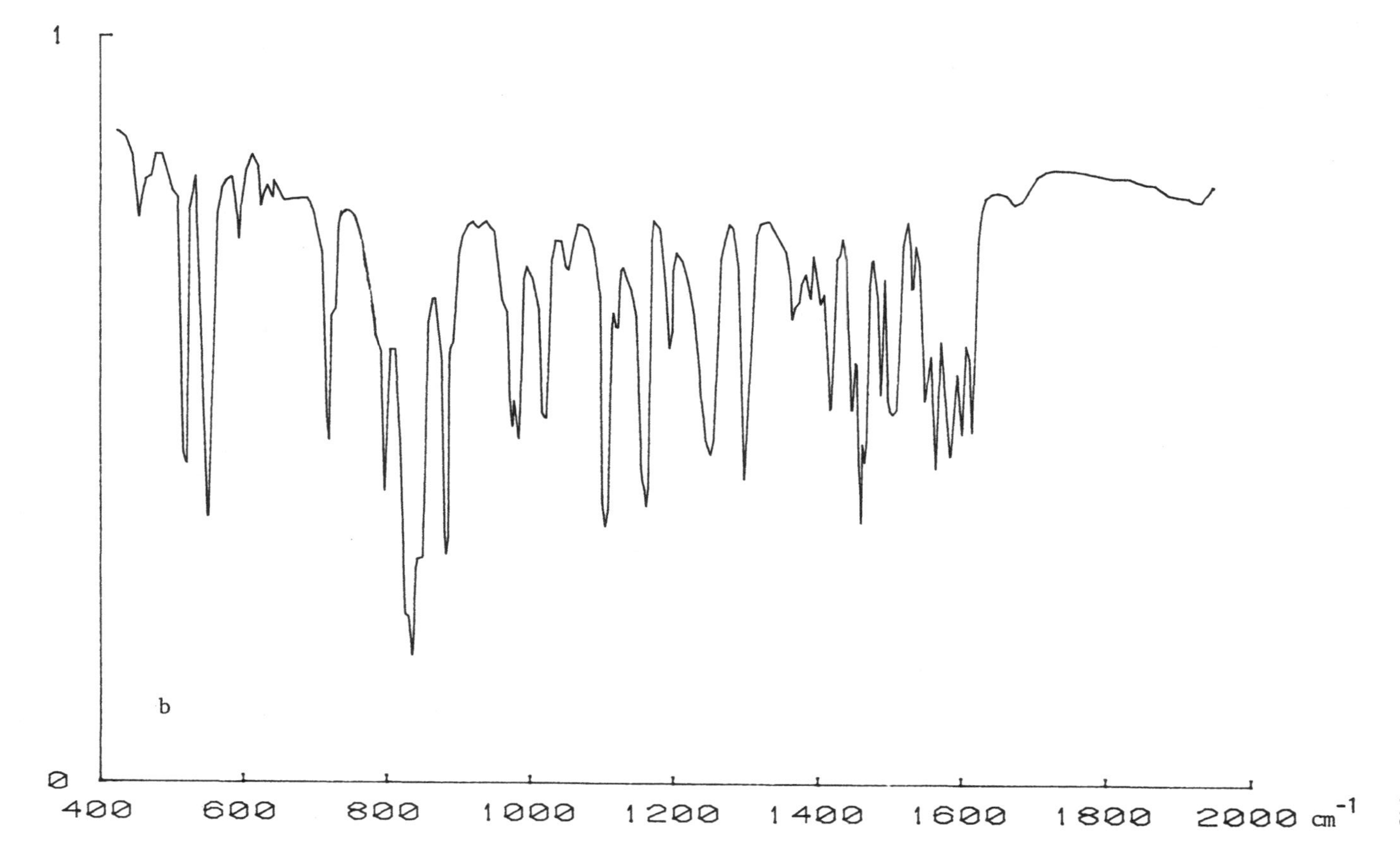

1
0
b
400
600
800
1000
1200
1400
1600
1800
2000 cm^{-1}

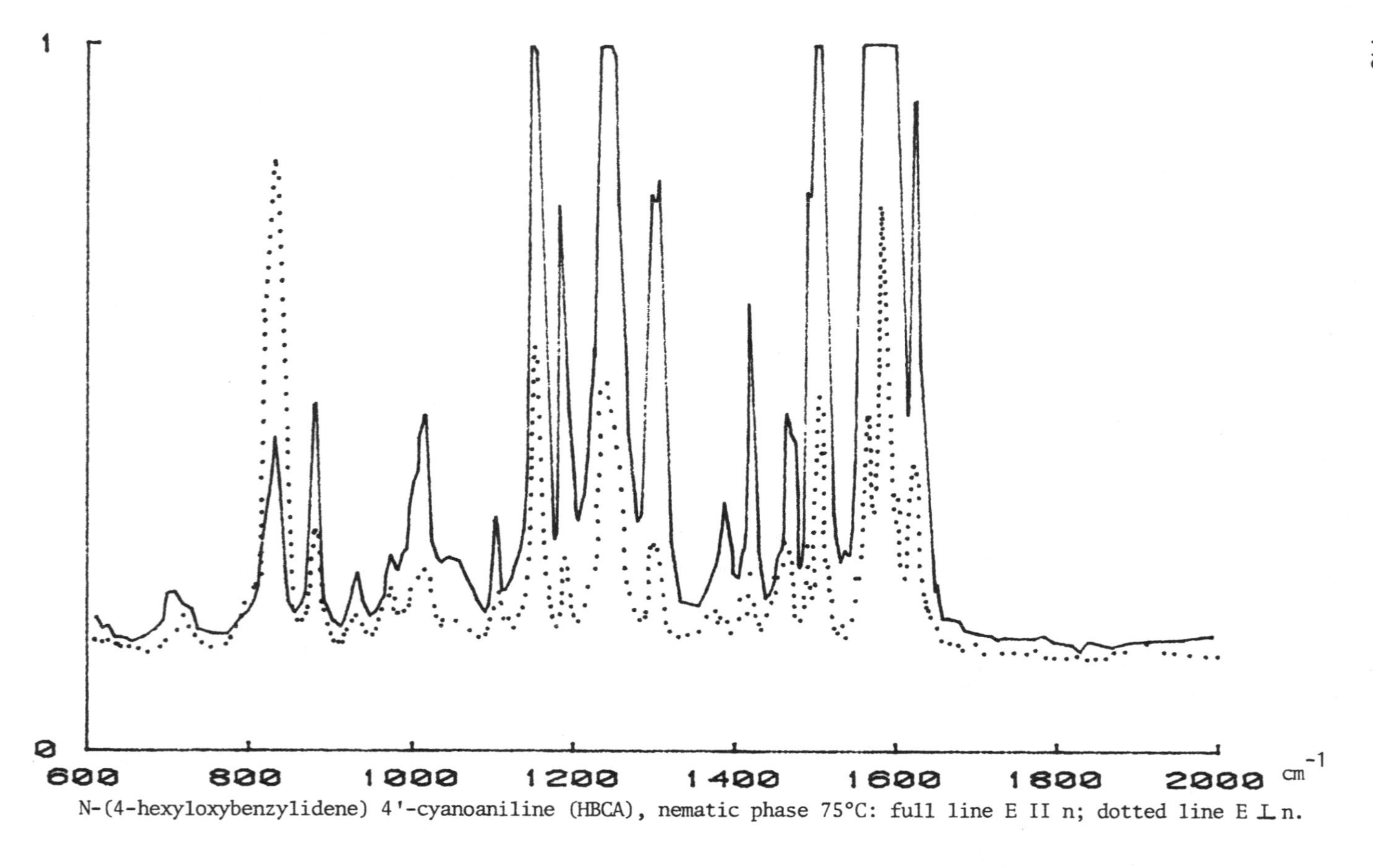

N-(4-hexyloxybenzylidene) 4'-cyanoaniline (HBCA), nematic phase 75°C: full line E II n; dotted line E ⊥ n.

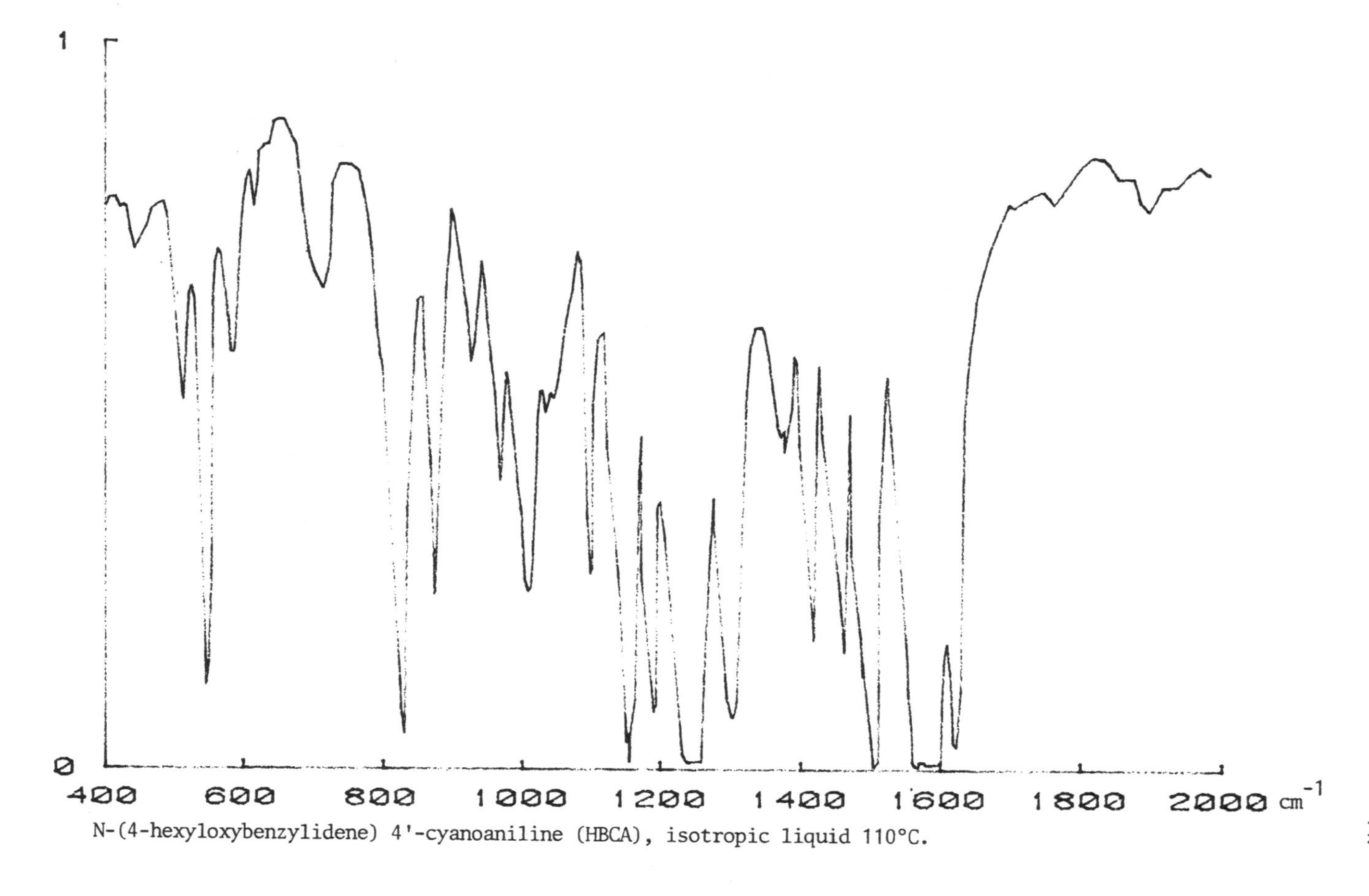

N-(4-hexyloxybenzylidene) 4'-cyanoaniline (HBCA), isotropic liquid 110°C.

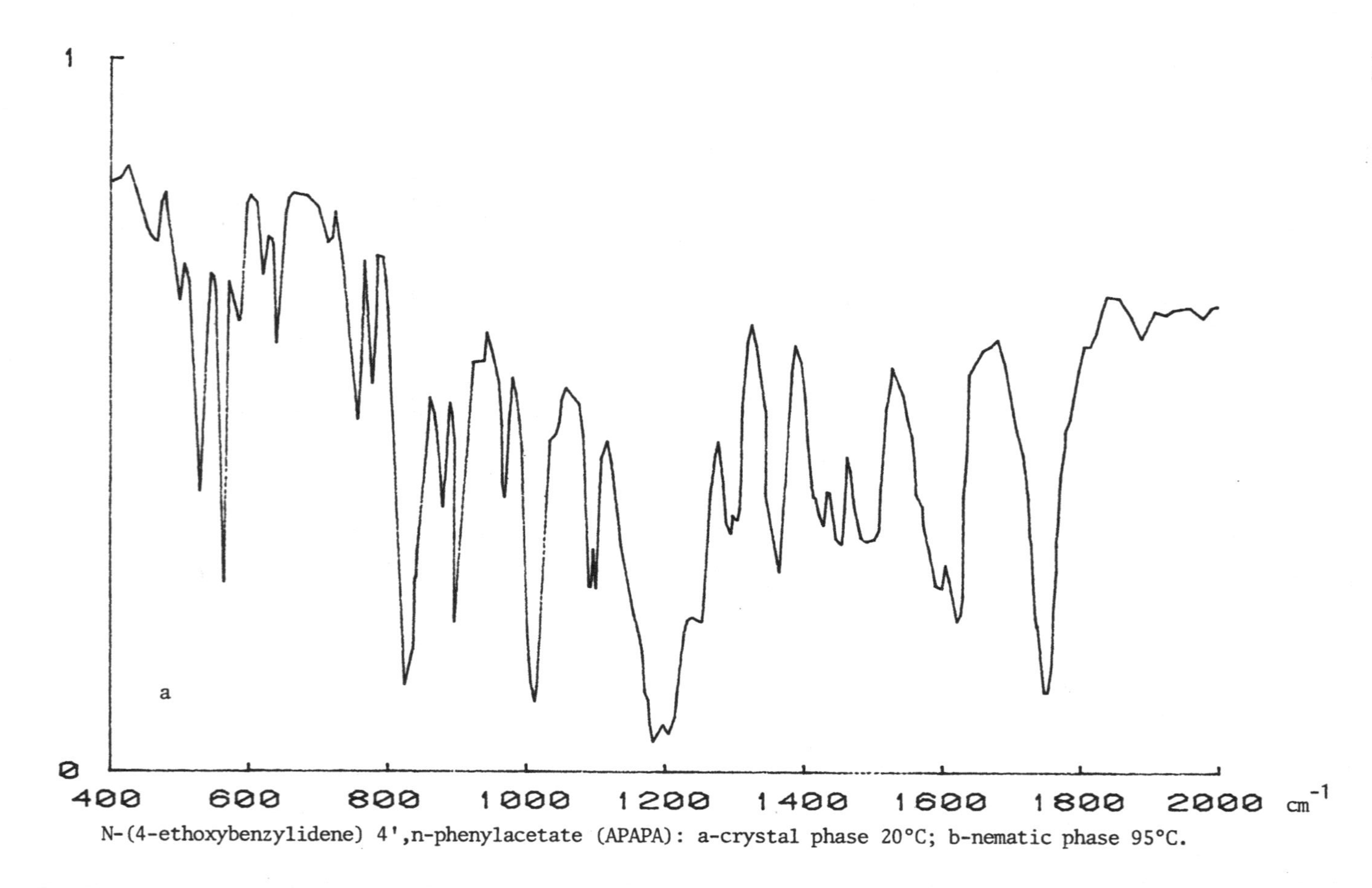

N-(4-ethoxybenzylidene) 4',n-phenylacetate (APAPA): a-crystal phase 20°C; b-nematic phase 95°C.

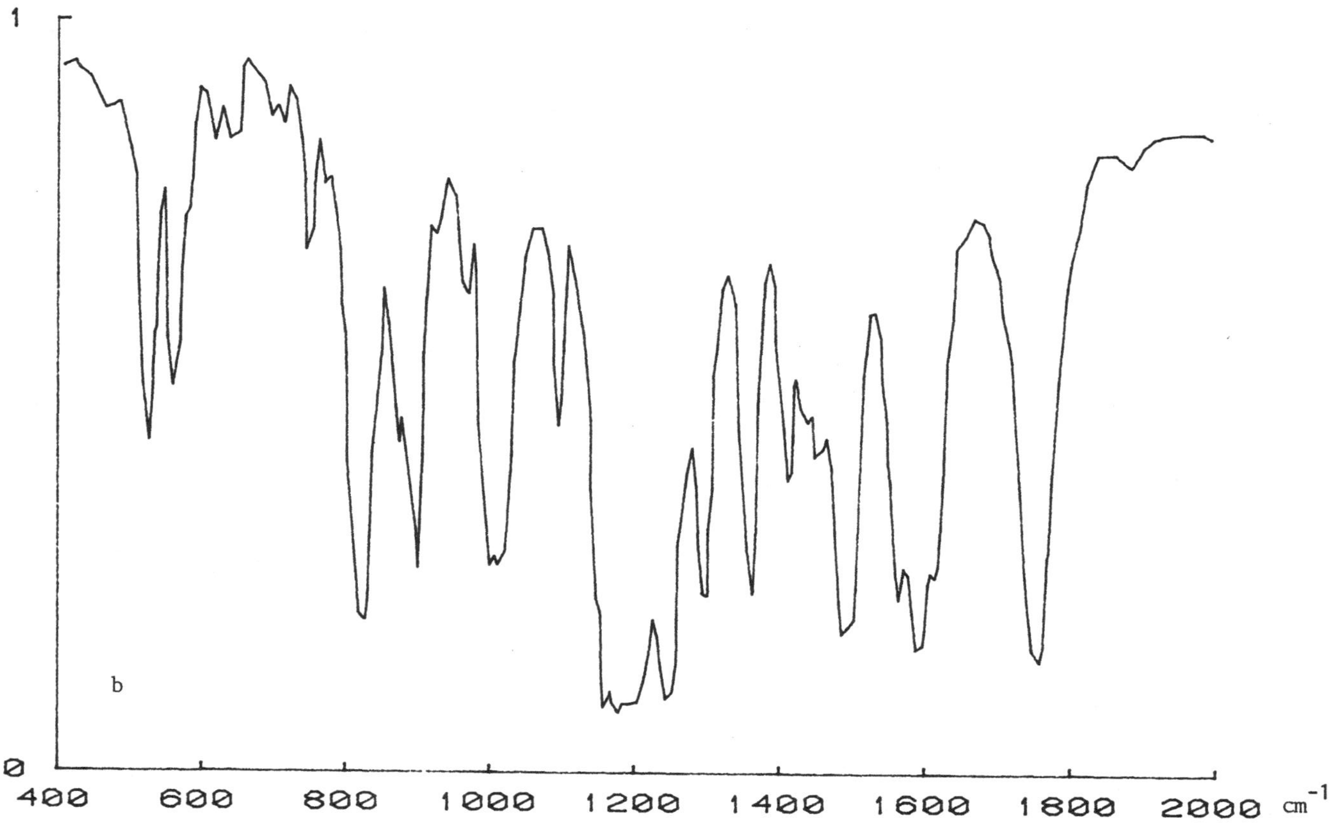

1

0

b

400 600 800 1000 1200 1400 1600 1800 2000 cm^{-1}

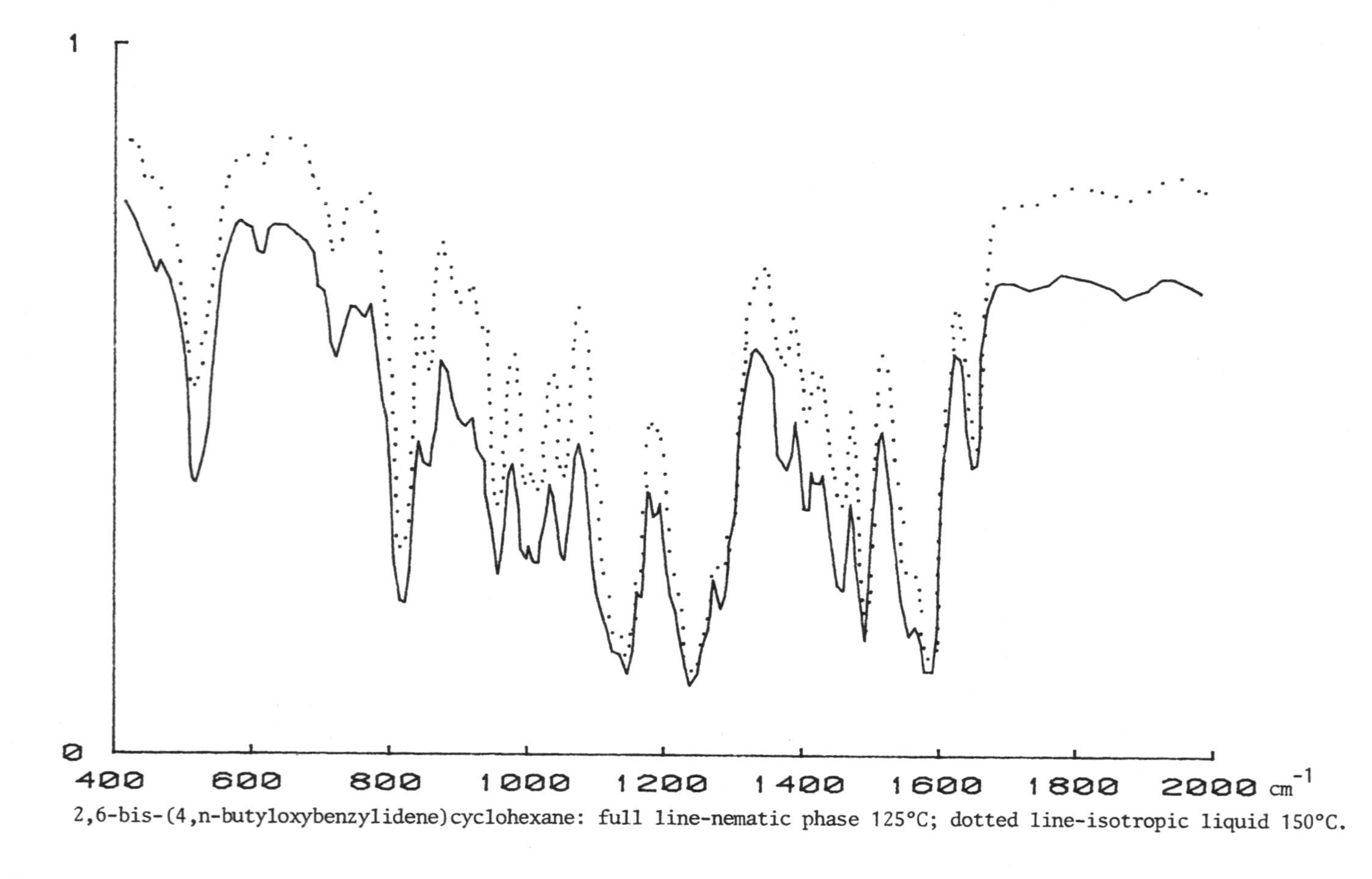

2,6-bis-(4,n-butyloxybenzylidene)cyclohexane: full line-nematic phase 125°C; dotted line-isotropic liquid 150°C.

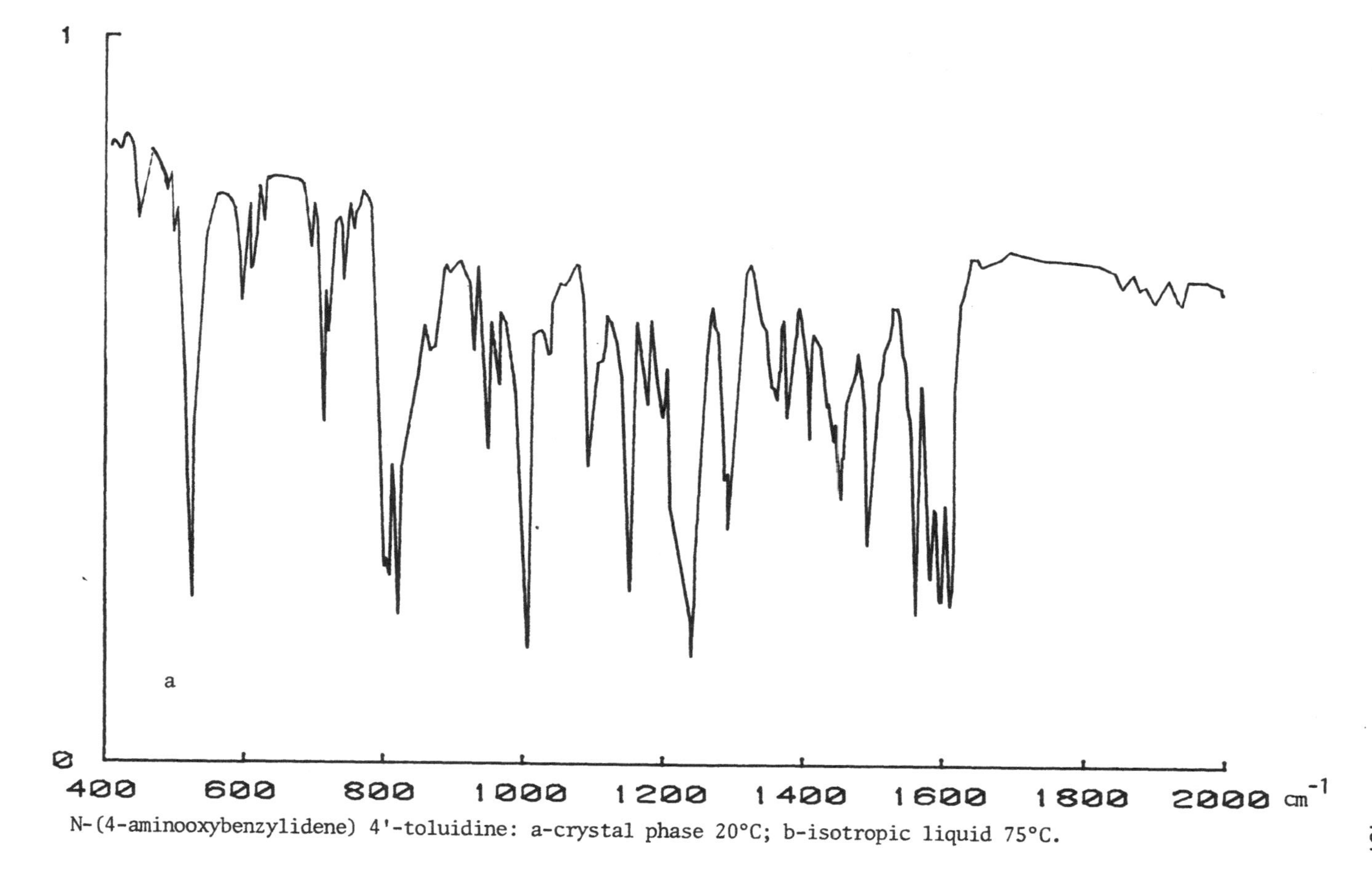

N-(4-aminooxybenzylidene) 4'-toluidine: a-crystal phase 20°C; b-isotropic liquid 75°C.

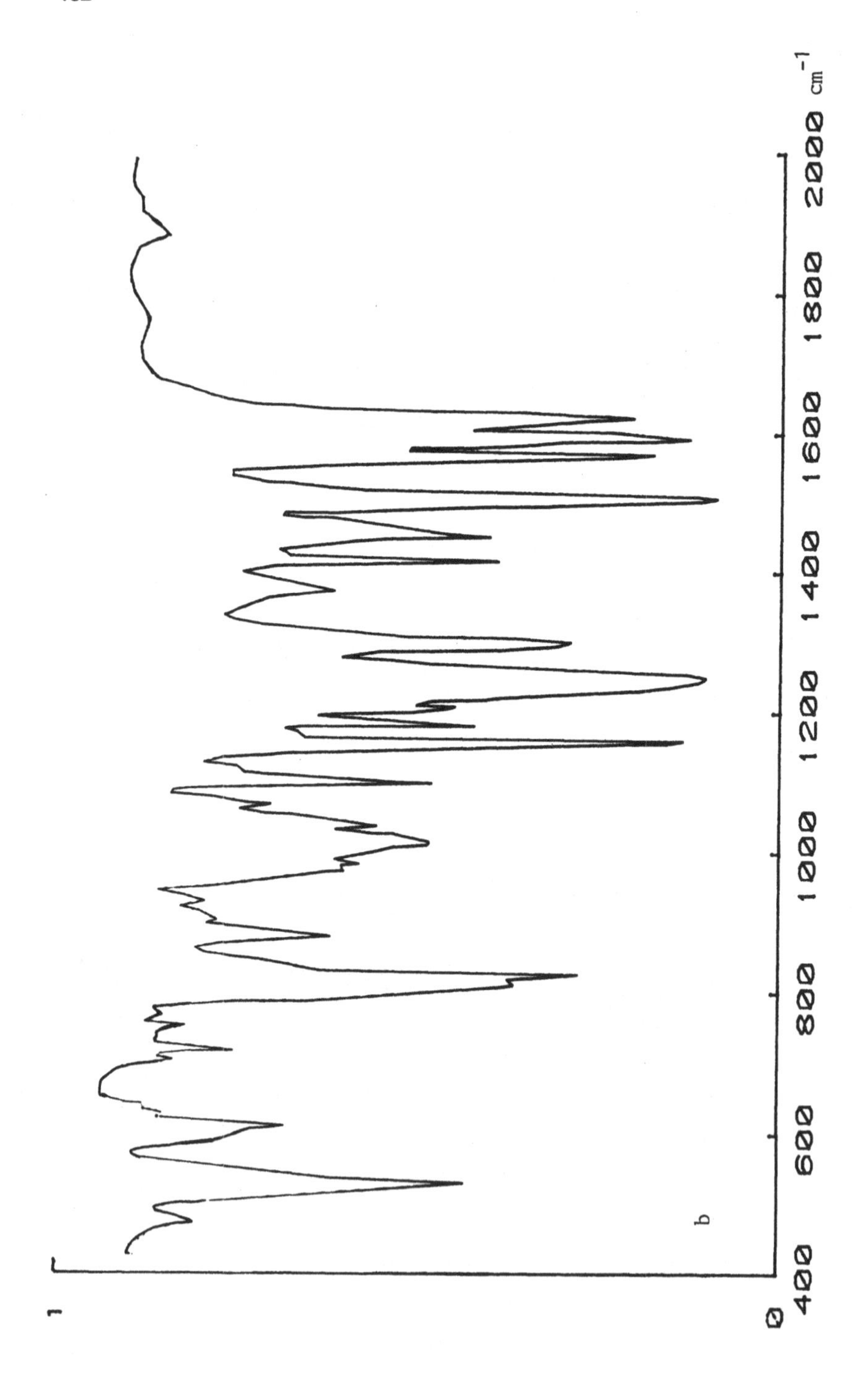
b
0
1
400
600
800
1000
1200
1400
1600
1800
2000
cm-1

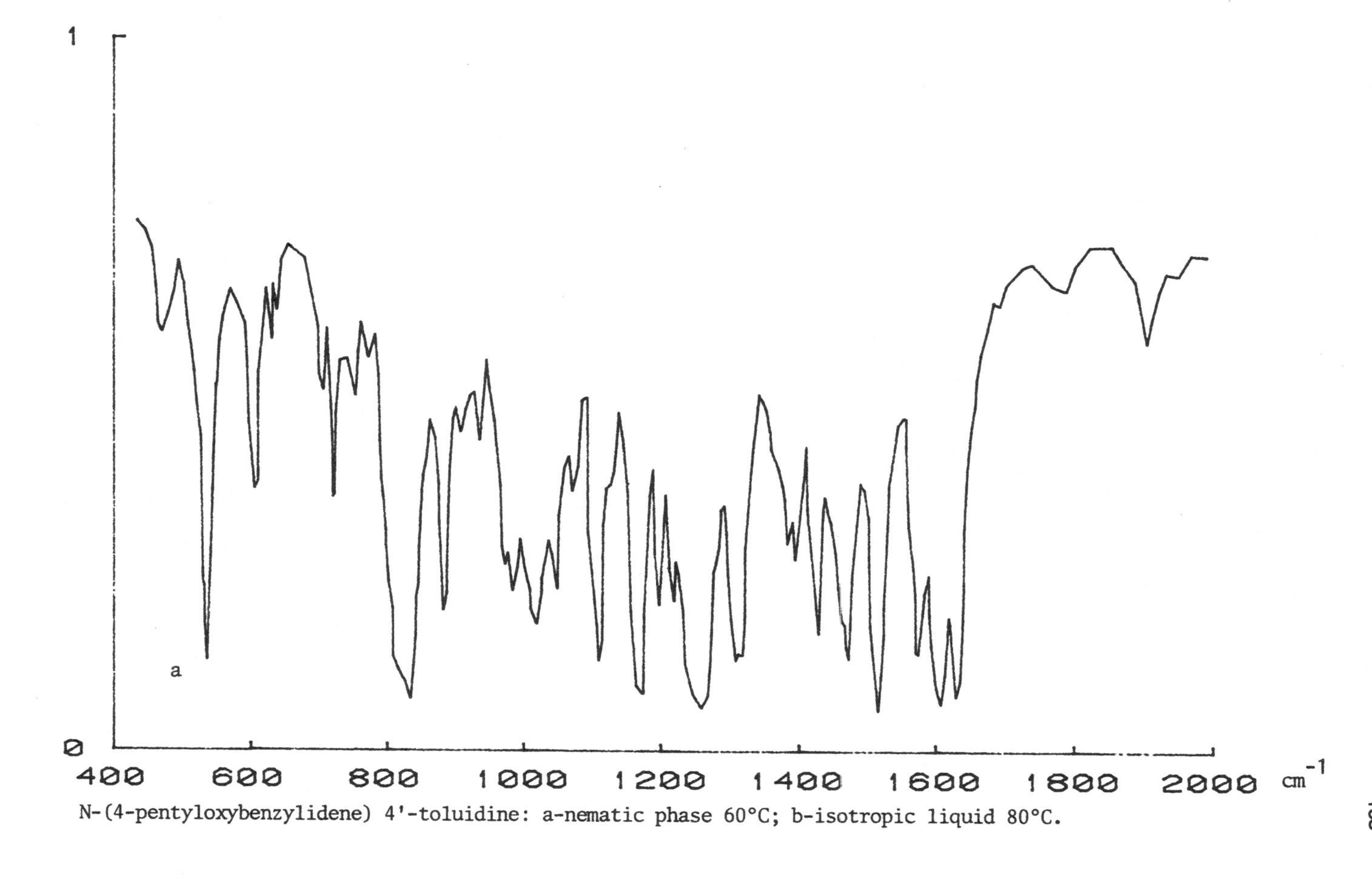

N-(4-pentyloxybenzylidene) 4'-toluidine: a-nematic phase 60°C; b-isotropic liquid 80°C.

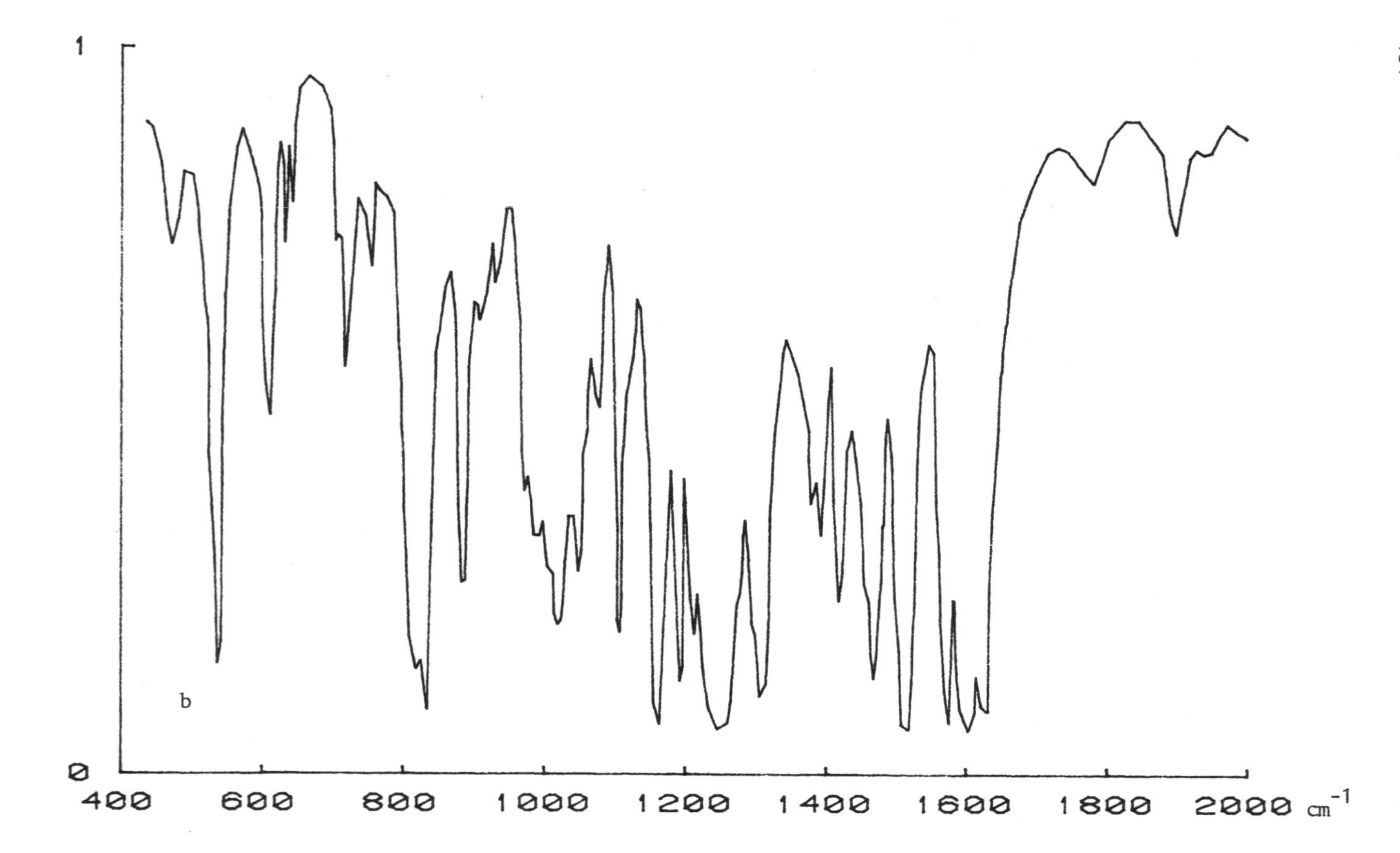

1
0
b
400
600
800
1000
1200
1400
1600
1800
2000
cm^{-1}

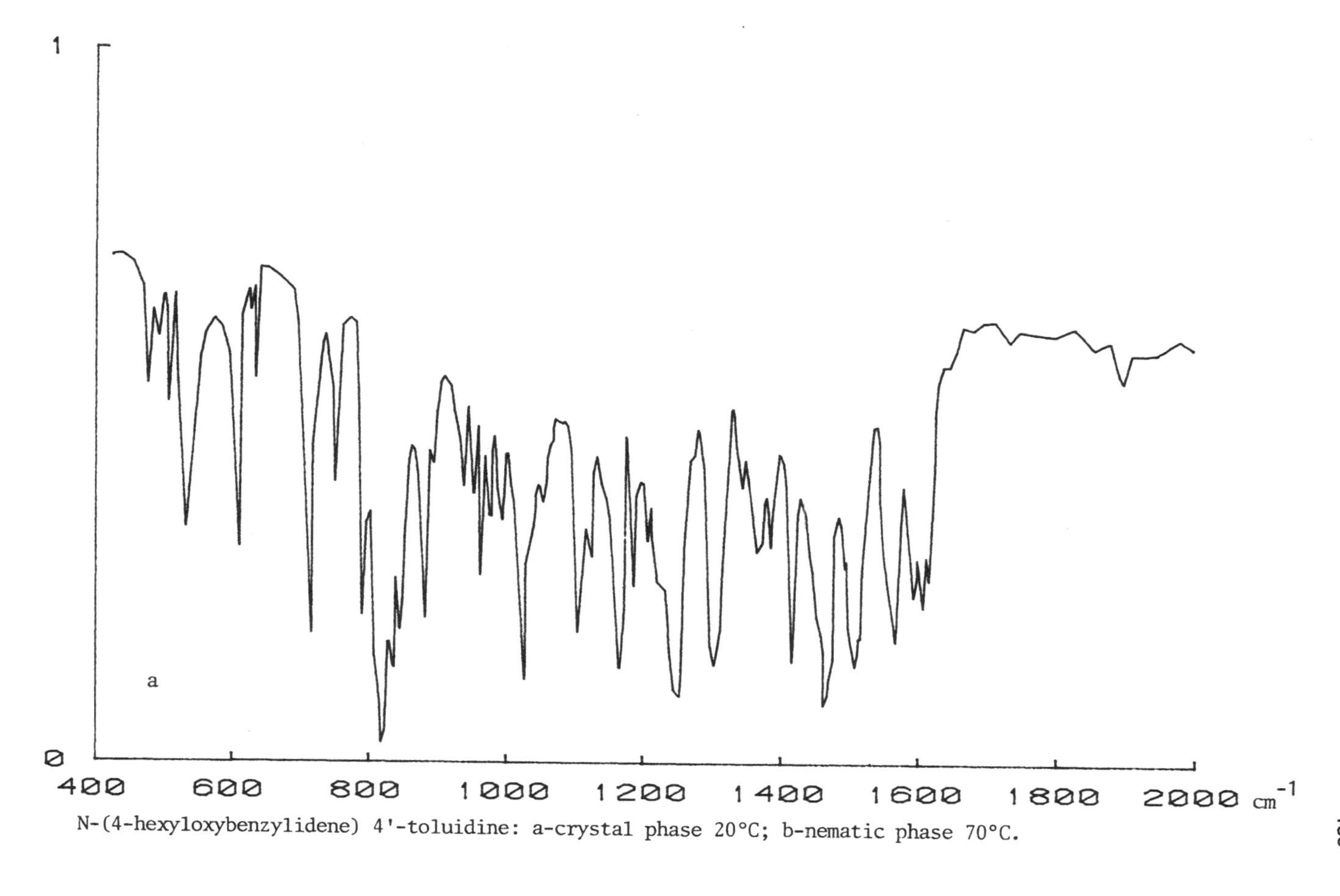

N-(4-hexyloxybenzylidene) 4'-toluidine: a-crystal phase 20°C; b-nematic phase 70°C.

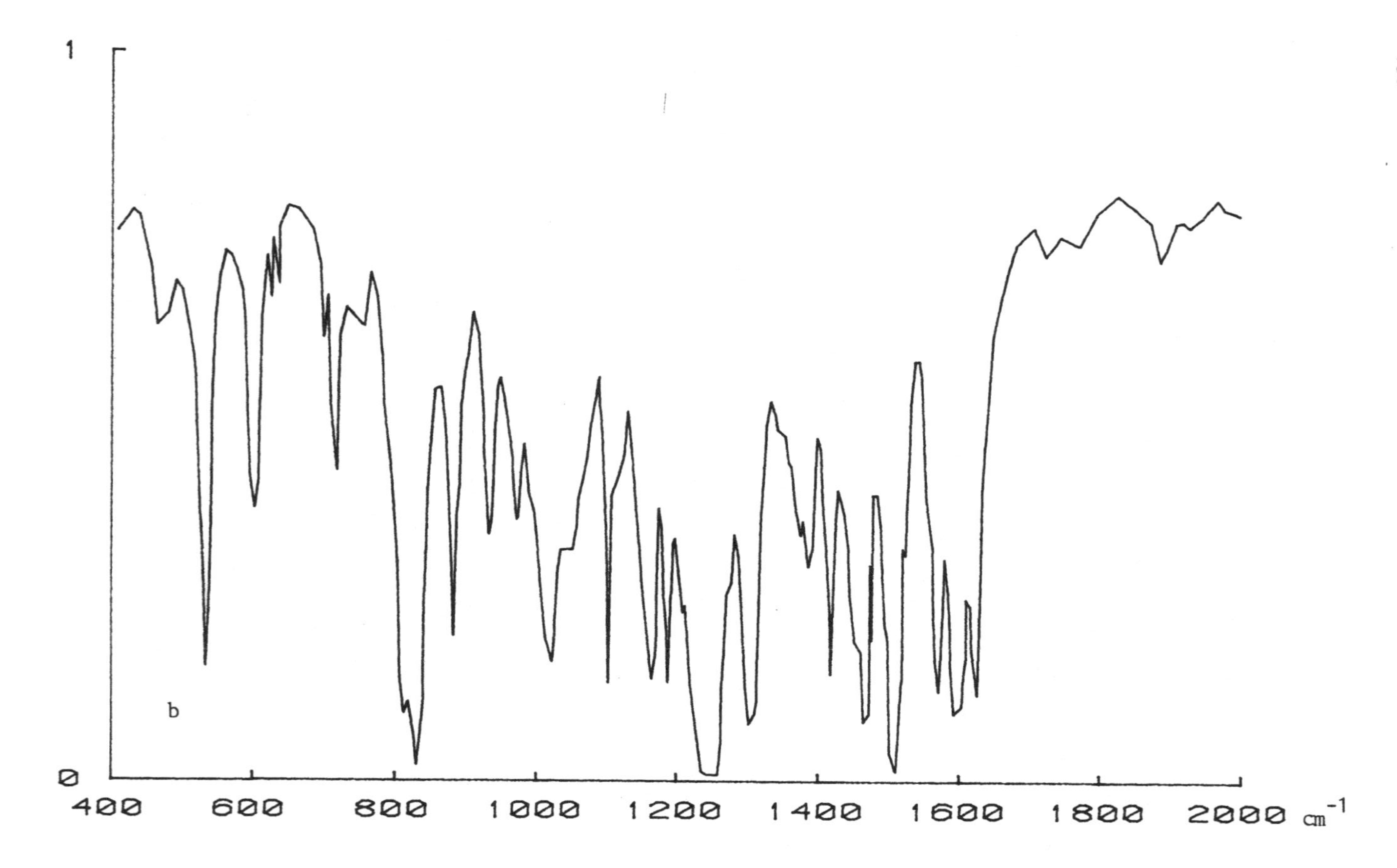

1
0
400
600
800
1000
1200
1400
1600
1800
2000 cm-1
b

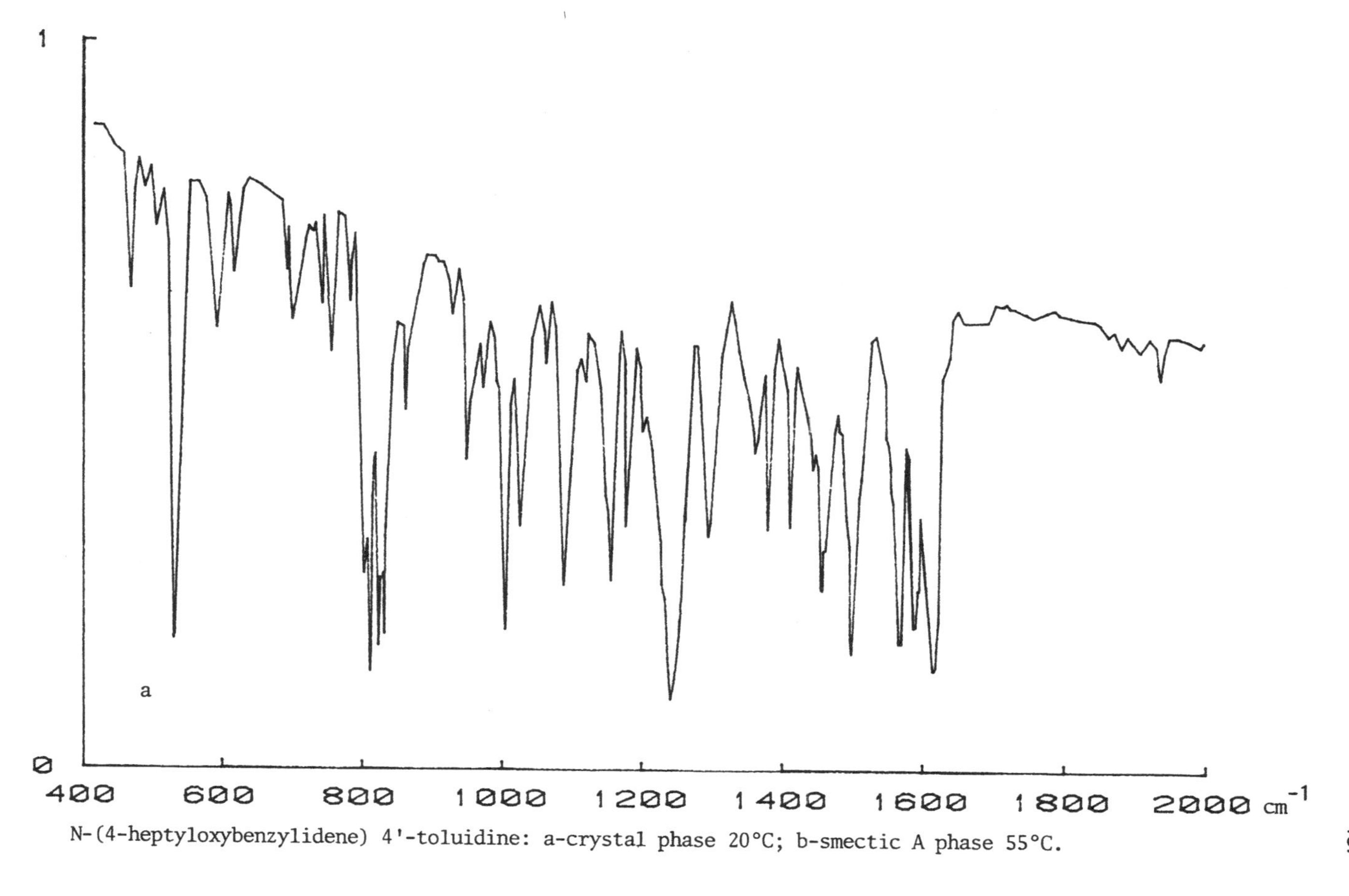

N-(4-heptyloxybenzylidene) 4'-toluidine: a-crystal phase 20°C; b-smectic A phase 55°C.

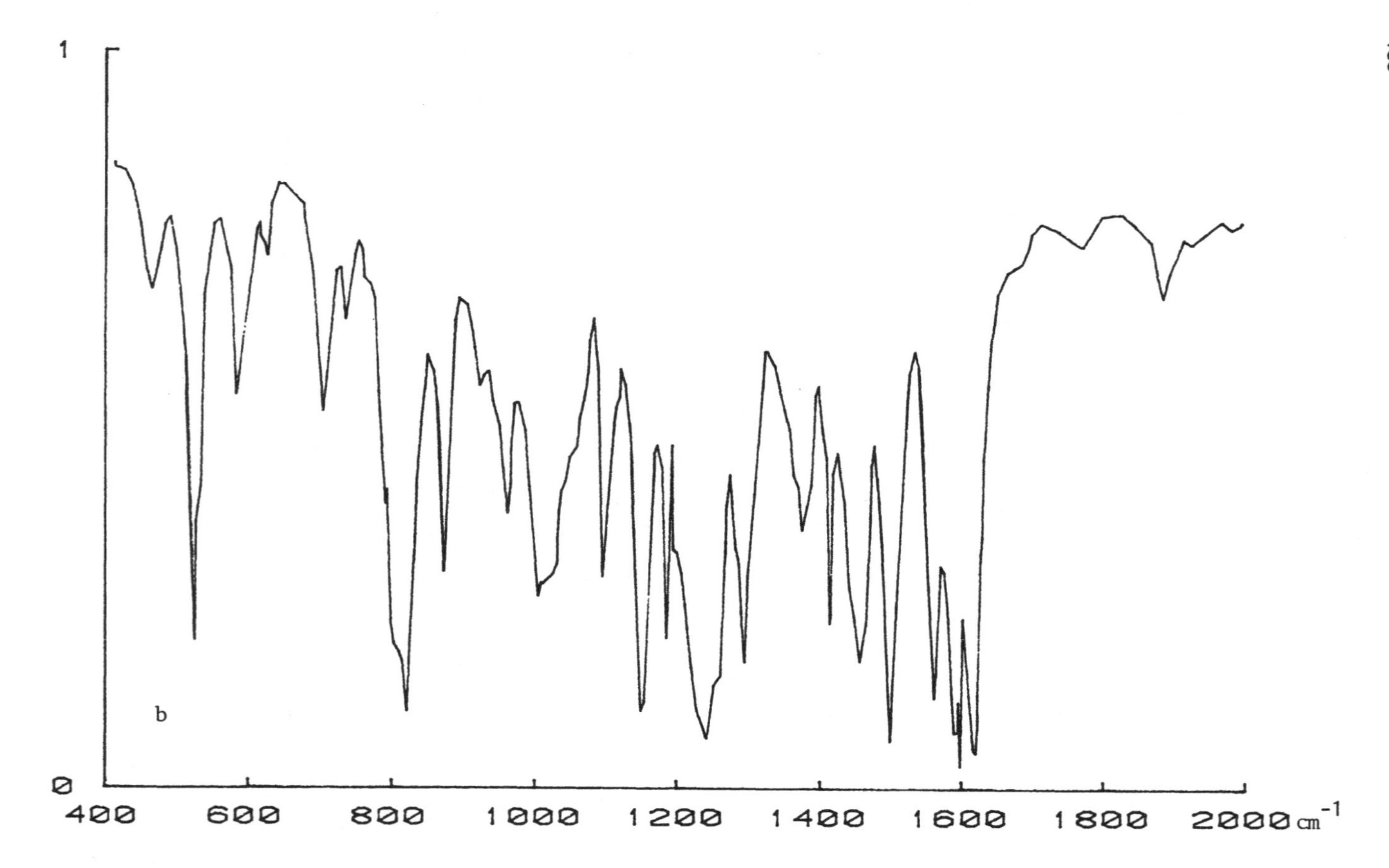

1
0
b
400
600
800
1000
1200
1400
1600
1800
2000 cm⁻¹

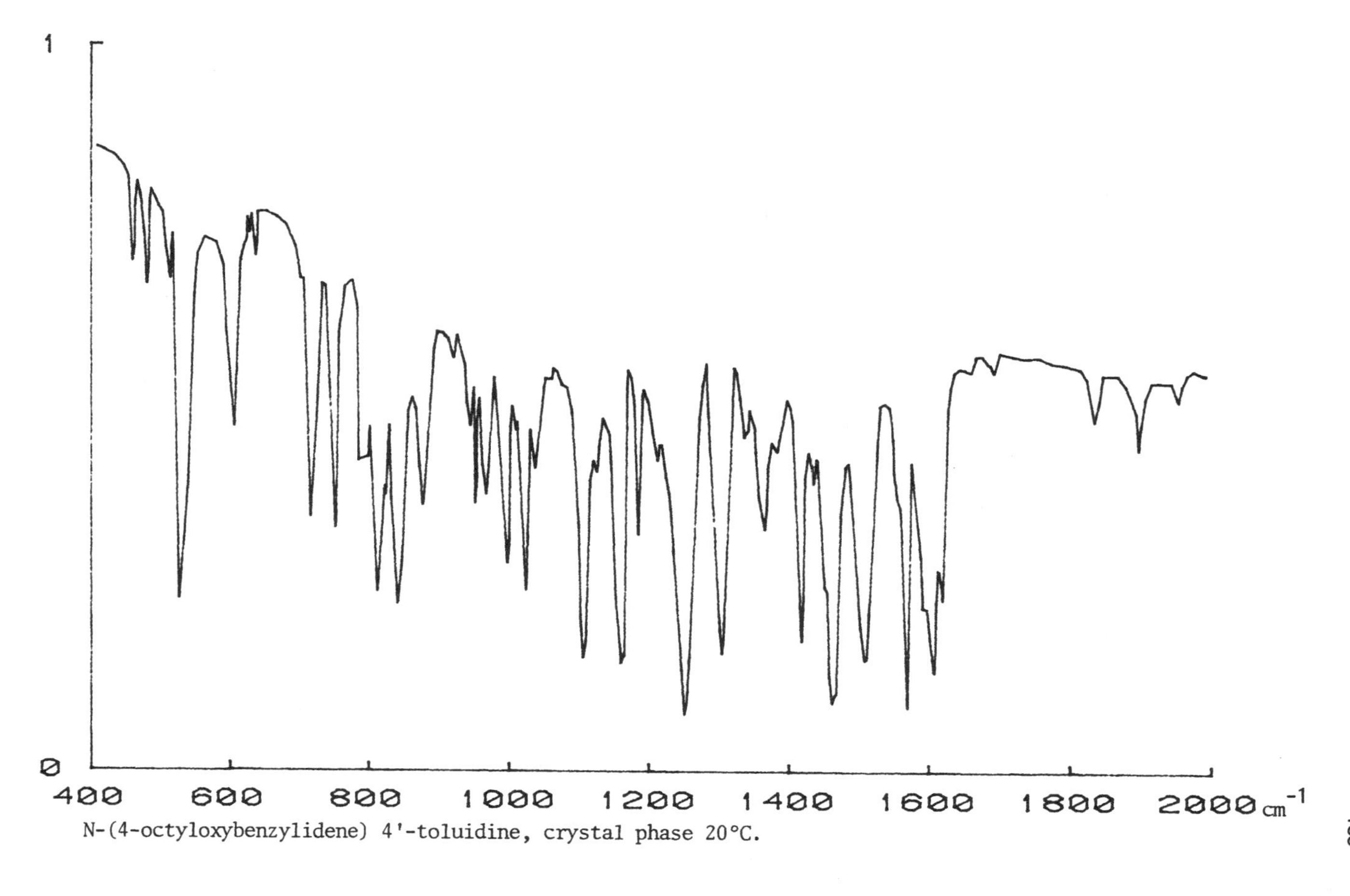

N-(4-octyloxybenzylidene) 4'-toluidine, crystal phase 20°C.

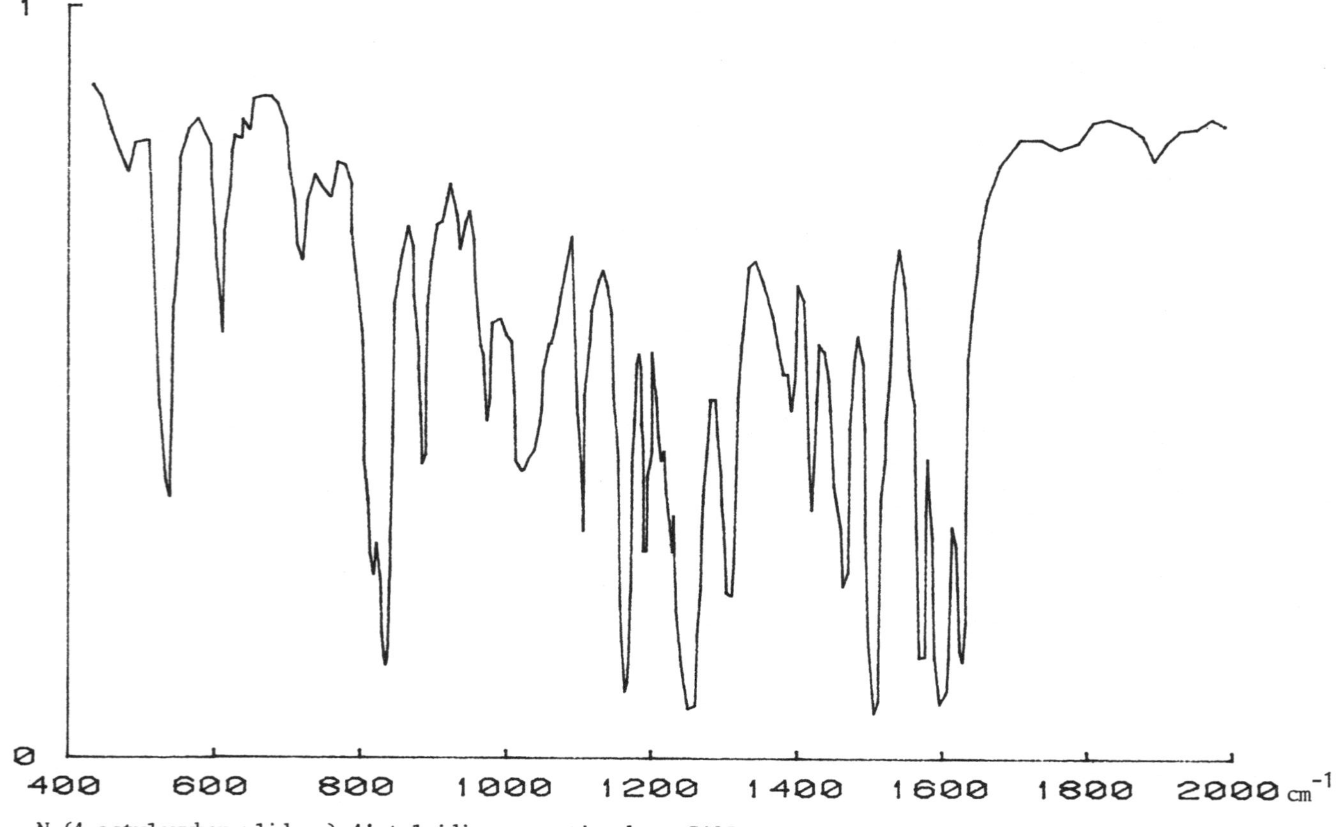

N-(4-octyloxybenzylidene) 4'-toluidine, nematic phase 74°C.

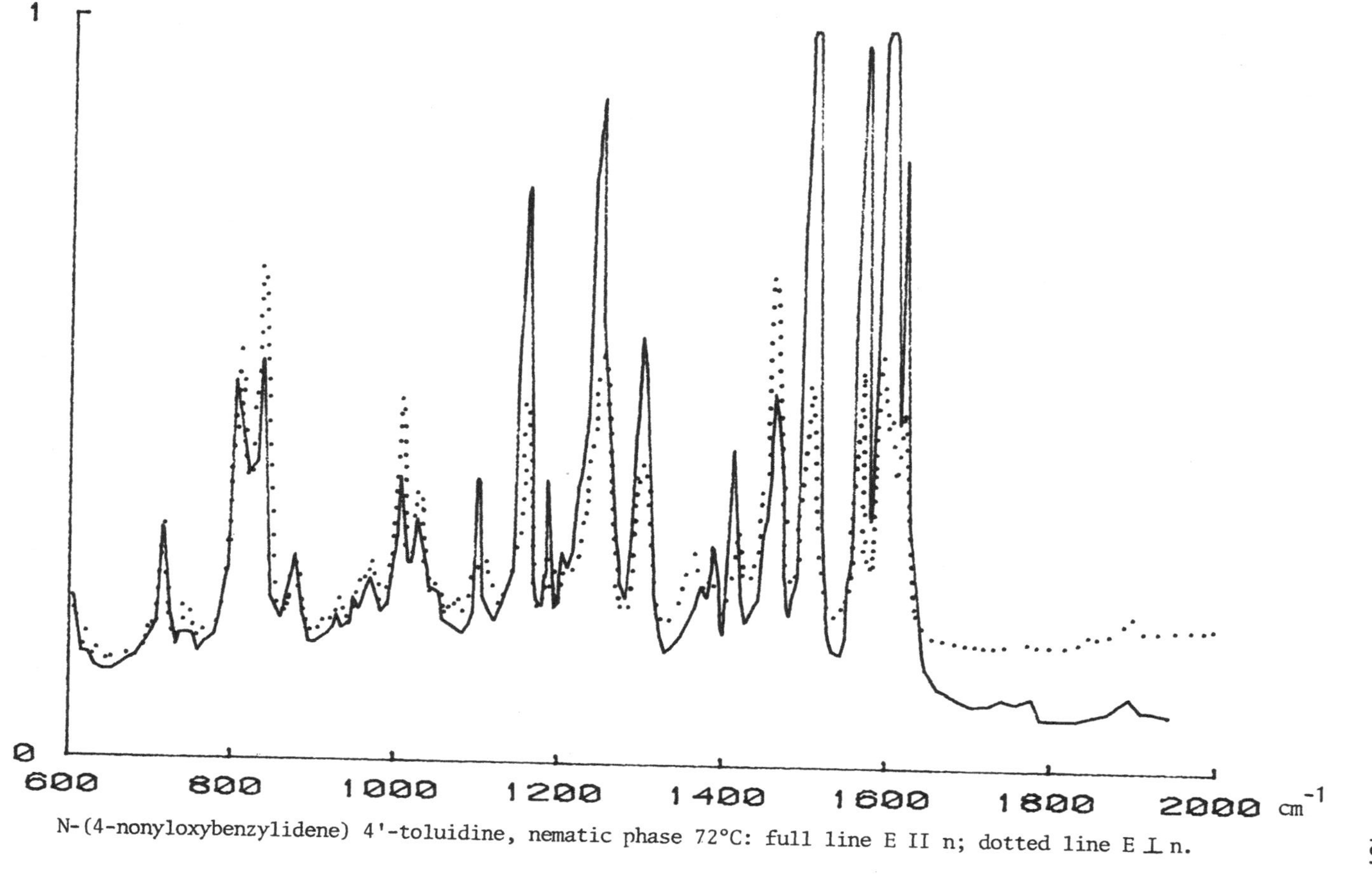

N-(4-nonyloxybenzylidene) 4'-toluidine, nematic phase 72°C: full line E II n; dotted line E ⊥ n.

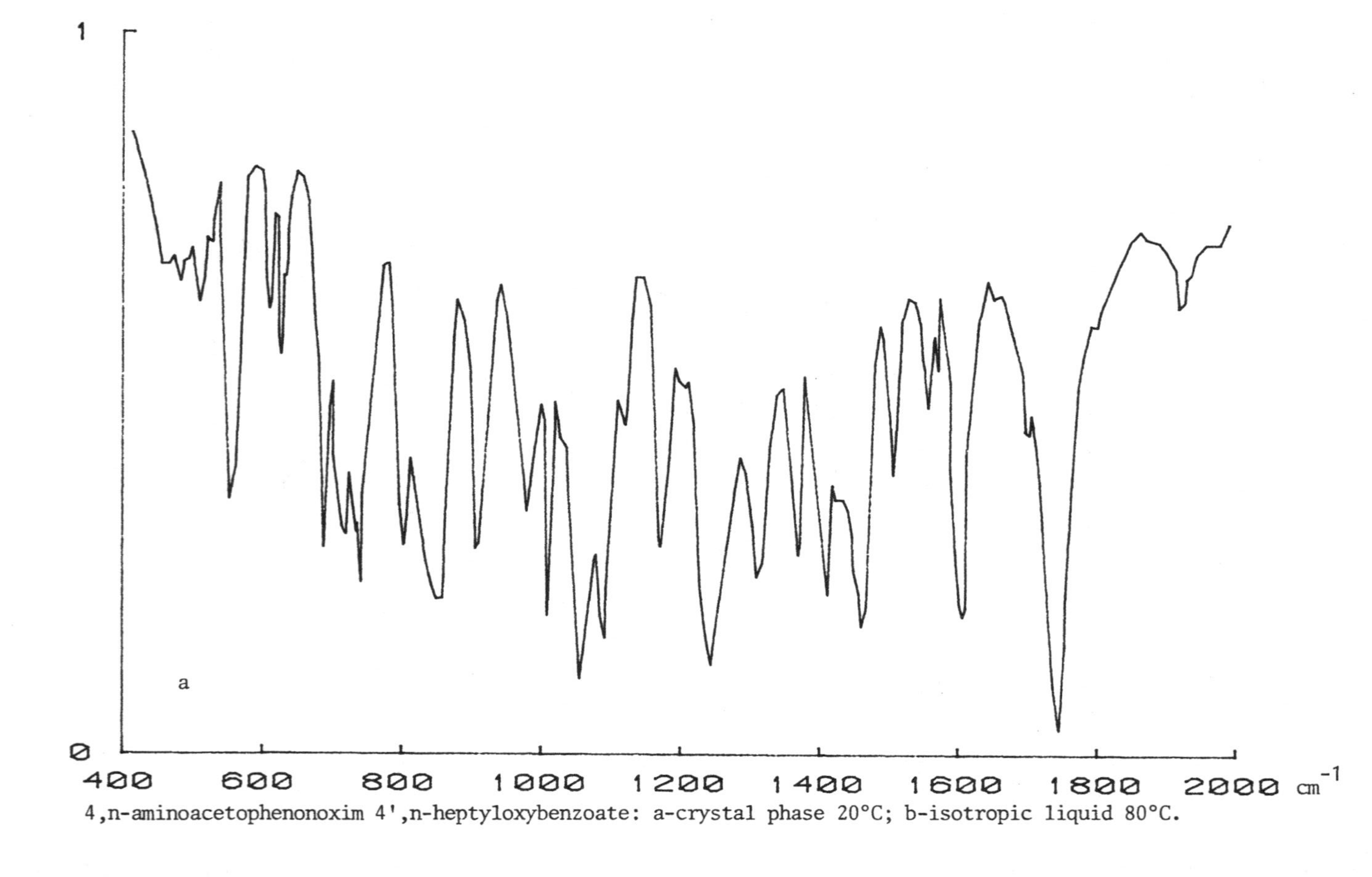

4,n-aminoacetophenonoxim 4',n-heptyloxybenzoate: a-crystal phase 20°C; b-isotropic liquid 80°C.

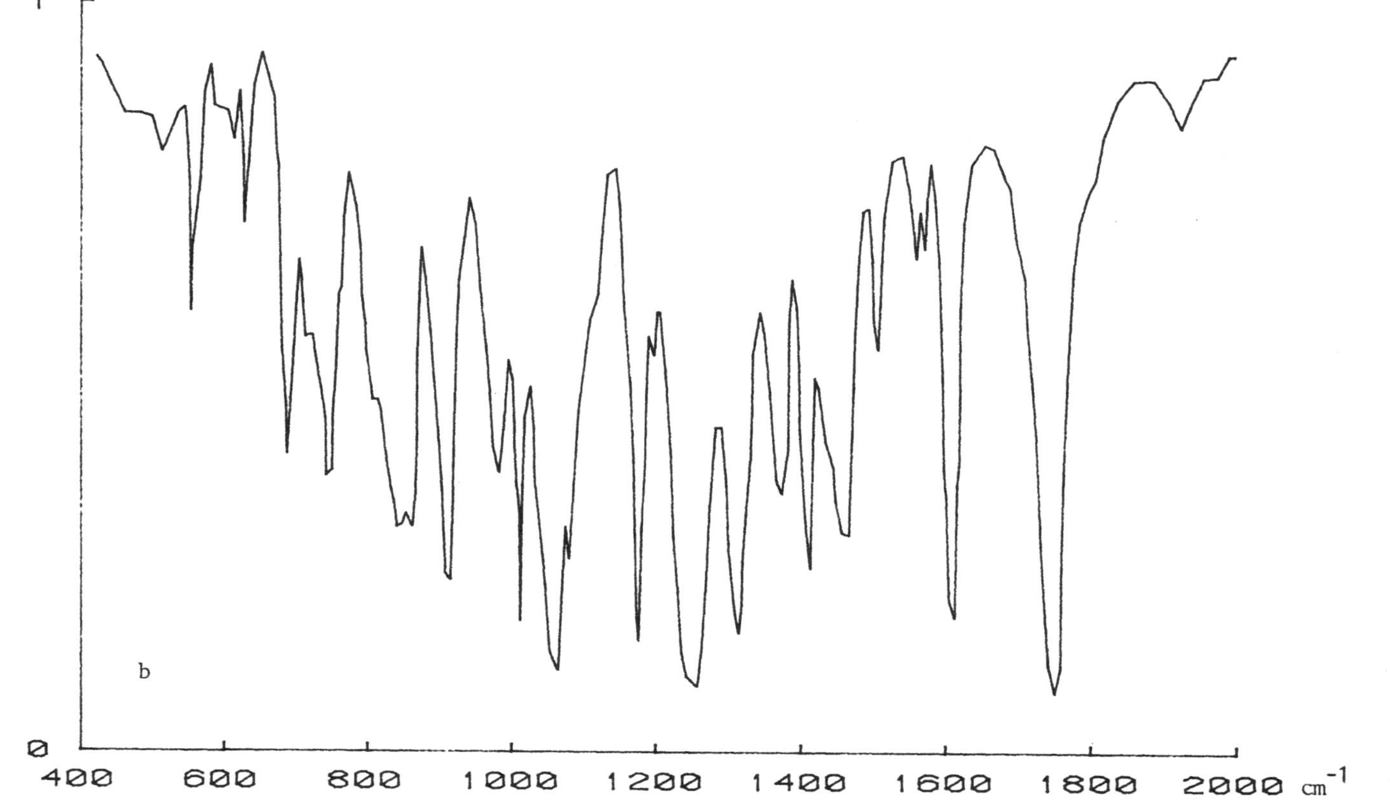

1
0
400
600
800
1000
1200
1400
1600
1800
2000 cm-1
b

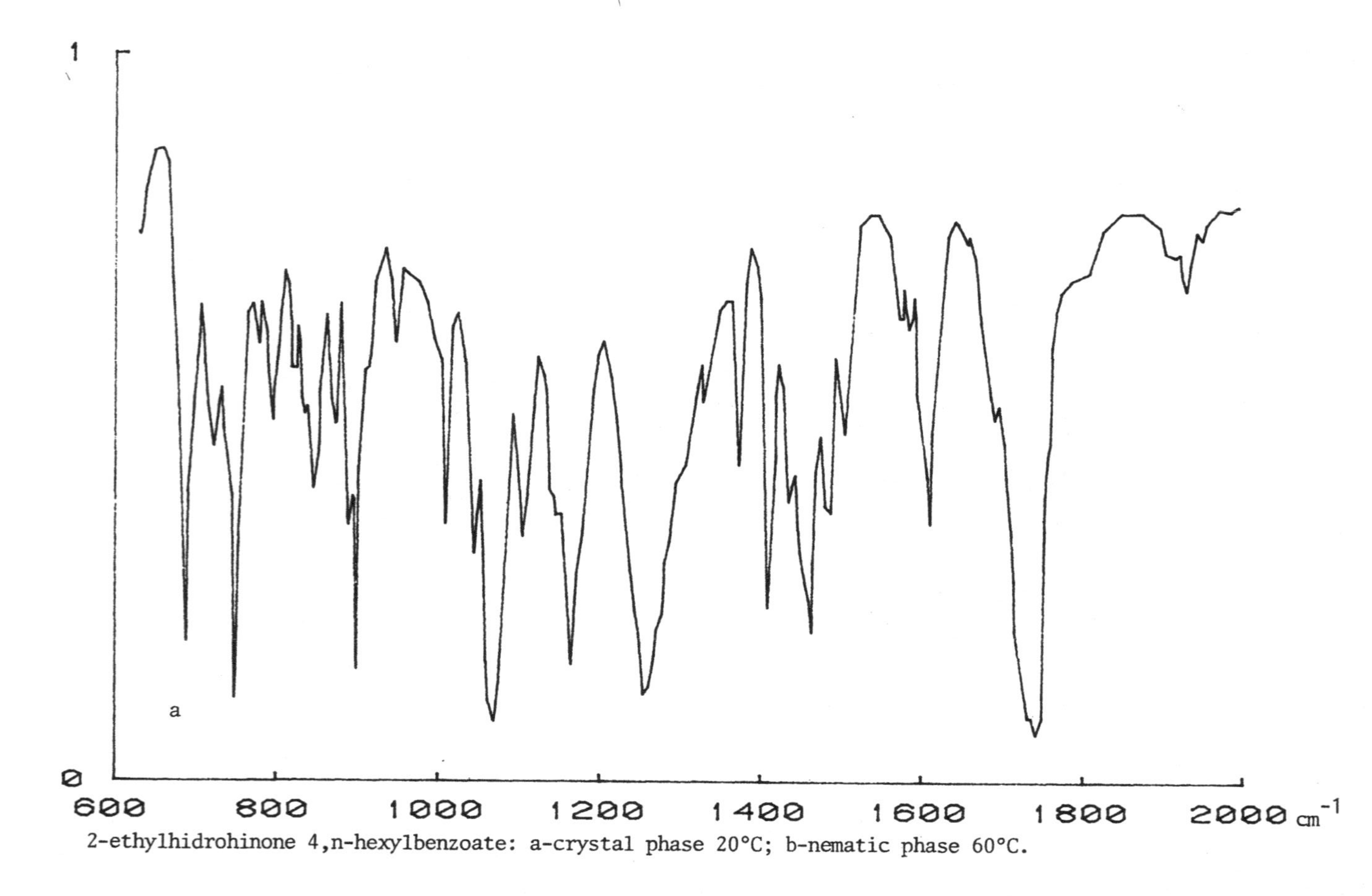

2-ethylhidrohinone 4,n-hexylbenzoate: a-crystal phase 20°C; b-nematic phase 60°C.

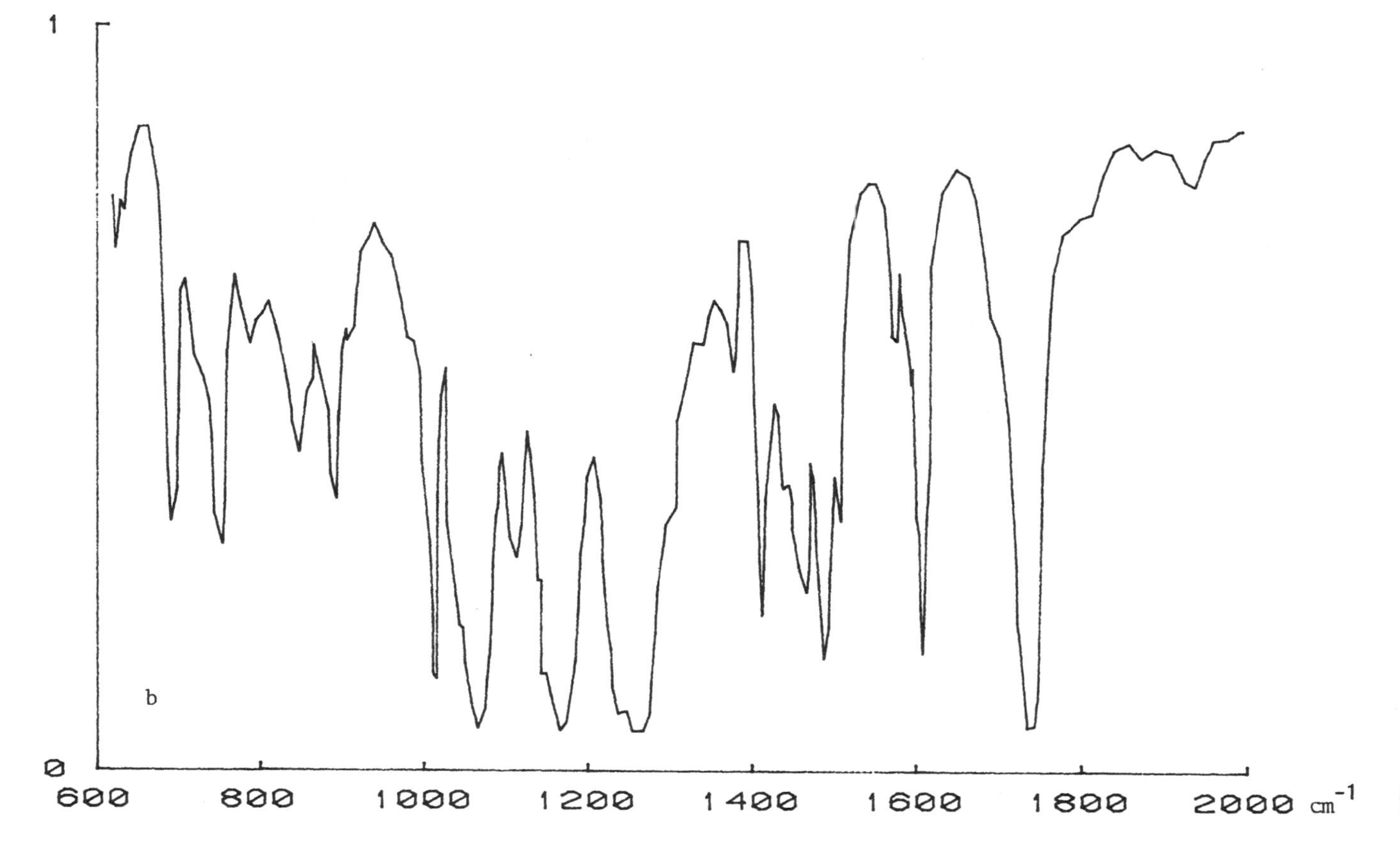
1
0
600
800
1000
1200
1400
1600
1800
2000
cm^{-1}
b

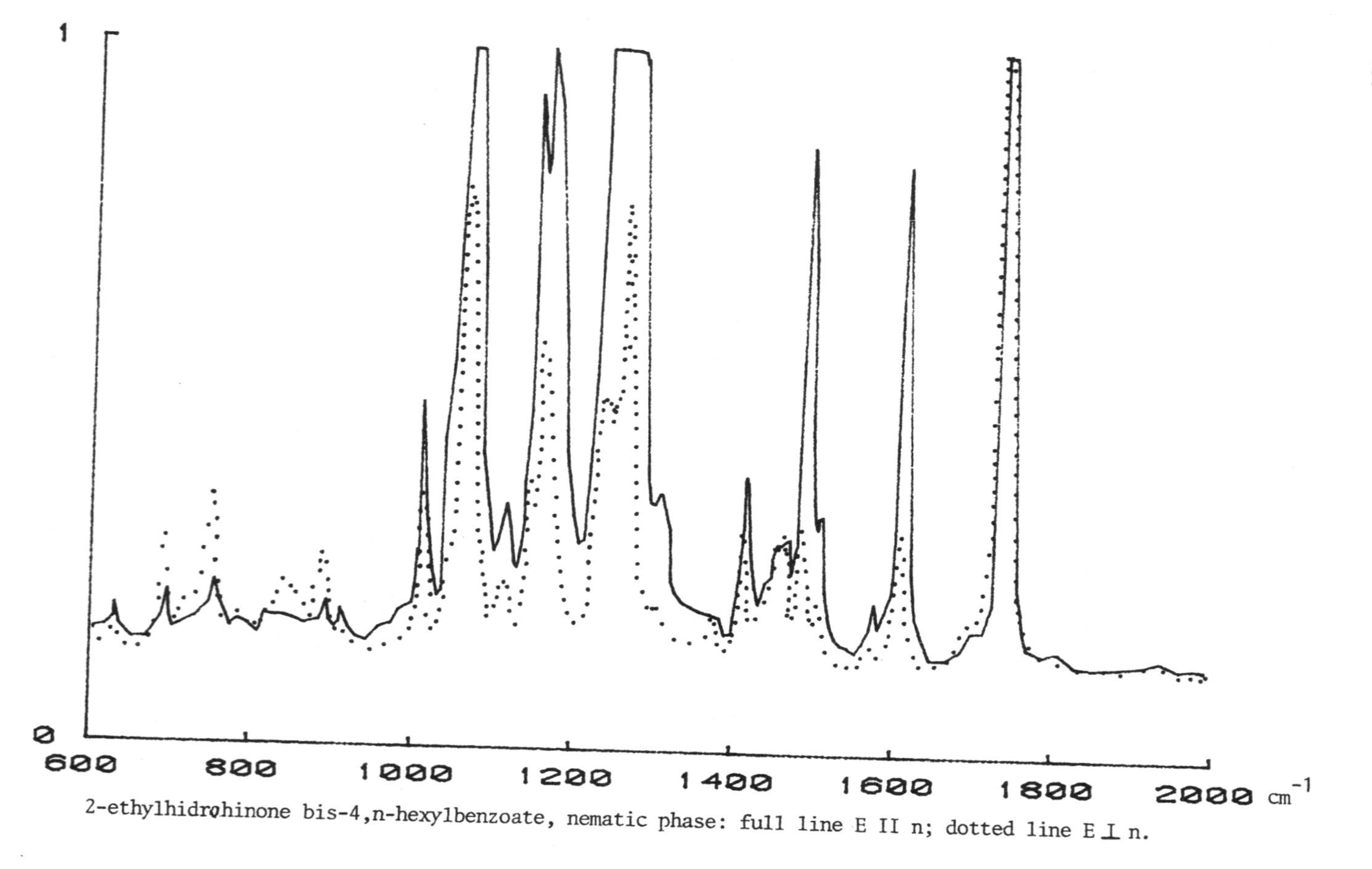

2-ethylhidrohinone bis-4,n-hexylbenzoate, nematic phase: full line E II n; dotted line E ⊥ n.

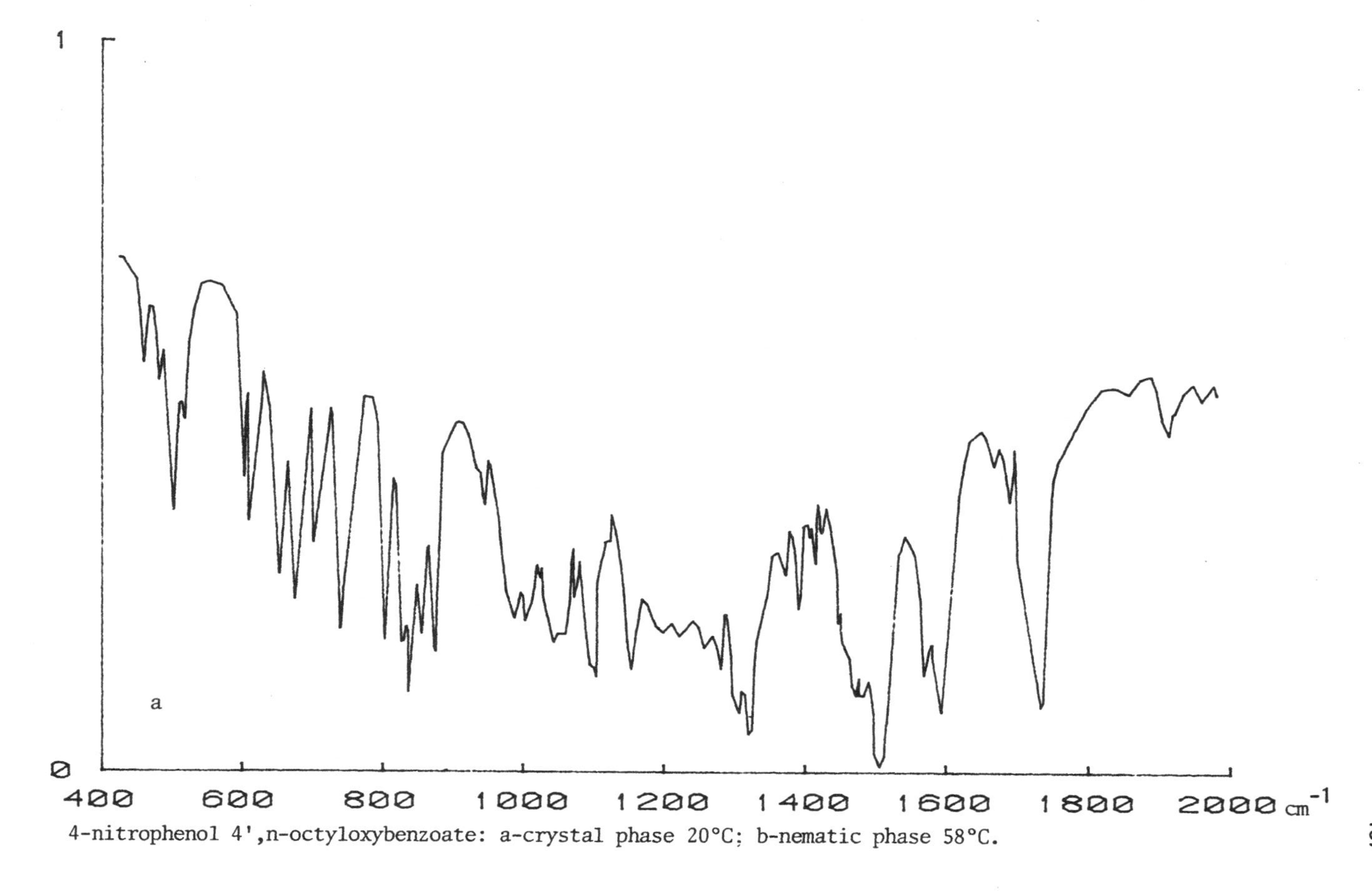

4-nitrophenol 4',n-octyloxybenzoate: a-crystal phase 20°C; b-nematic phase 58°C.

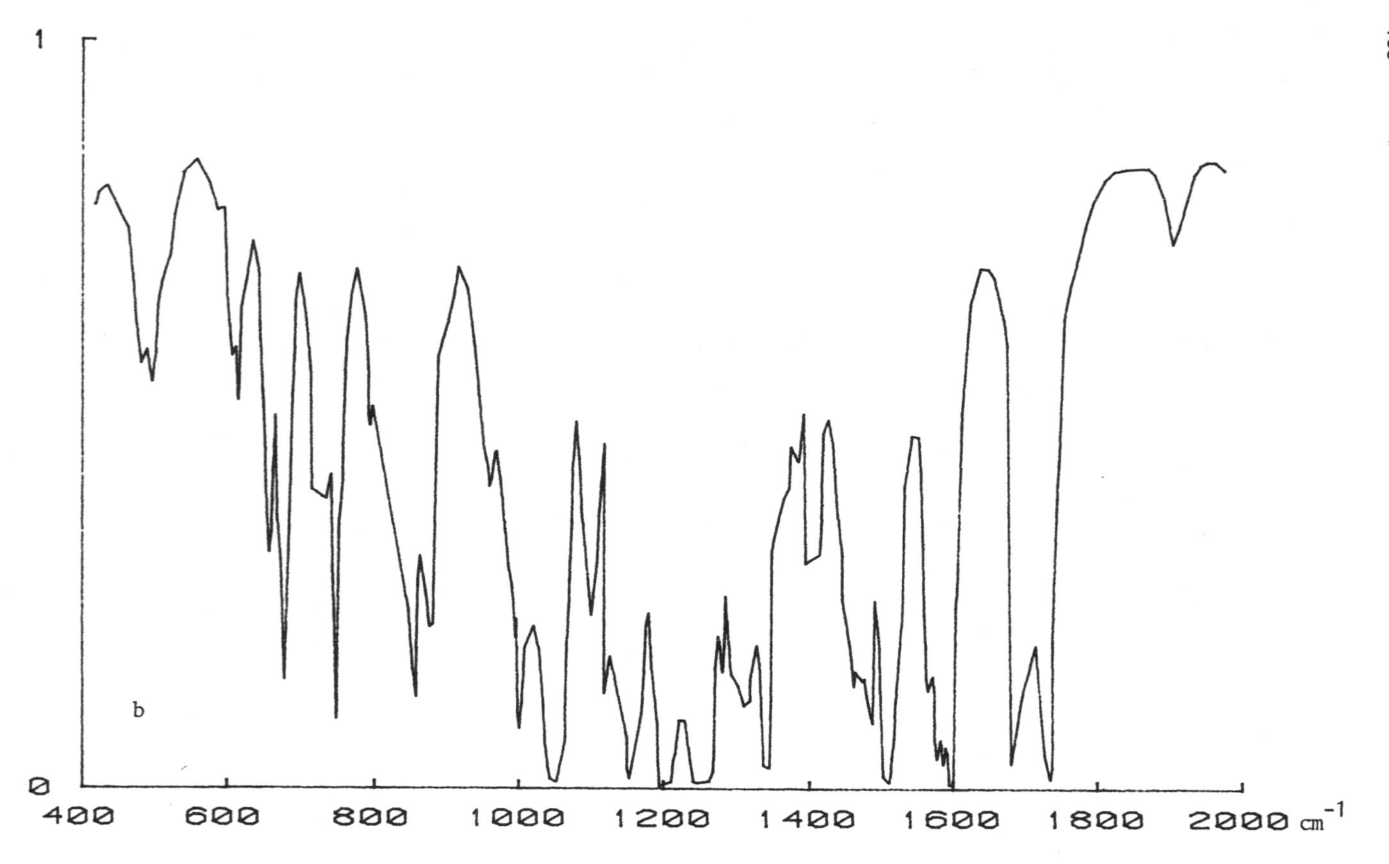

1
0
400
600
800
1000
1200
1400
1600
1800
2000 cm-1
b

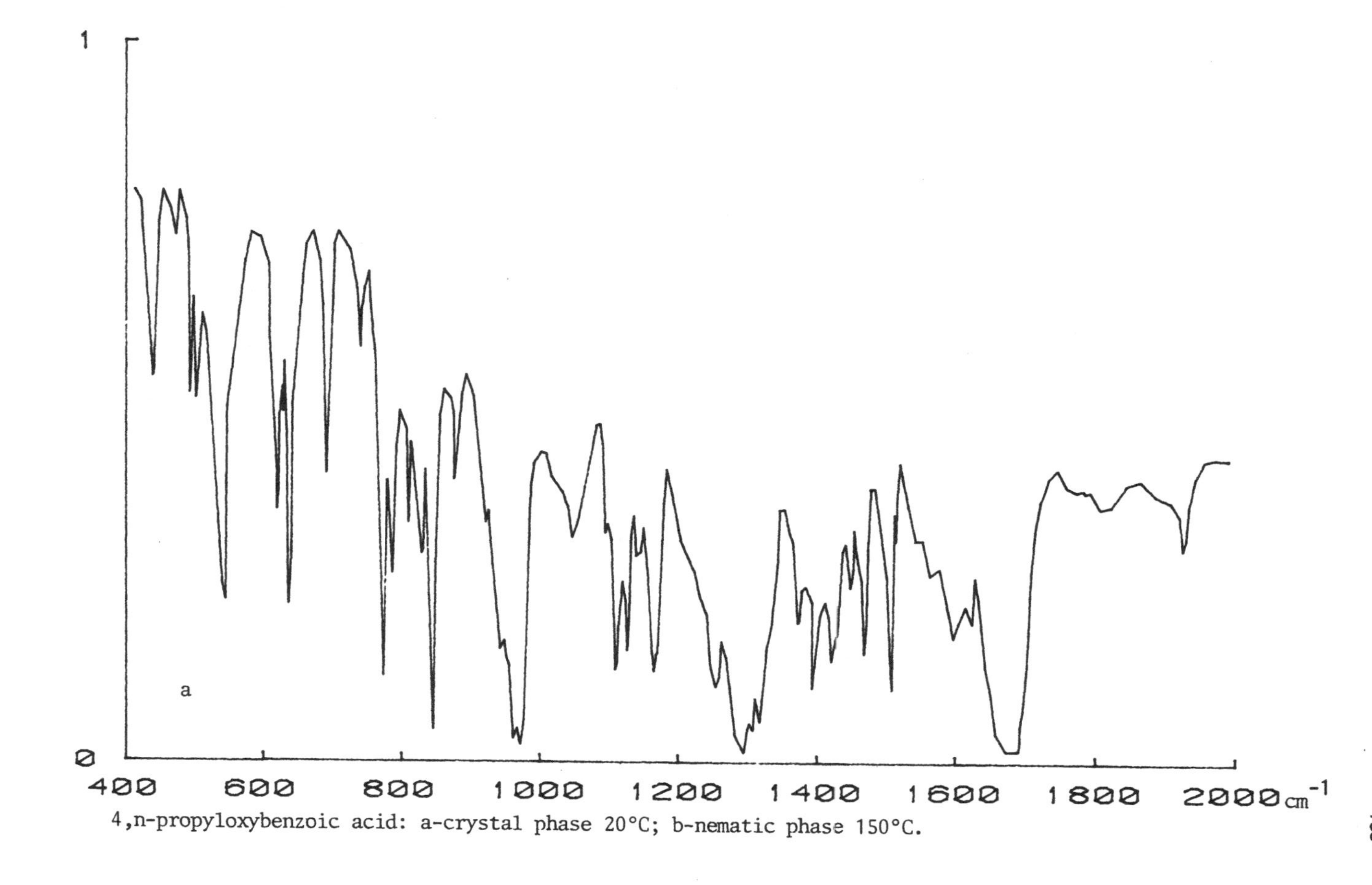

4,n-propyloxybenzoic acid: a-crystal phase 20°C; b-nematic phase 150°C.

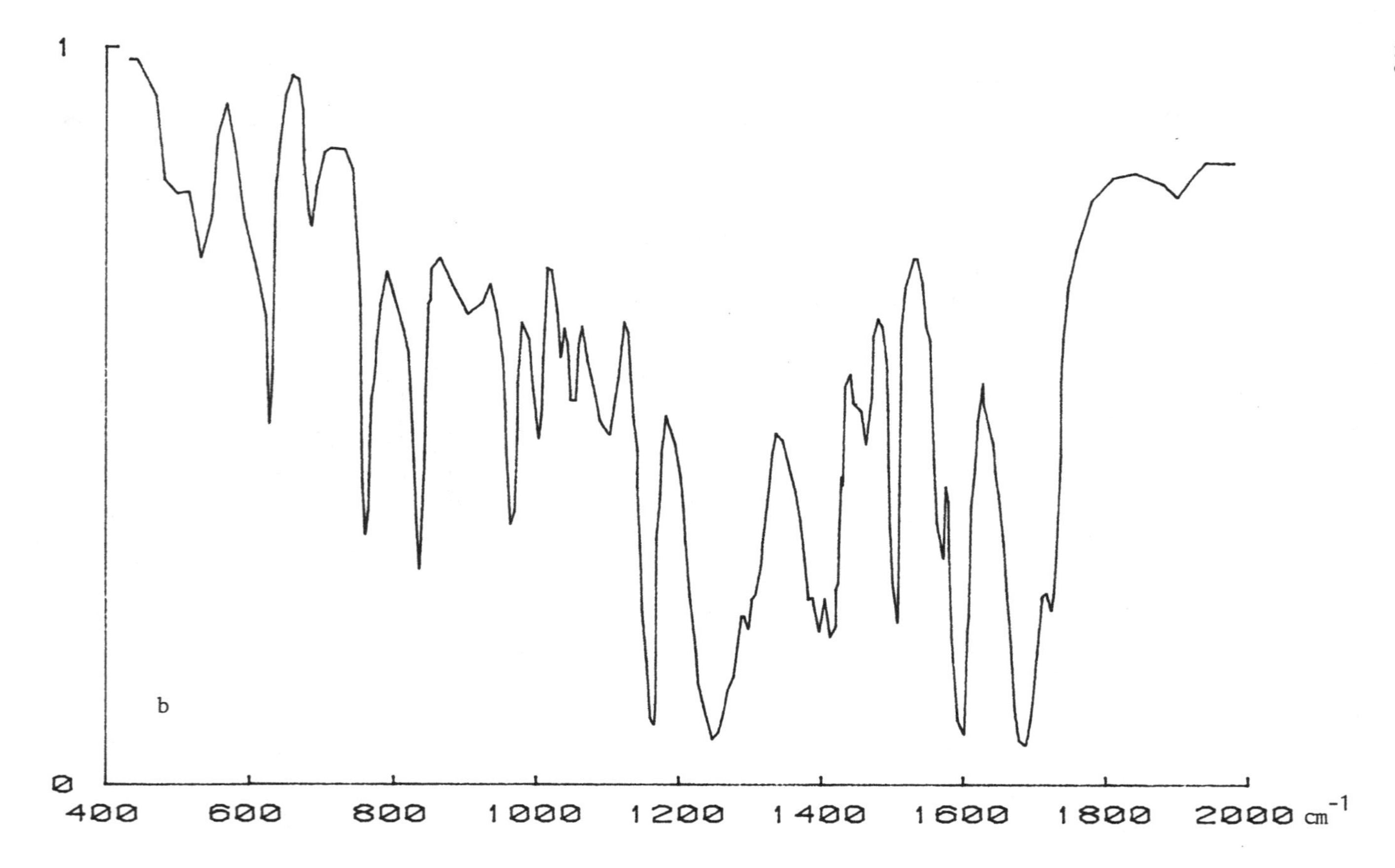
1
0
b
400
600
800
1000
1200
1400
1600
1800
2000 cm^{-1}

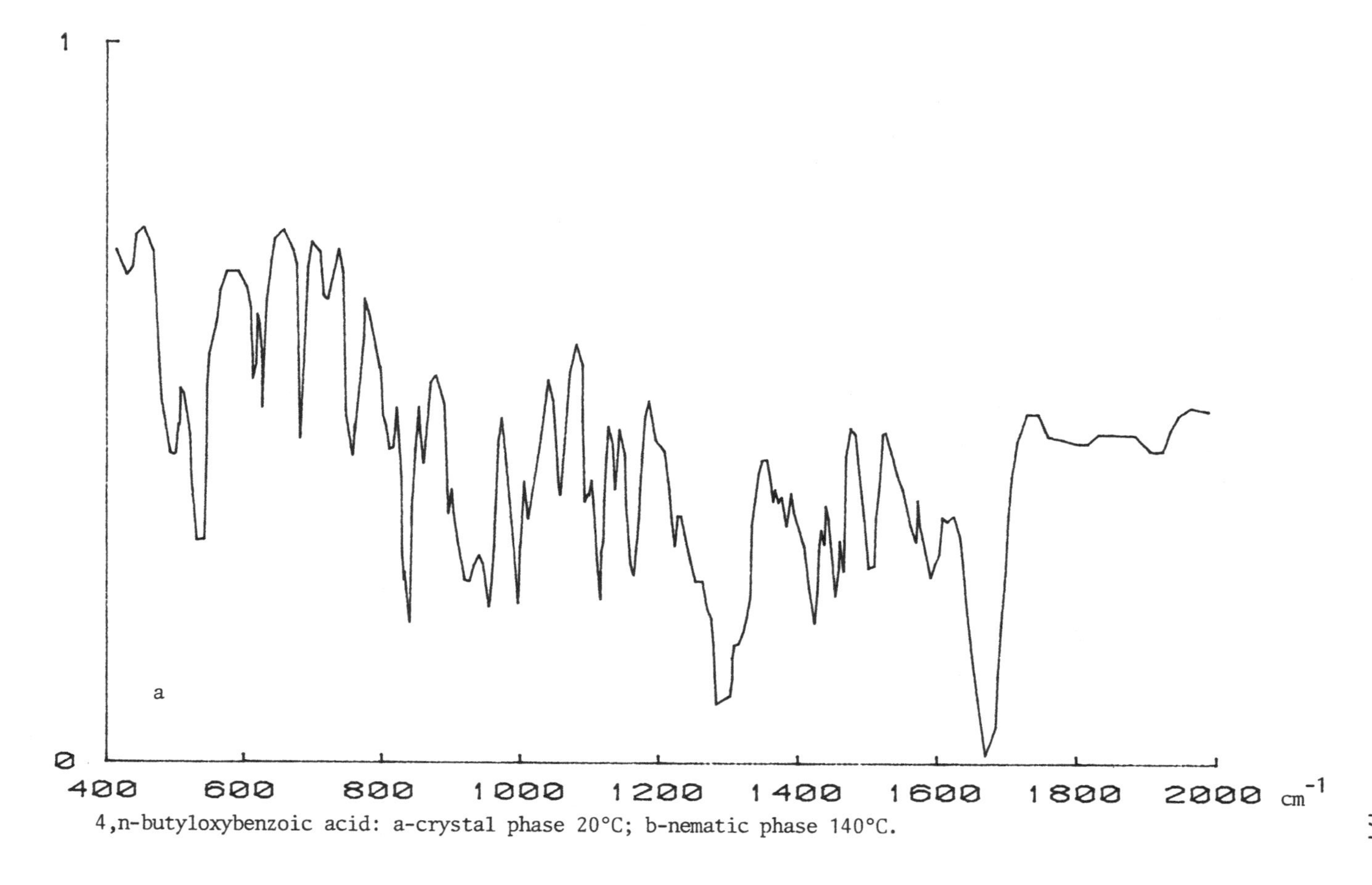

4,n-butyloxybenzoic acid: a-crystal phase 20°C; b-nematic phase 140°C.

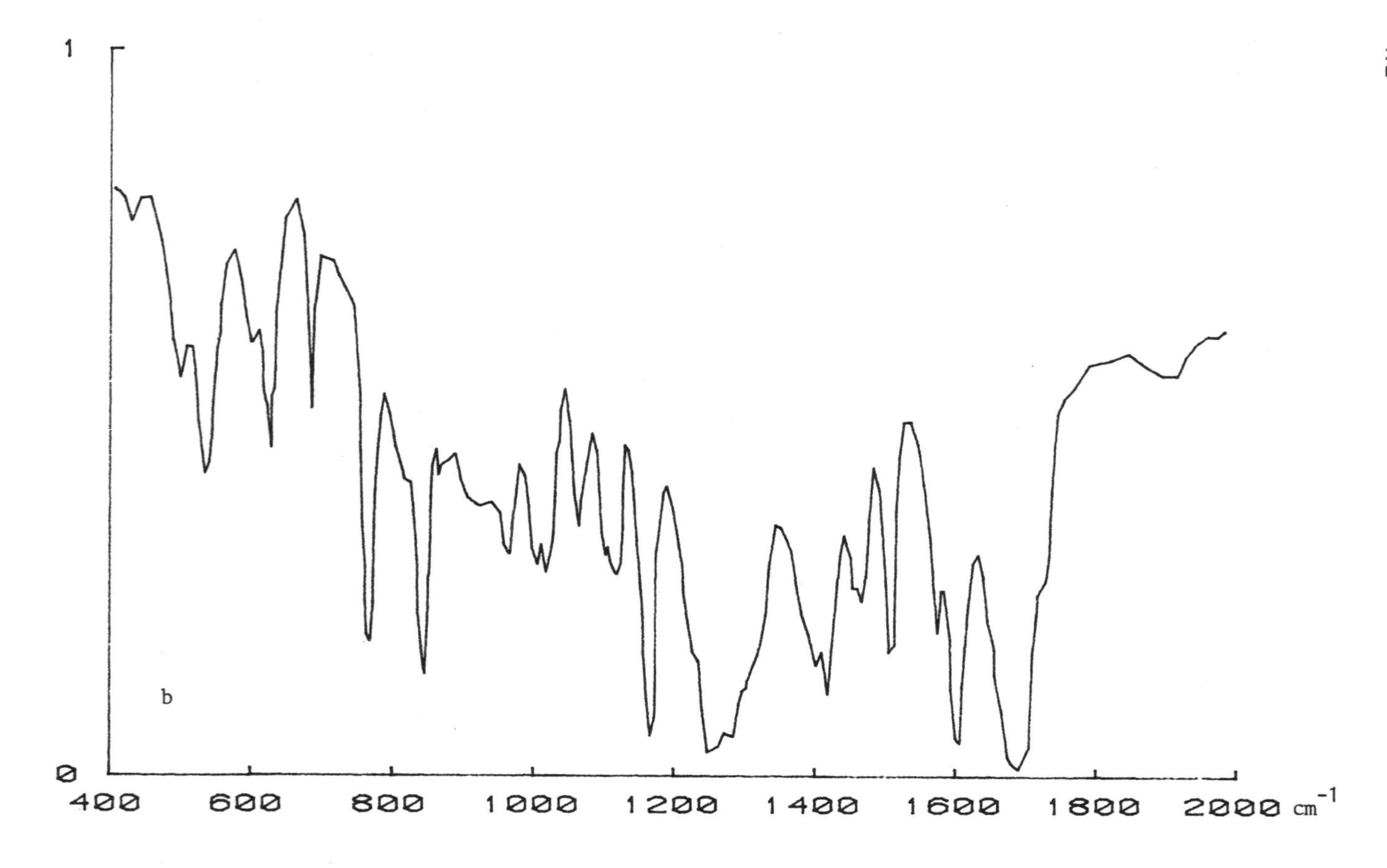

1
0
b
400
600
800
1000
1200
1400
1600
1800
2000 cm-1

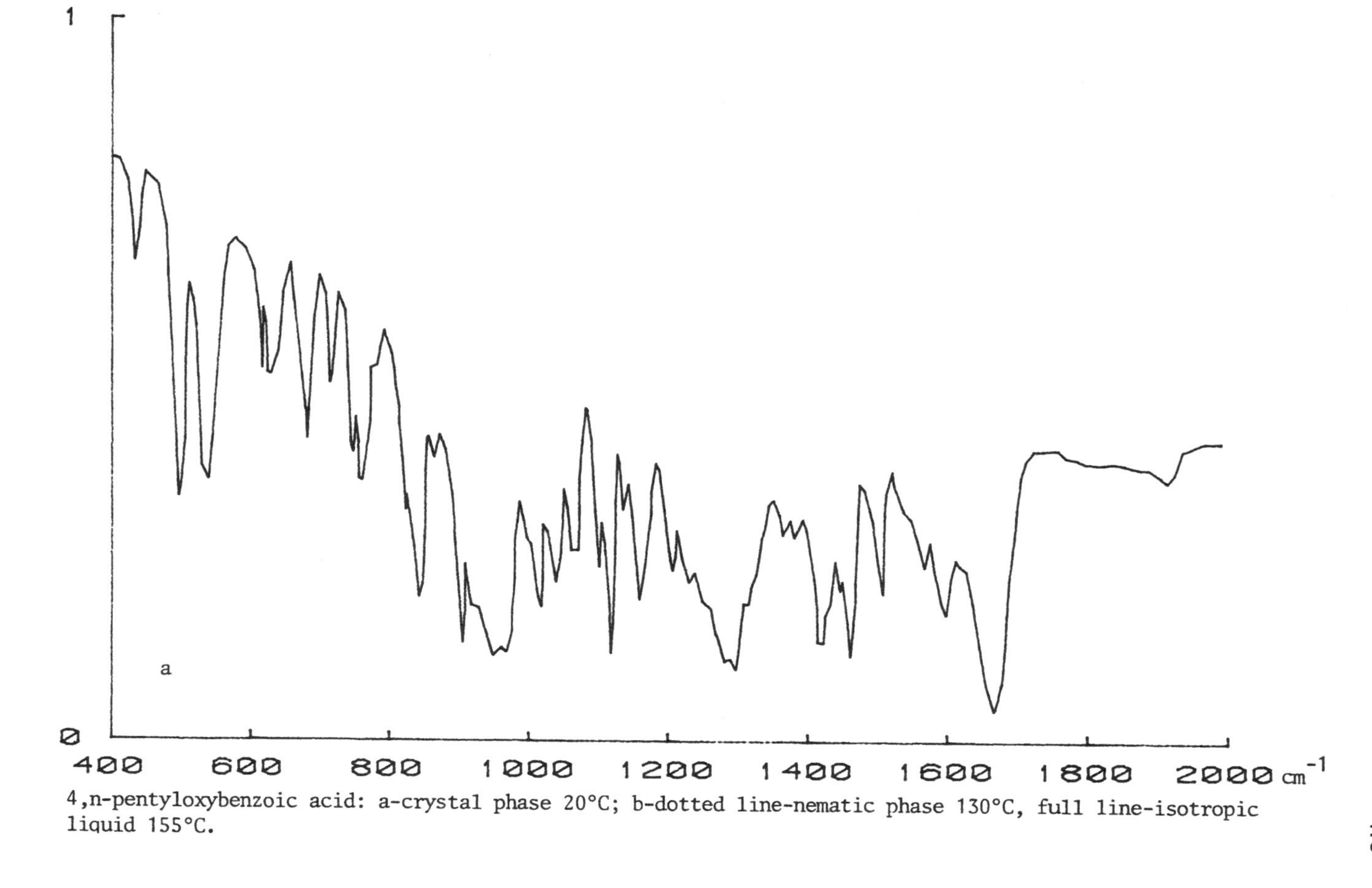

4,n-pentyloxybenzoic acid: a-crystal phase 20°C; b-dotted line-nematic phase 130°C, full line-isotropic liquid 155°C.

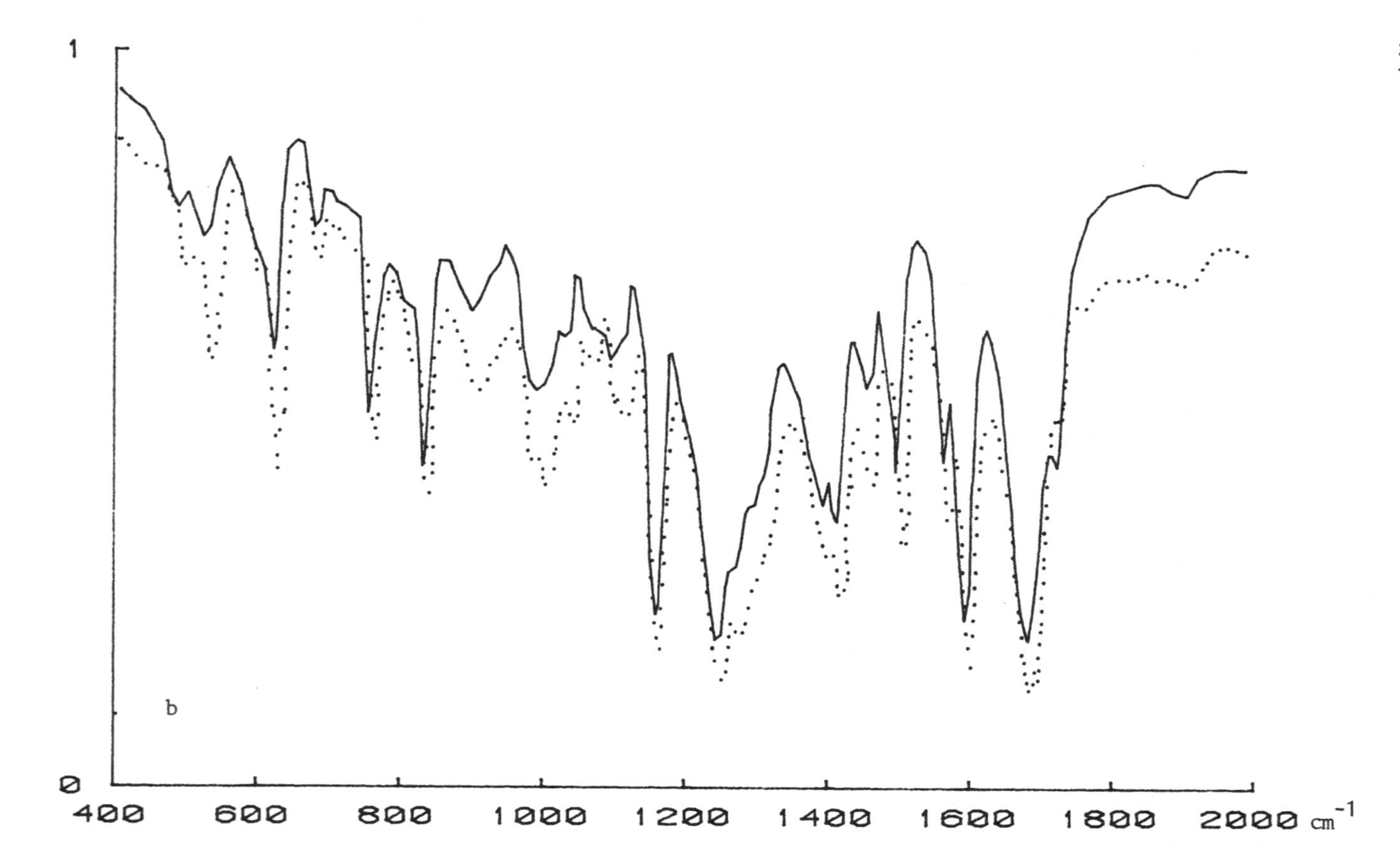

1
0
b
400
600
800
1000
1200
1400
1600
1800
2000 cm⁻¹

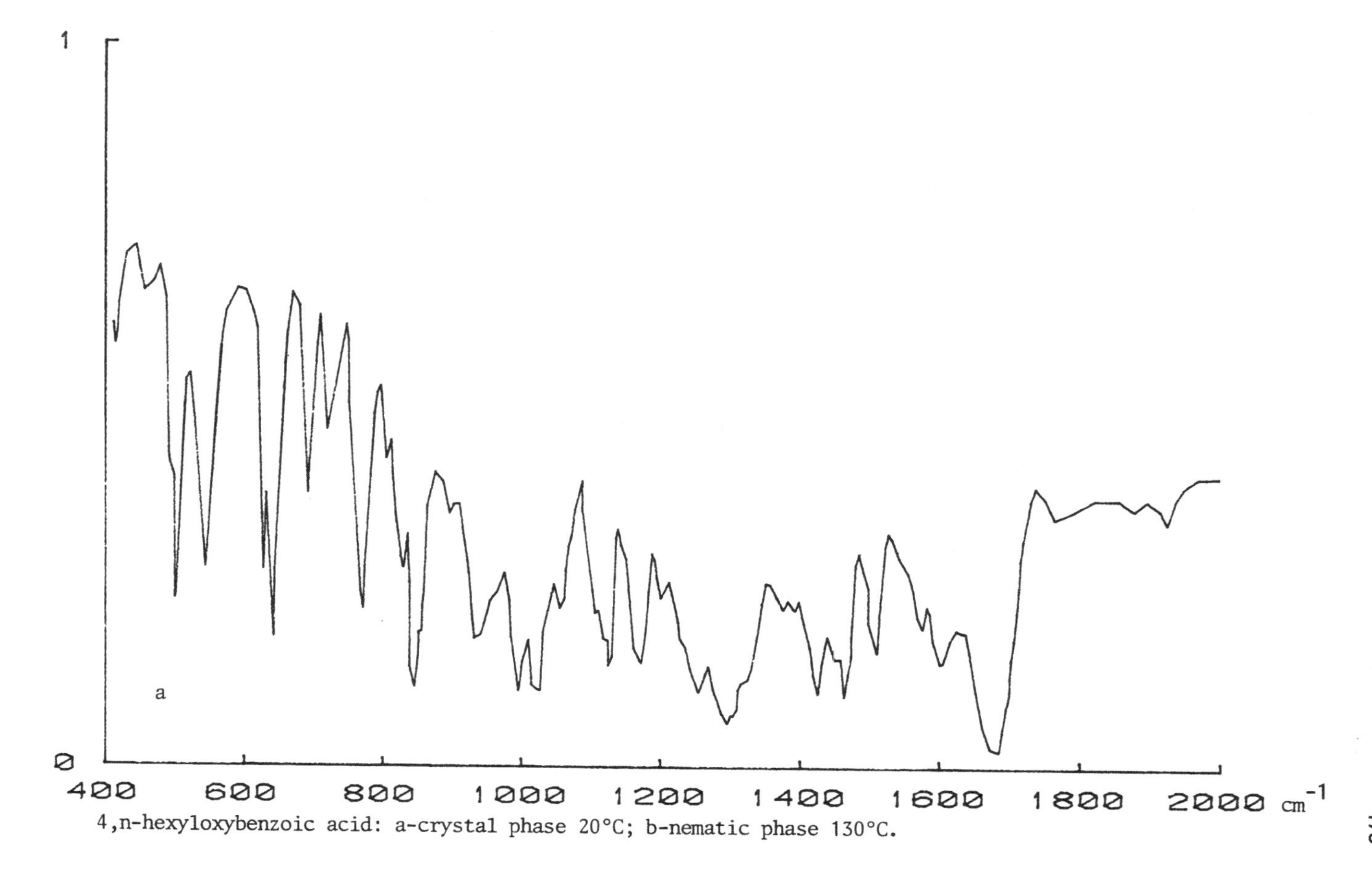

4,n-hexyloxybenzoic acid: a-crystal phase 20°C; b-nematic phase 130°C.

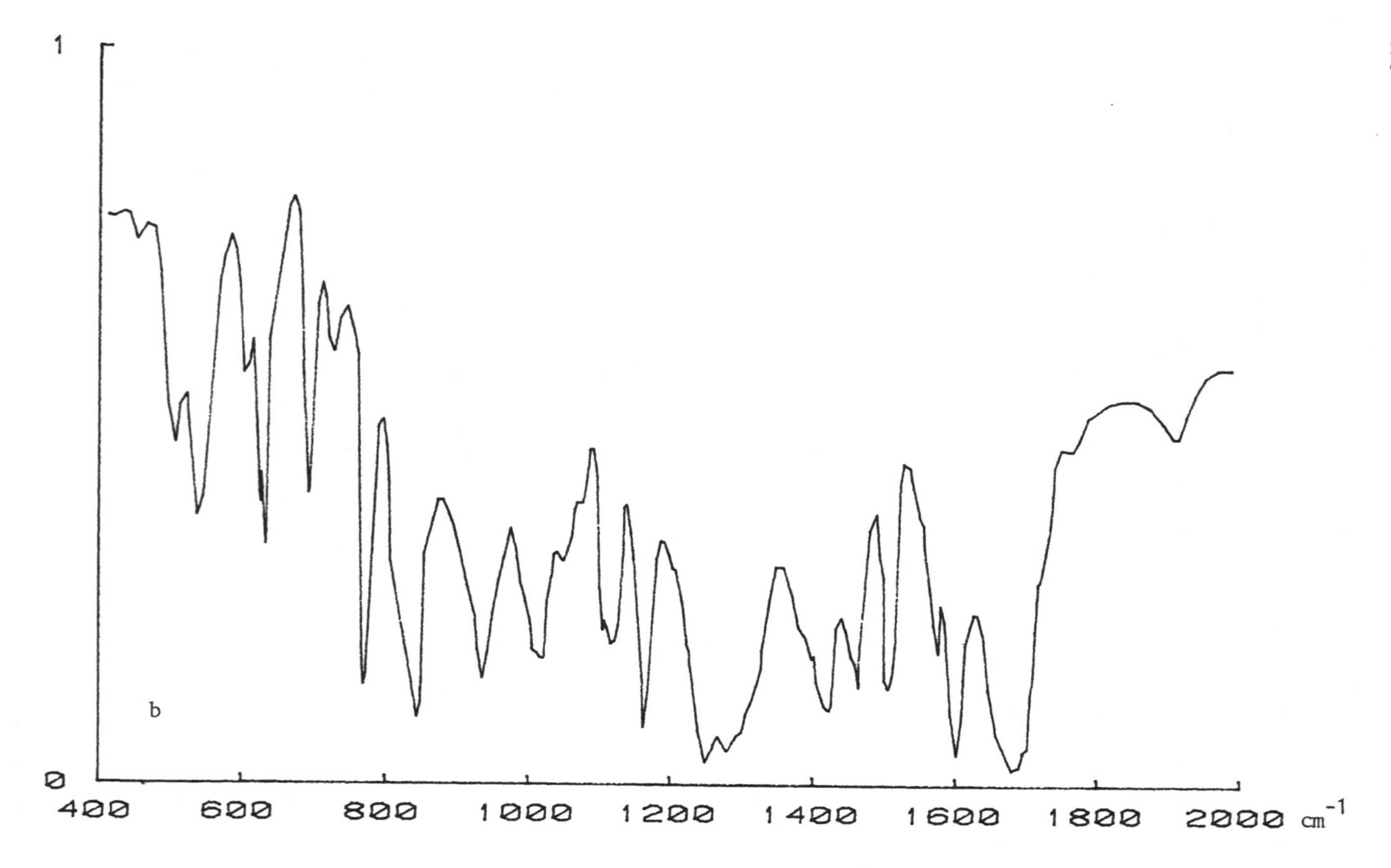
1
0
400
600
800
1000
1200
1400
1600
1800
2000 cm^{-1}
b

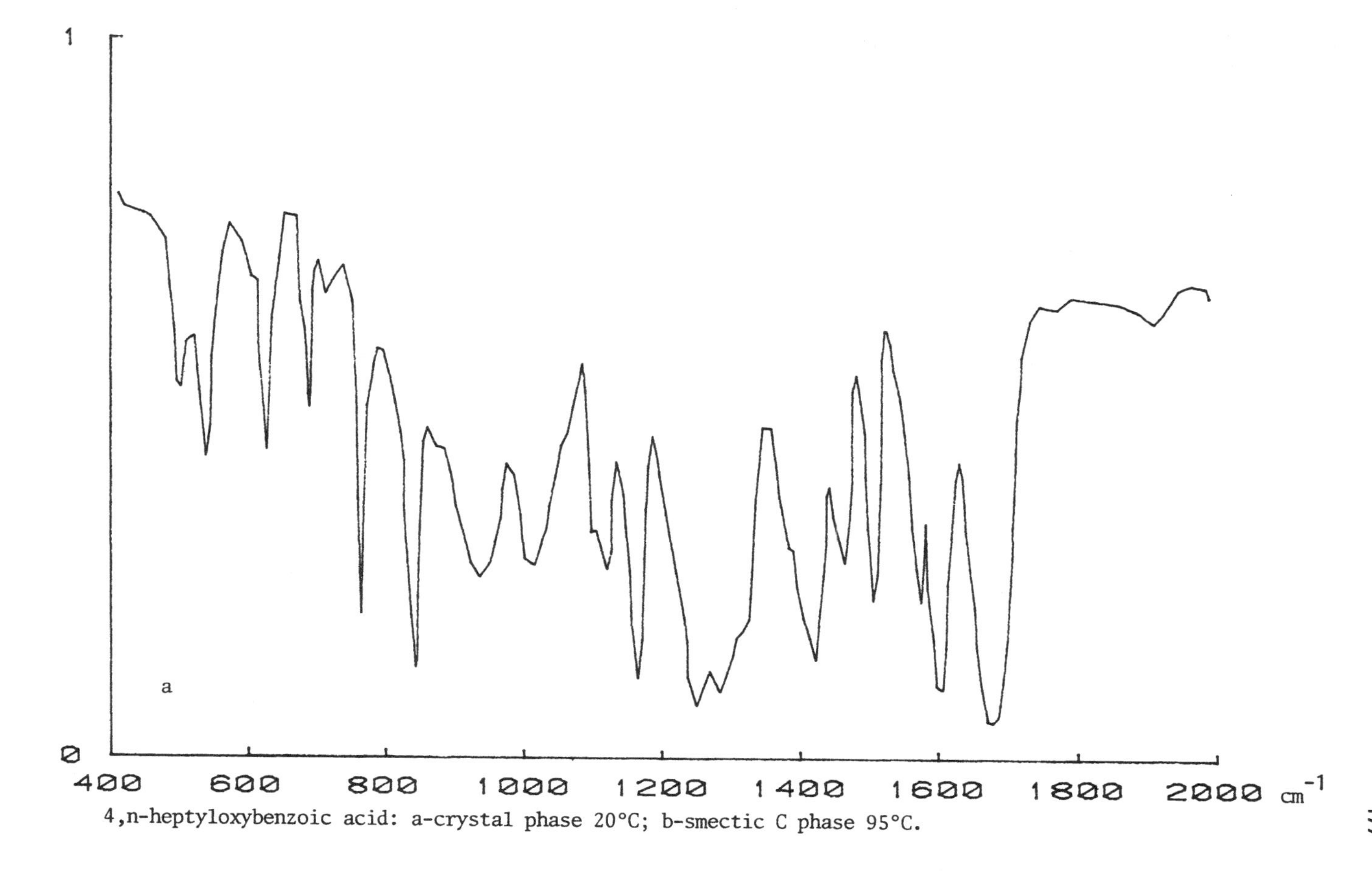

4,n-heptyloxybenzoic acid: a-crystal phase 20°C; b-smectic C phase 95°C.

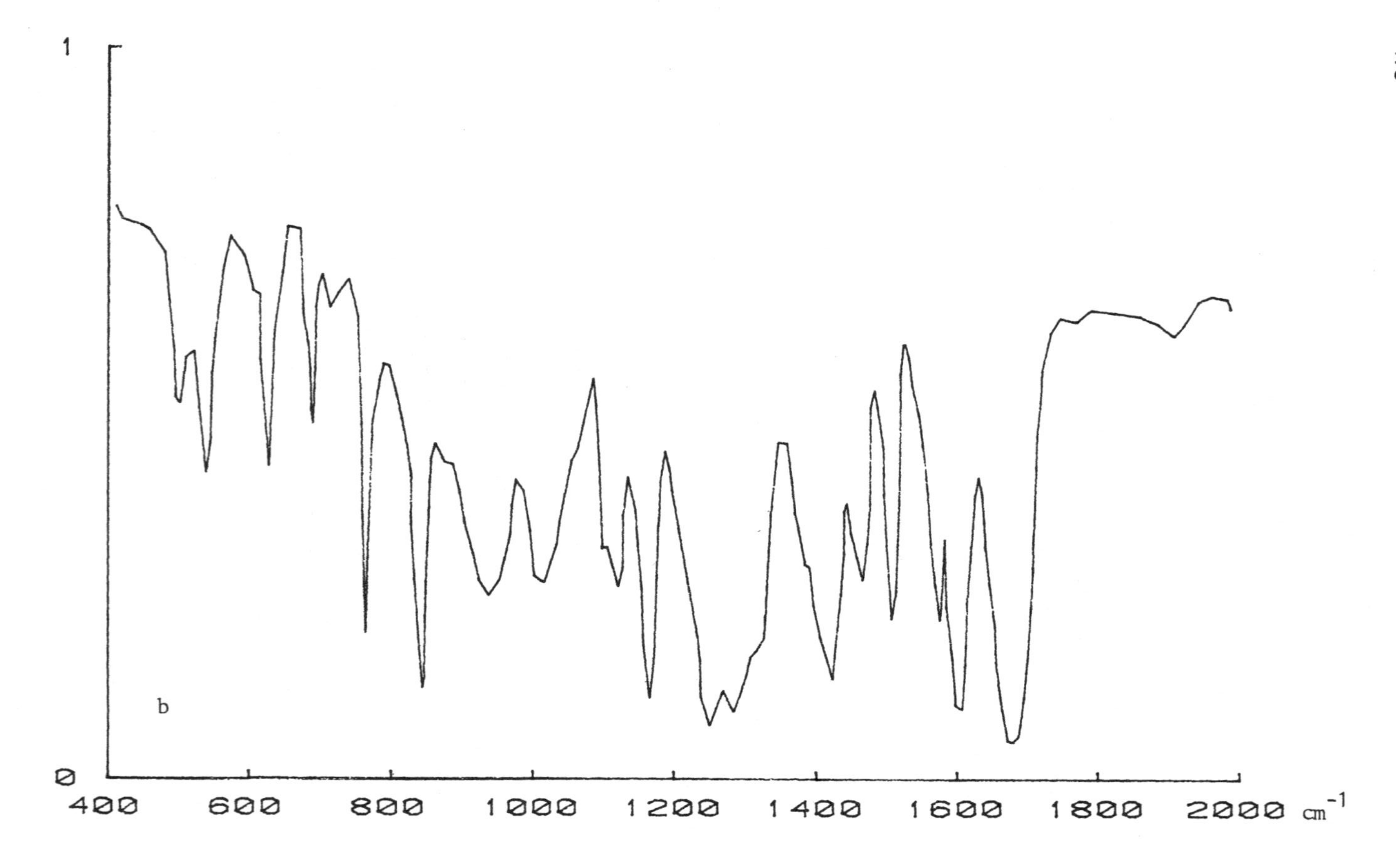

1
0
b
400
600
800
1000
1200
1400
1600
1800
2000 cm⁻¹

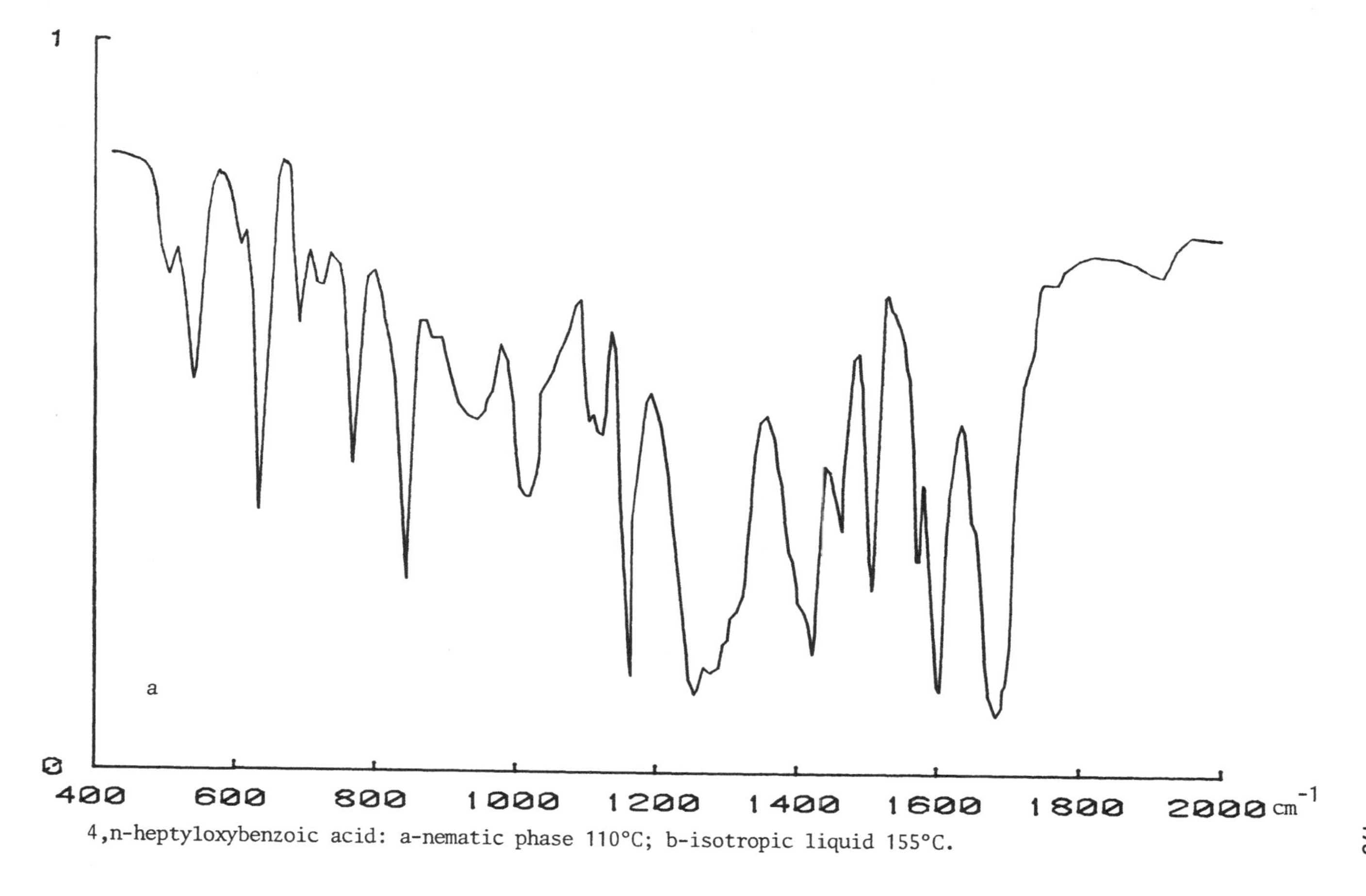

4,n-heptyloxybenzoic acid: a-nematic phase 110°C; b-isotropic liquid 155°C.

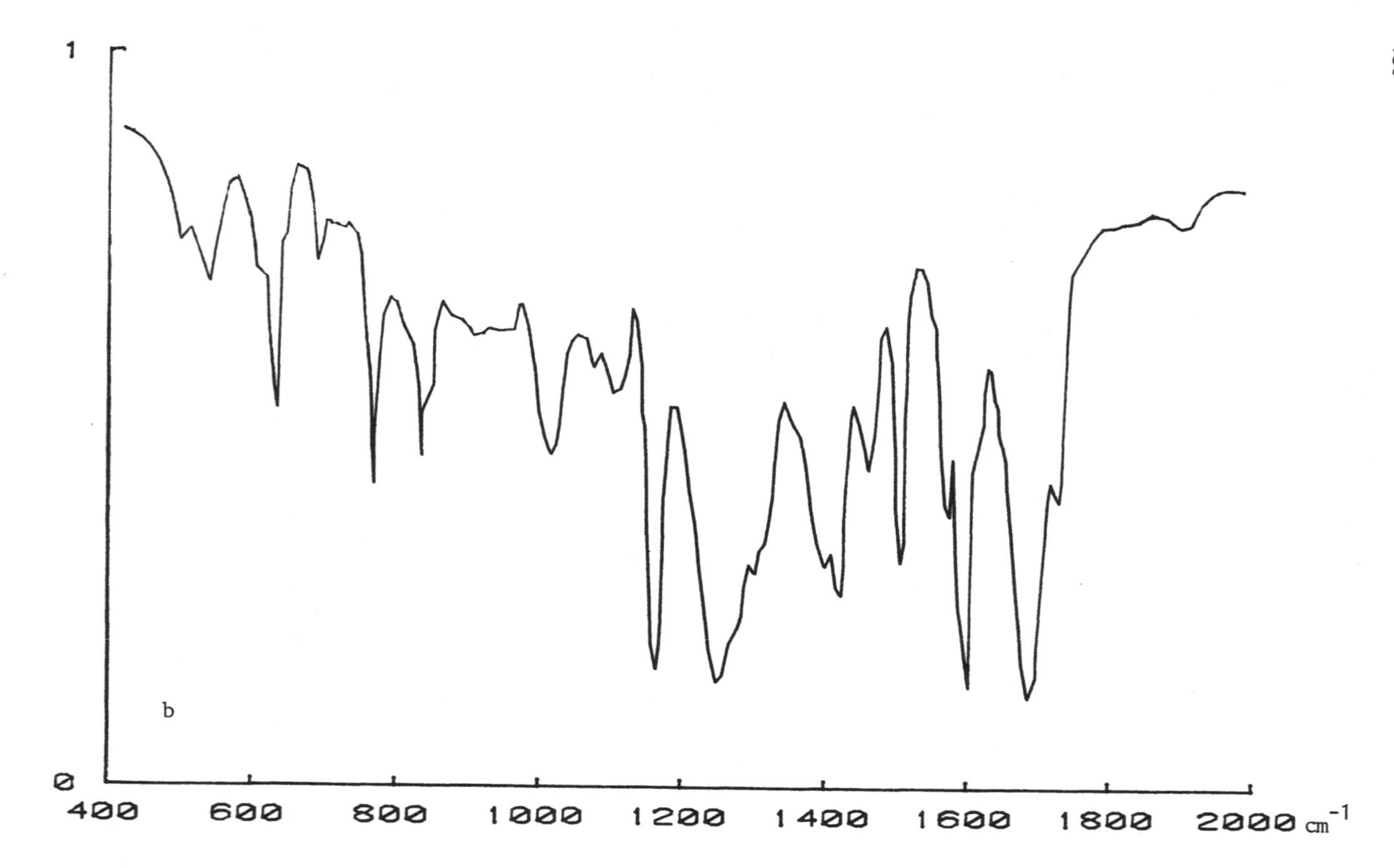
1
0
b
400
600
800
1000
1200
1400
1600
1800
2000 cm-1

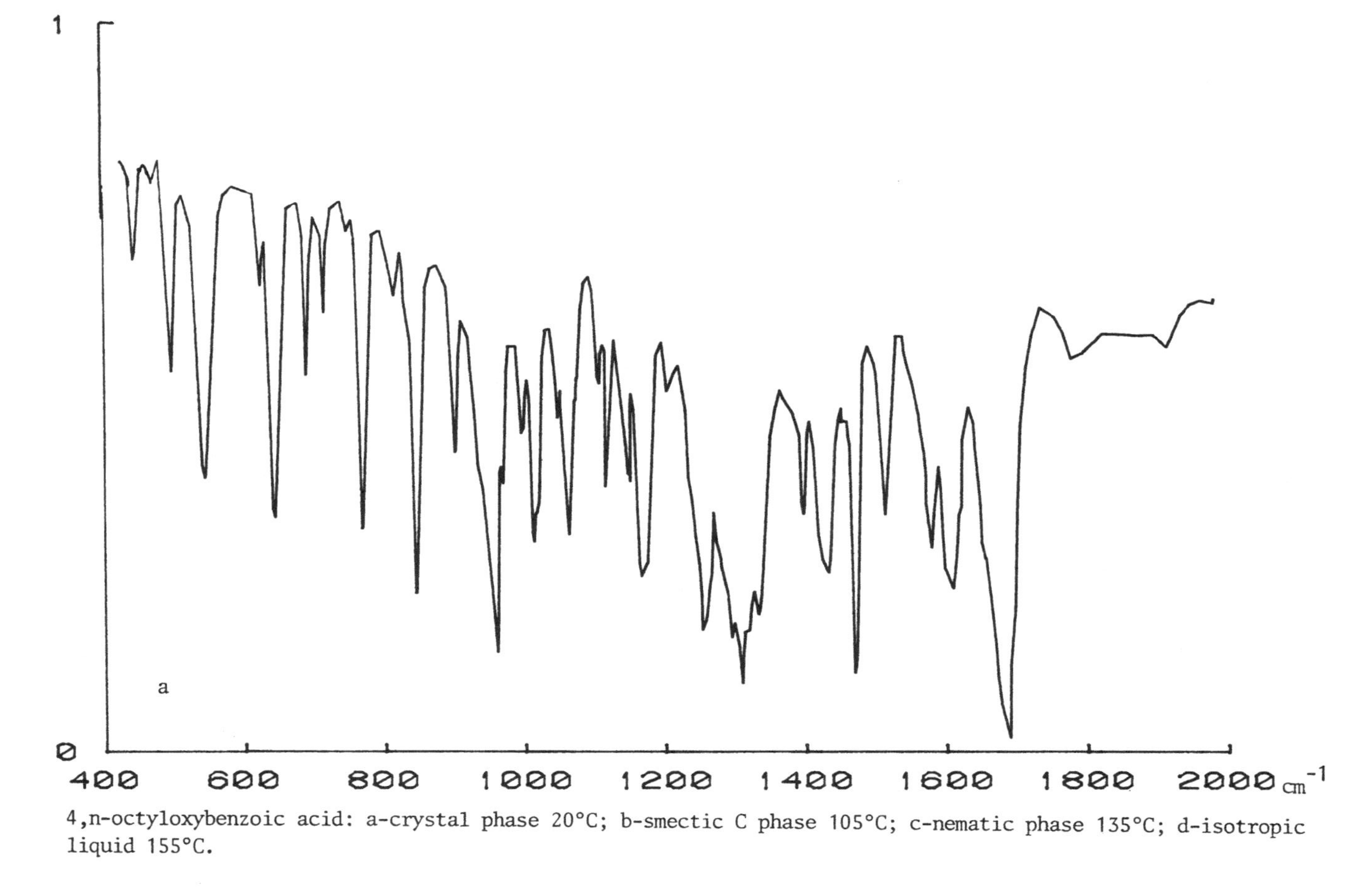

4,n-octyloxybenzoic acid: a-crystal phase 20°C; b-smectic C phase 105°C; c-nematic phase 135°C; d-isotropic liquid 155°C.

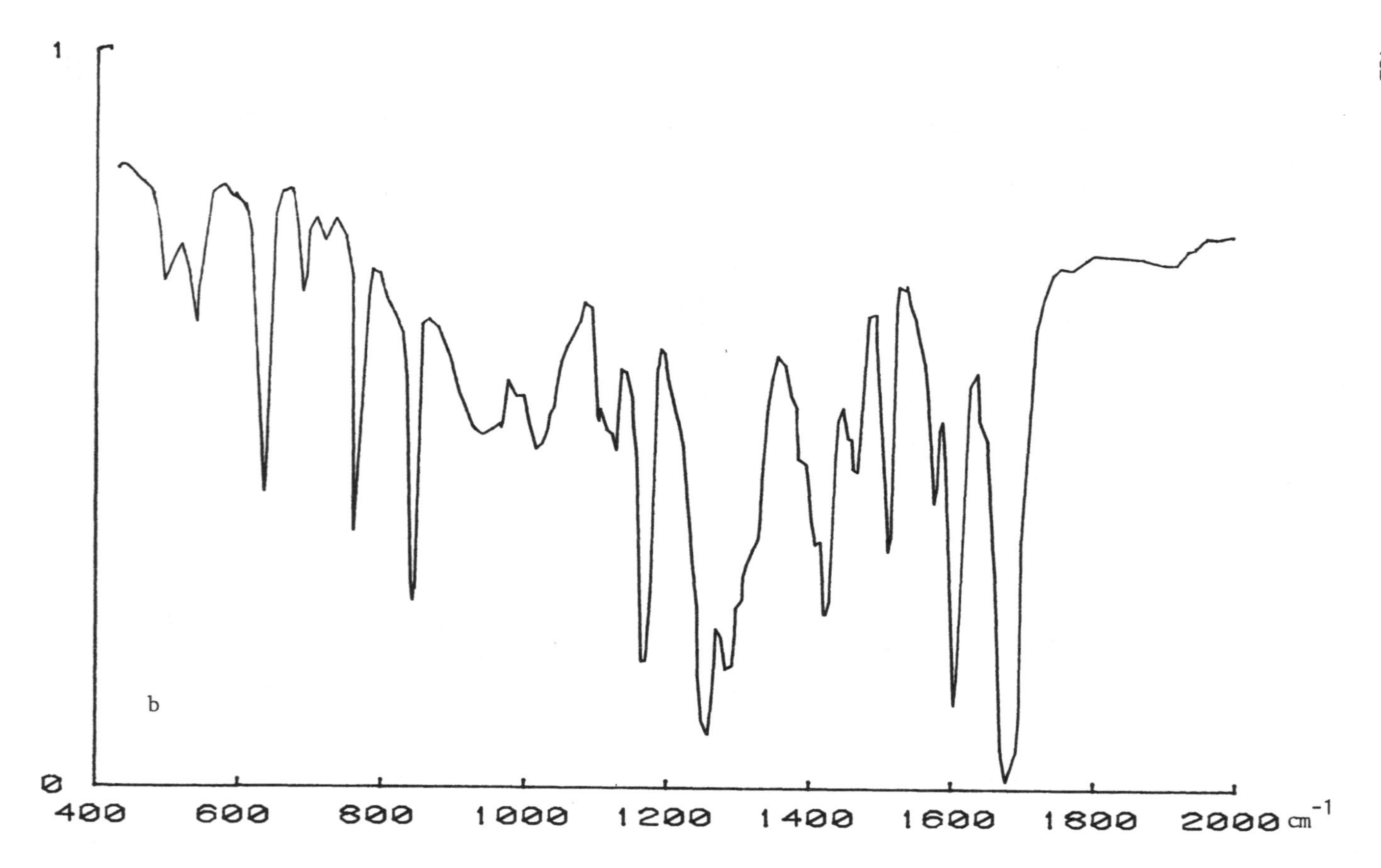

1
0
400
600
800
1000
1200
1400
1600
1800
2000
cm^{-1}
b

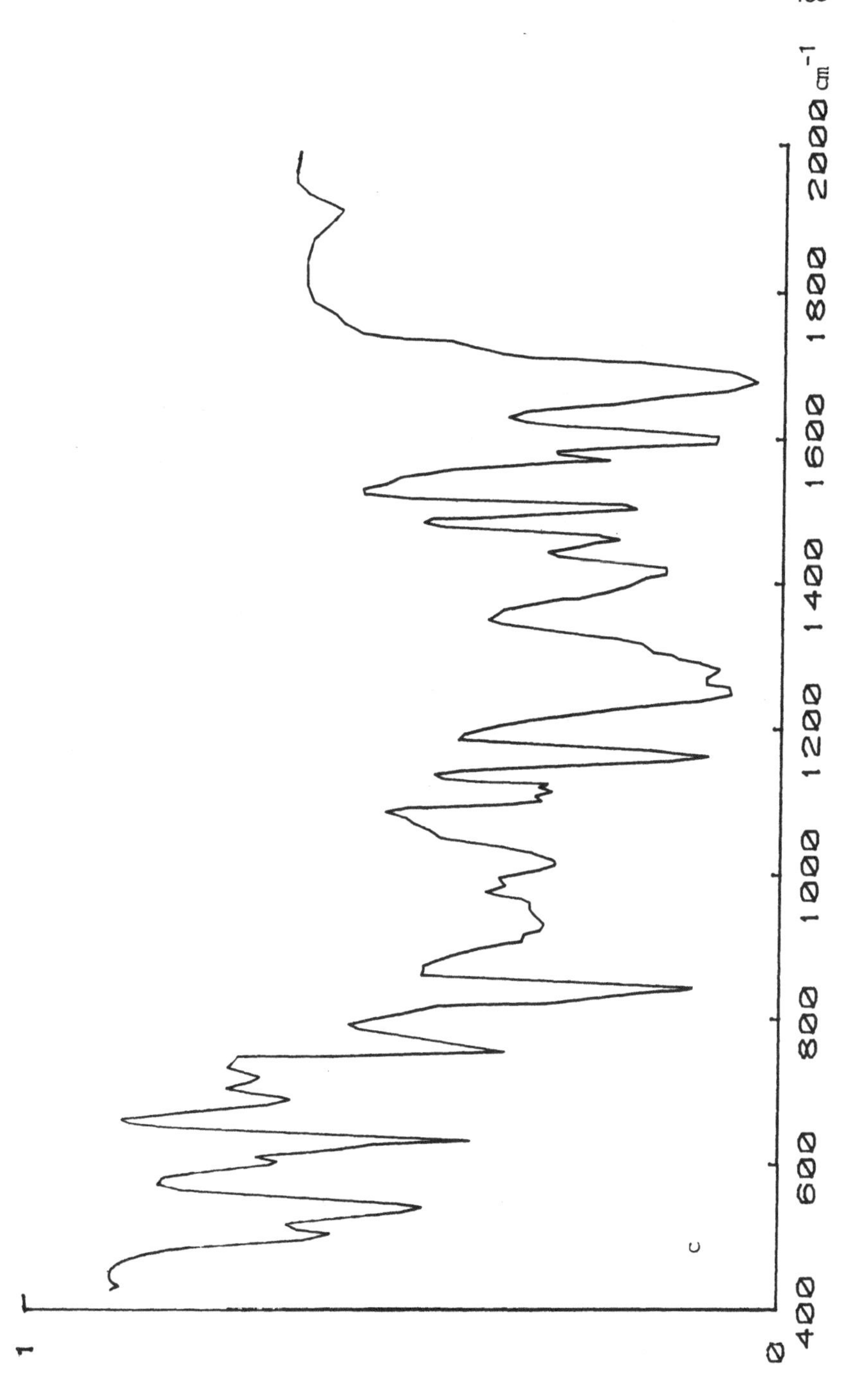

1
0
c
400
600
800
1000
1200
1400
1600
1800
2000 cm⁻¹

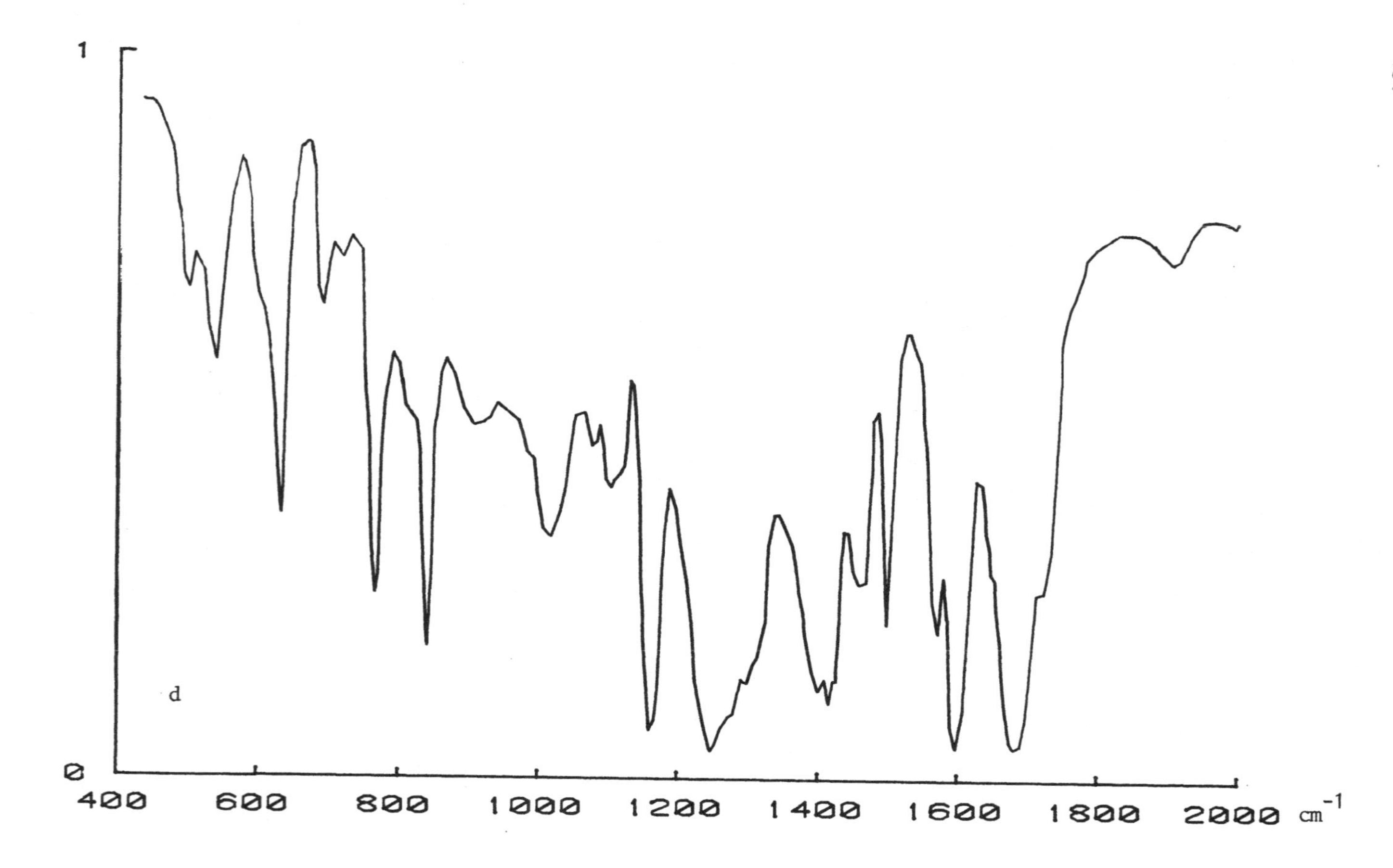

1
0
d
400
600
800
1000
1200
1400
1600
1800
2000
cm^{-1}

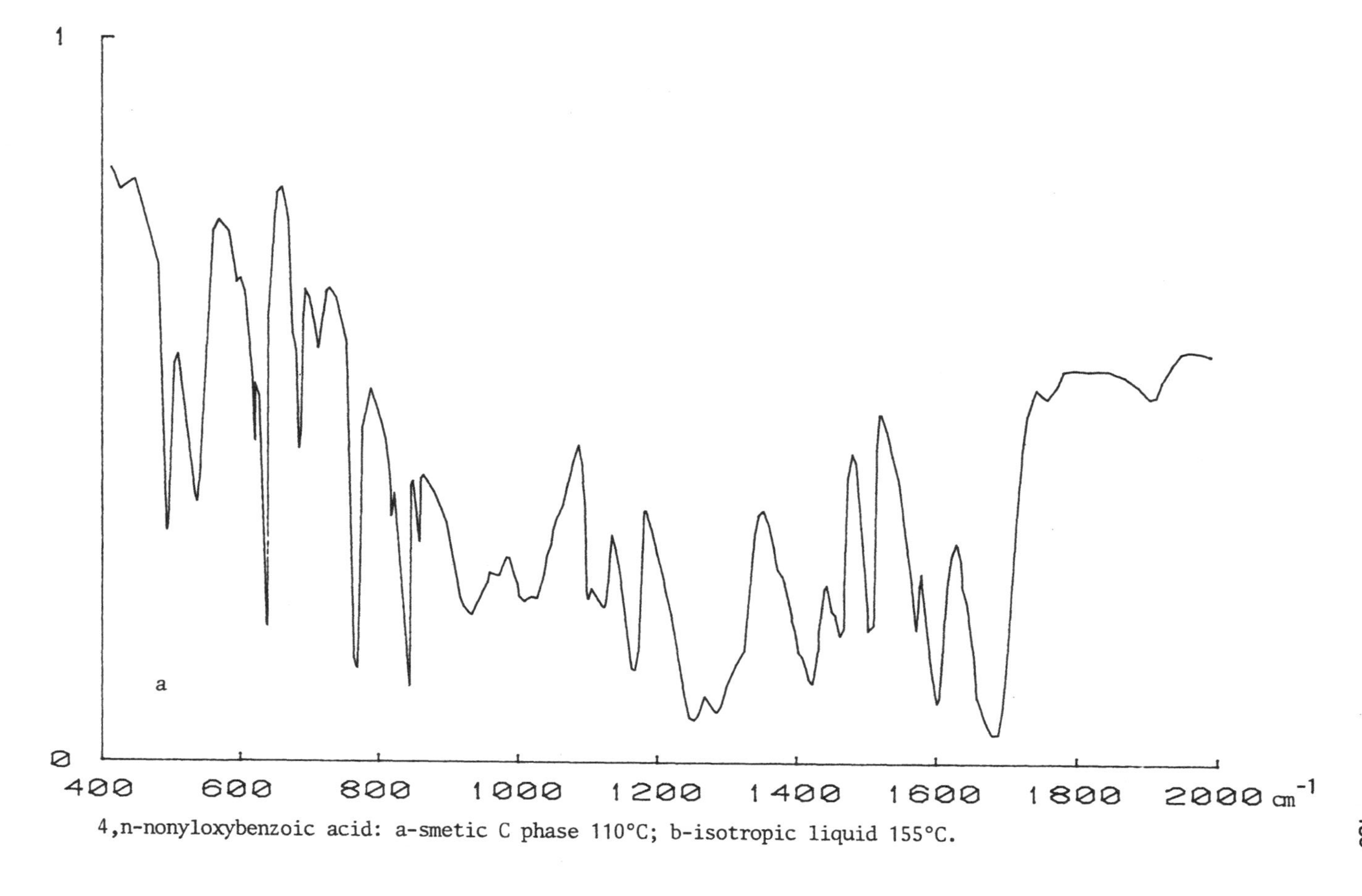

4,n-nonyloxybenzoic acid: a-smetic C phase 110°C; b-isotropic liquid 155°C.

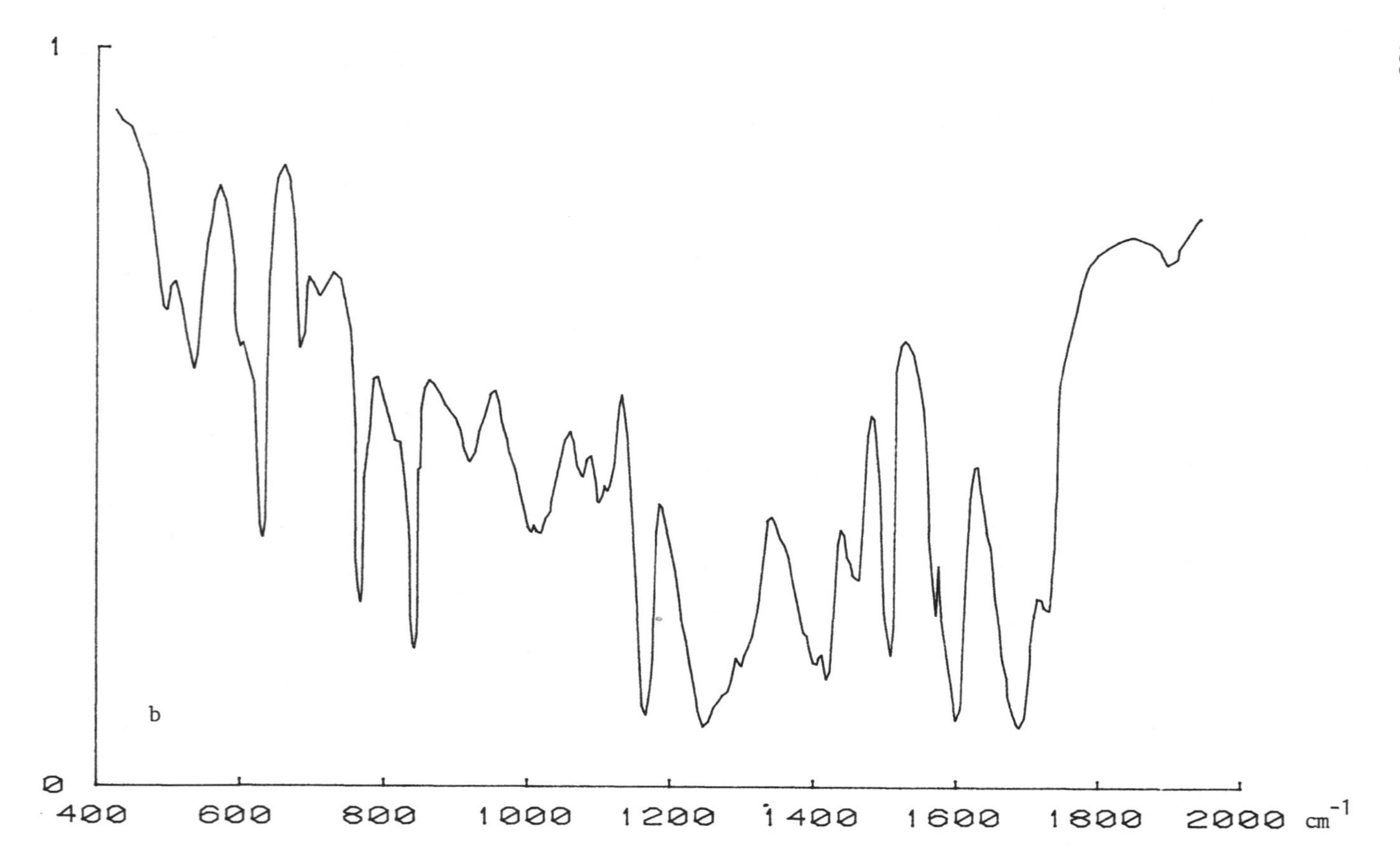

1
0
b
400
600
800
1000
1200
1400
1600
1800
2000 cm-1

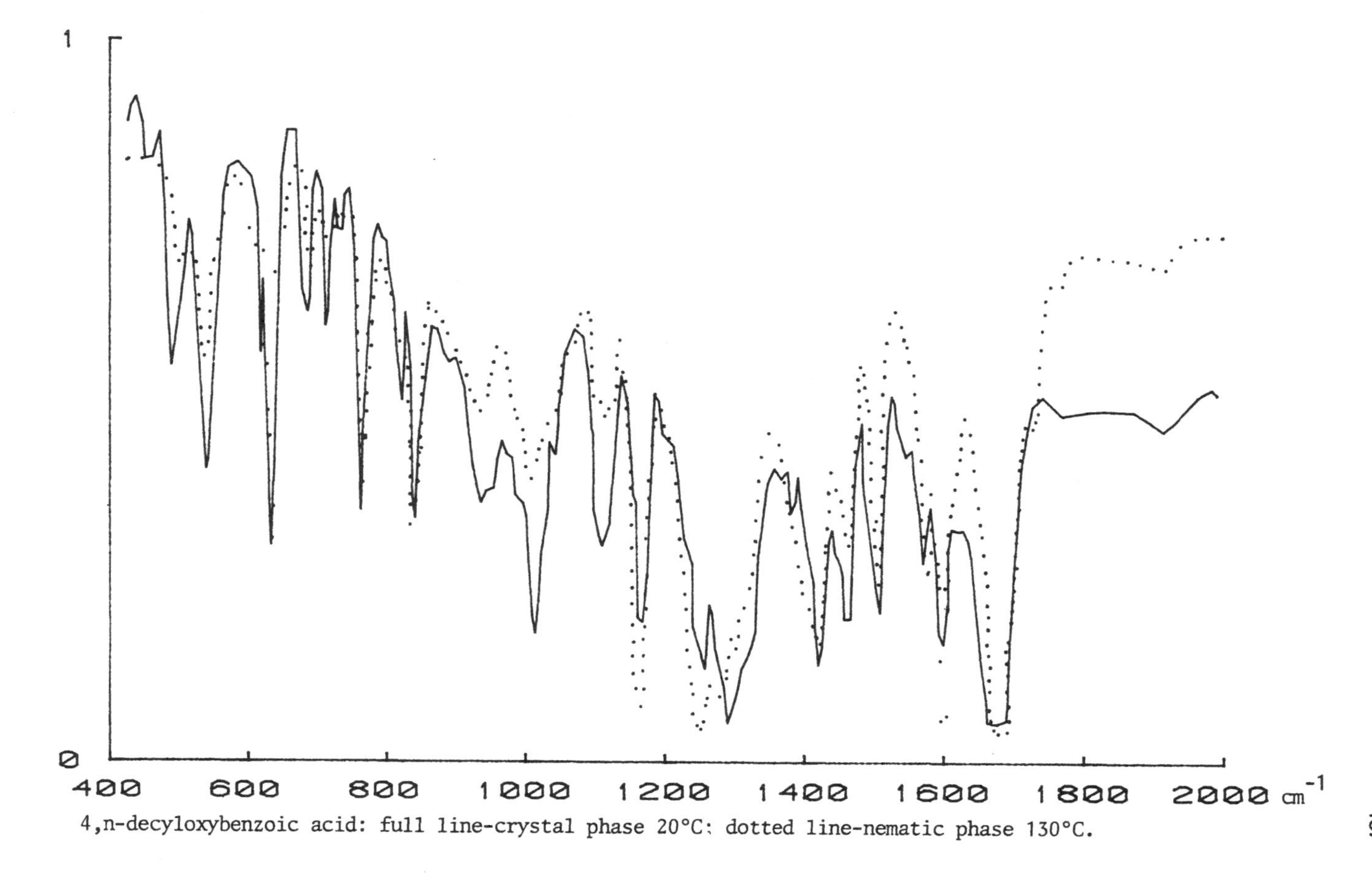

4,n-decyloxybenzoic acid: full line-crystal phase 20°C; dotted line-nematic phase 130°C.

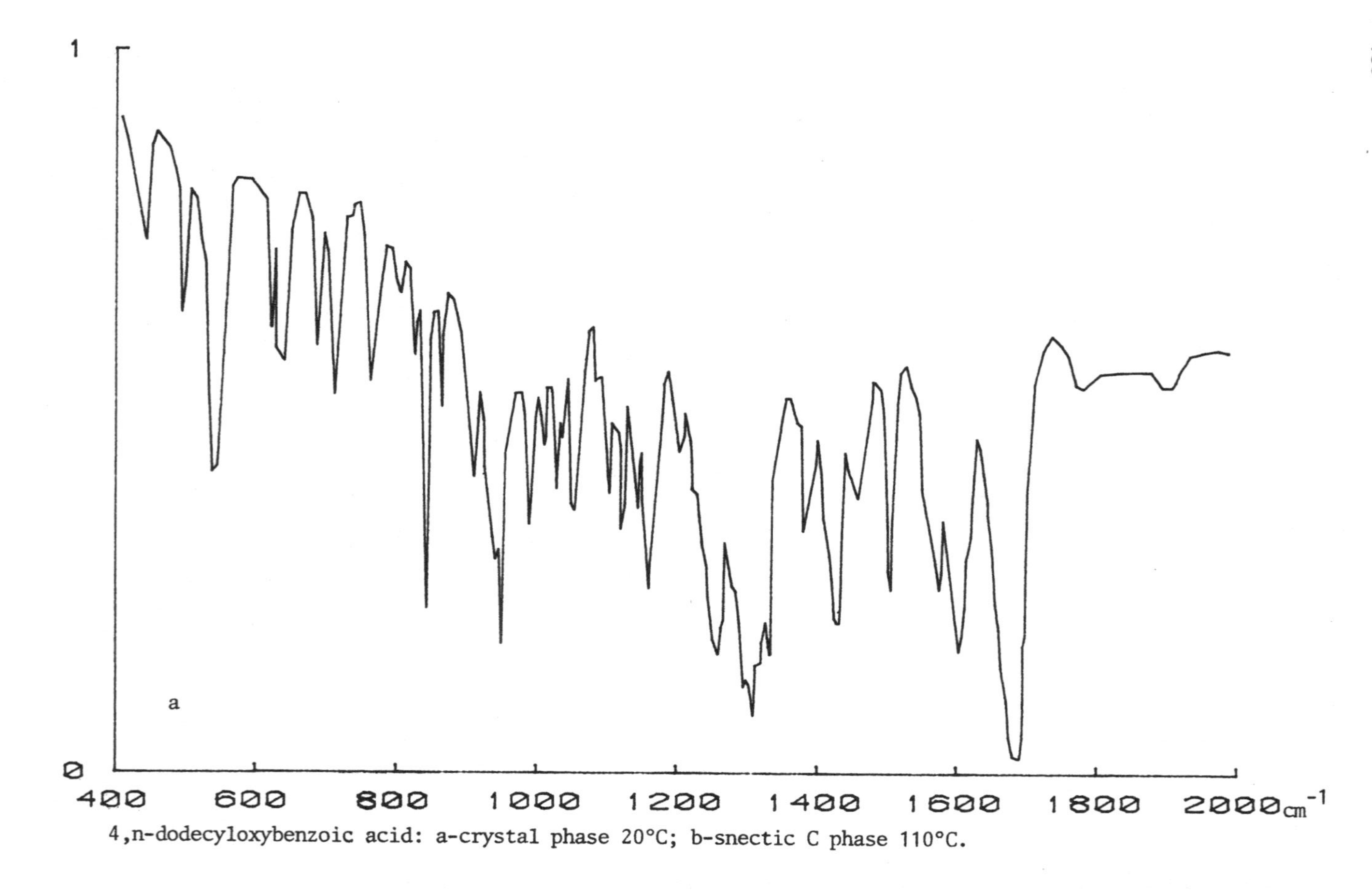

4,n-dodecyloxybenzoic acid: a-crystal phase 20°C; b-snectic C phase 110°C.

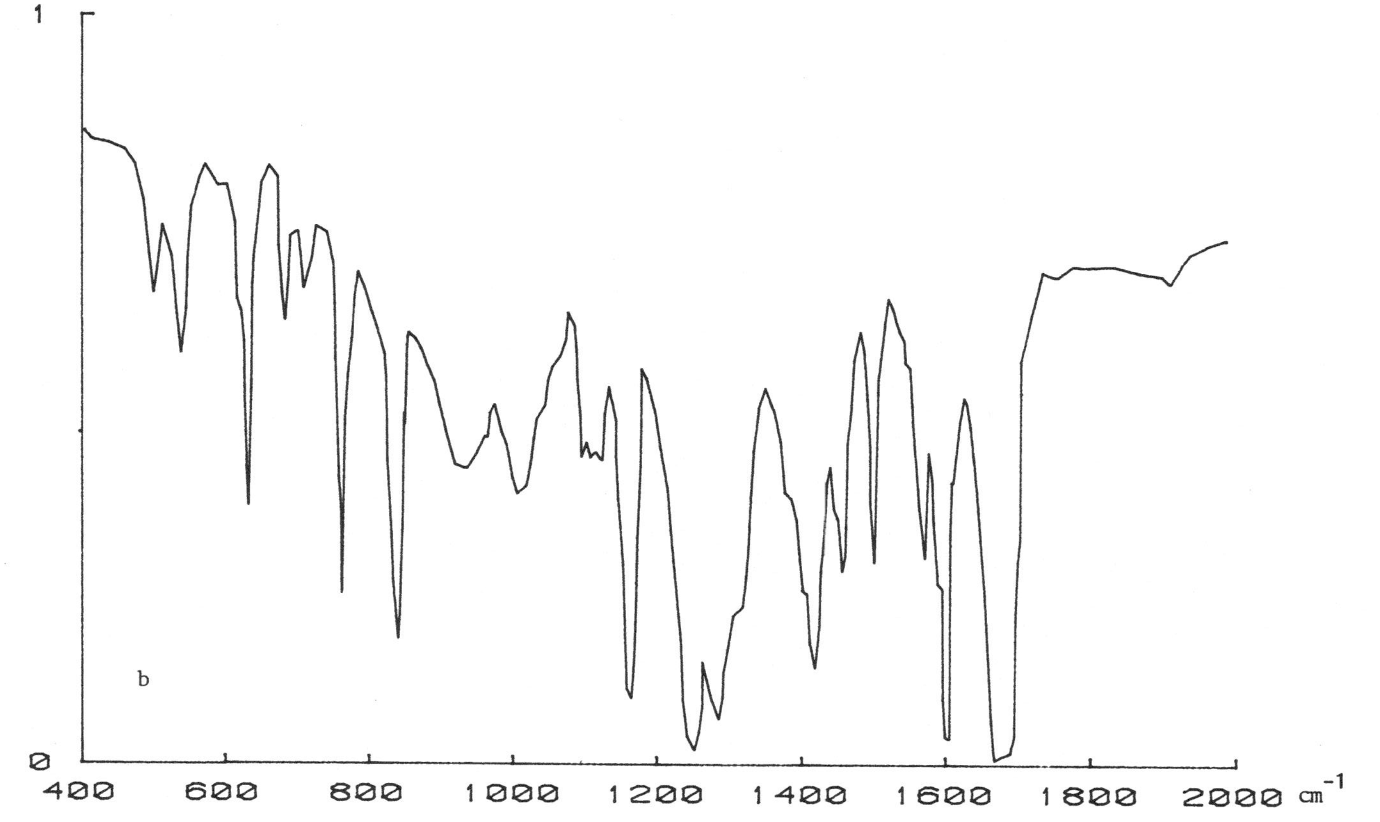
1
0
400
600
800
1000
1200
1400
1600
1800
2000 cm^{-1}
b

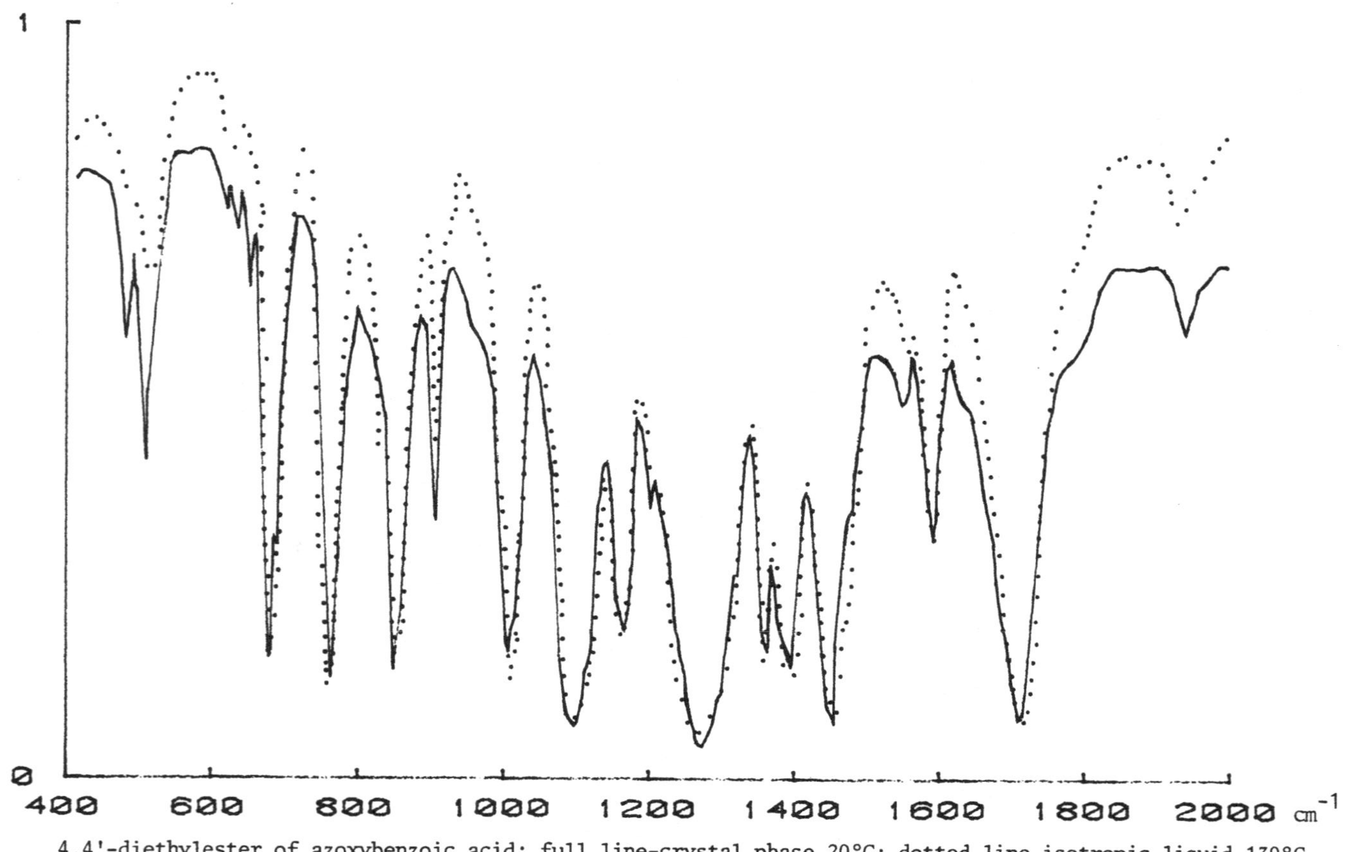

4,4'-diethylester of azoxybenzoic acid: full line-crystal phase 20°C; dotted line-isotropic liquid 130°C.

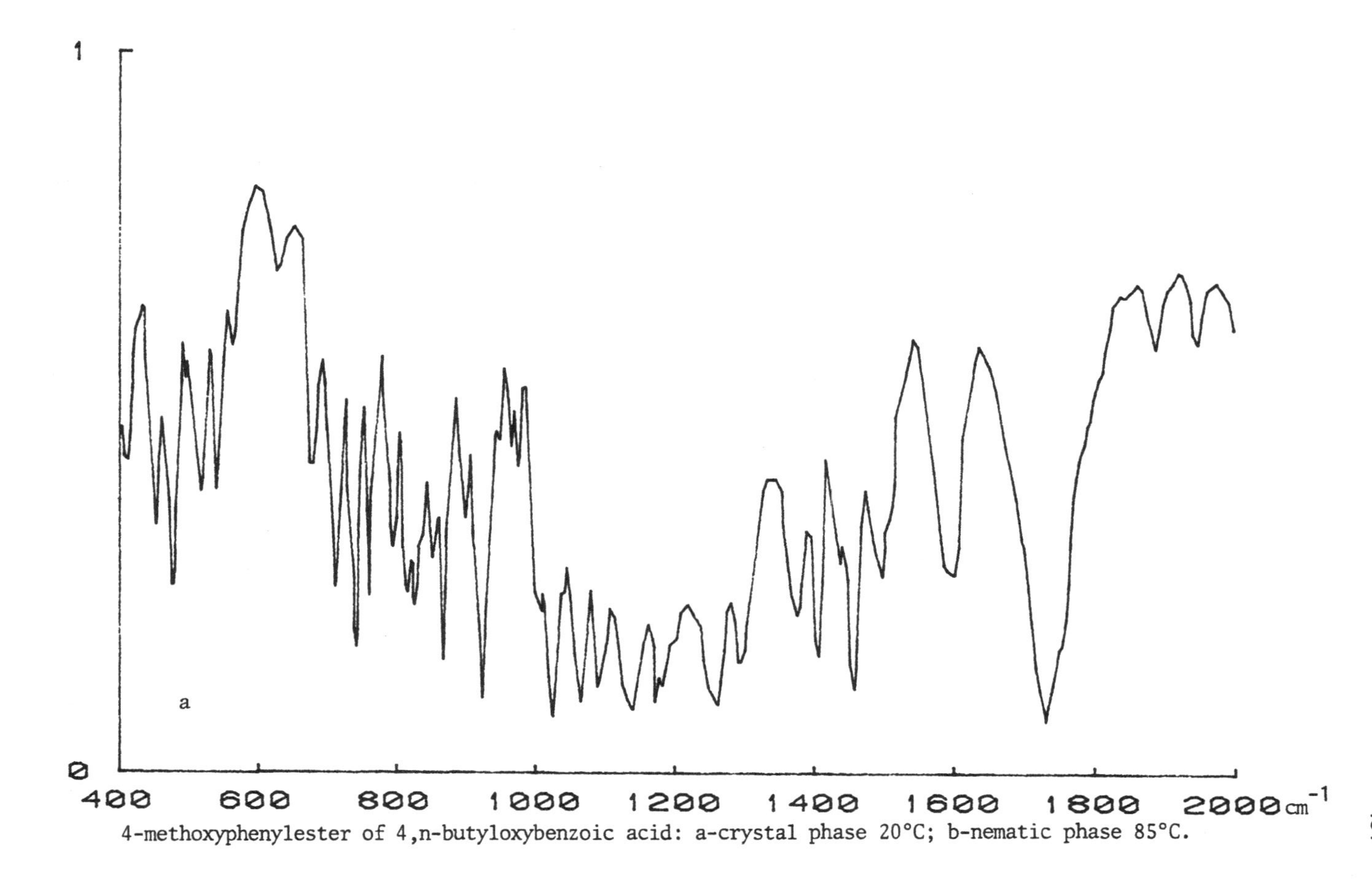

4-methoxyphenylester of 4,n-butyloxybenzoic acid: a-crystal phase 20°C; b-nematic phase 85°C.

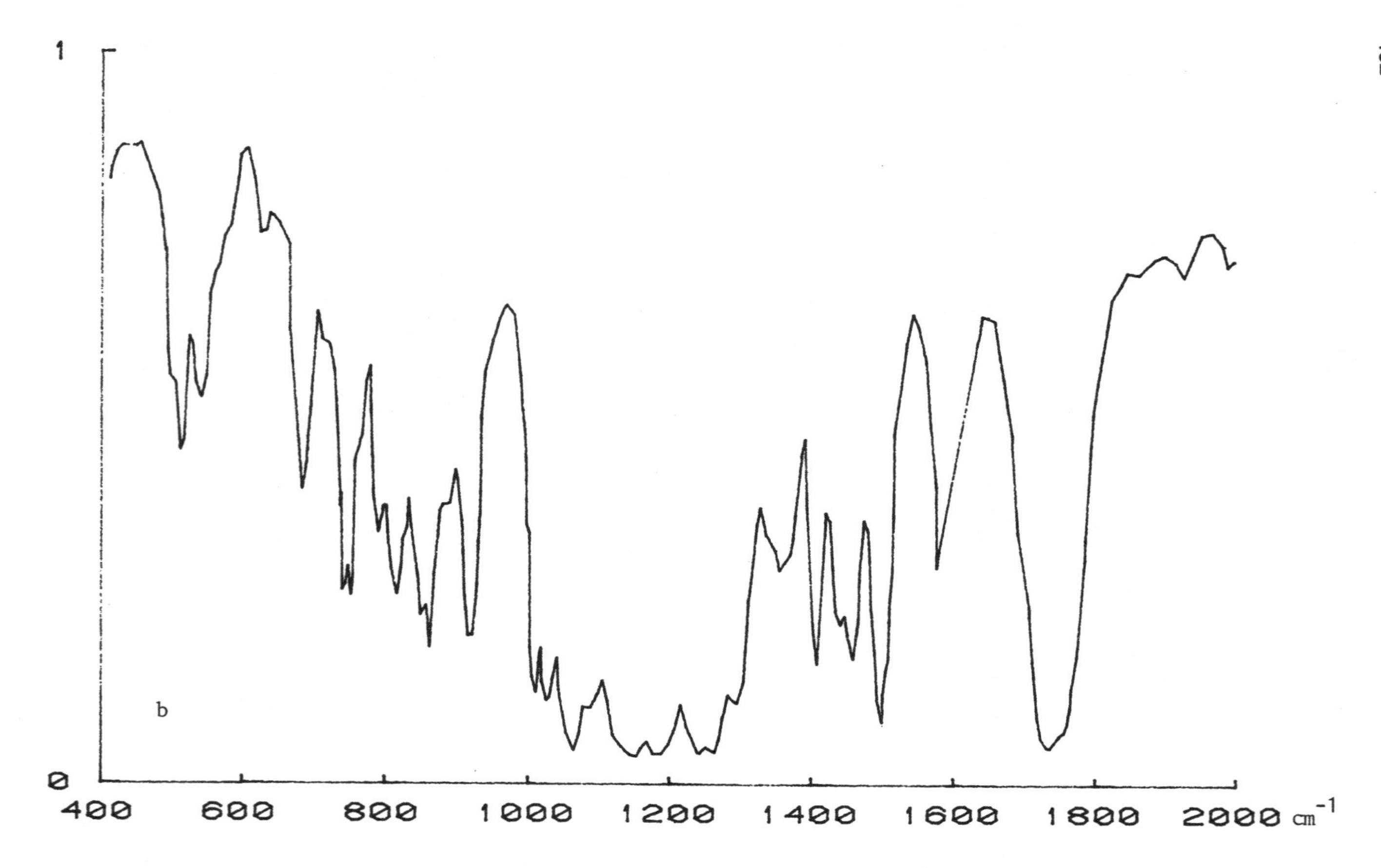
1
0
b
400
600
800
1000
1200
1400
1600
1800
2000 cm-1

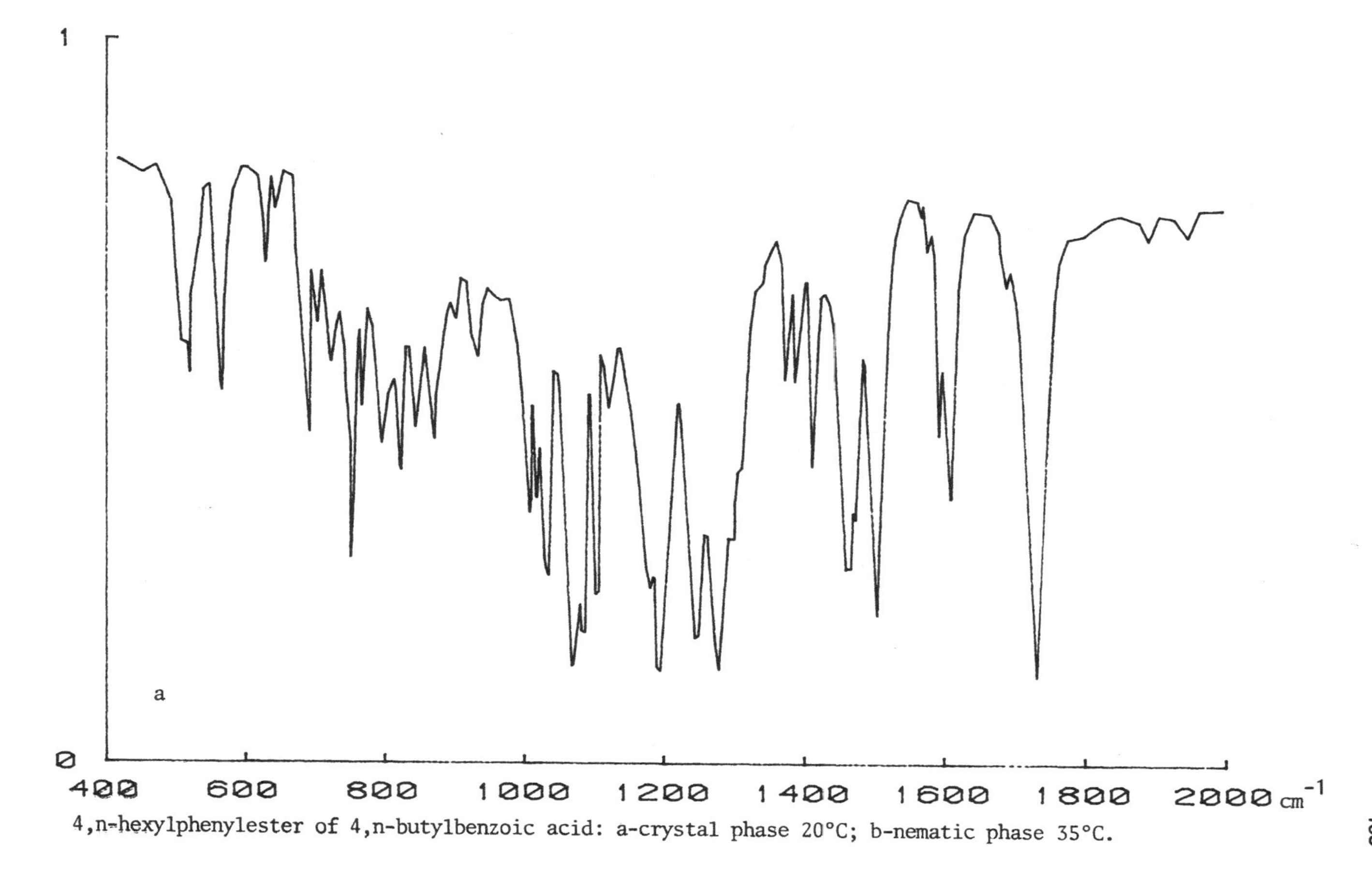

4,n-hexylphenylester of 4,n-butylbenzoic acid: a-crystal phase 20°C; b-nematic phase 35°C.

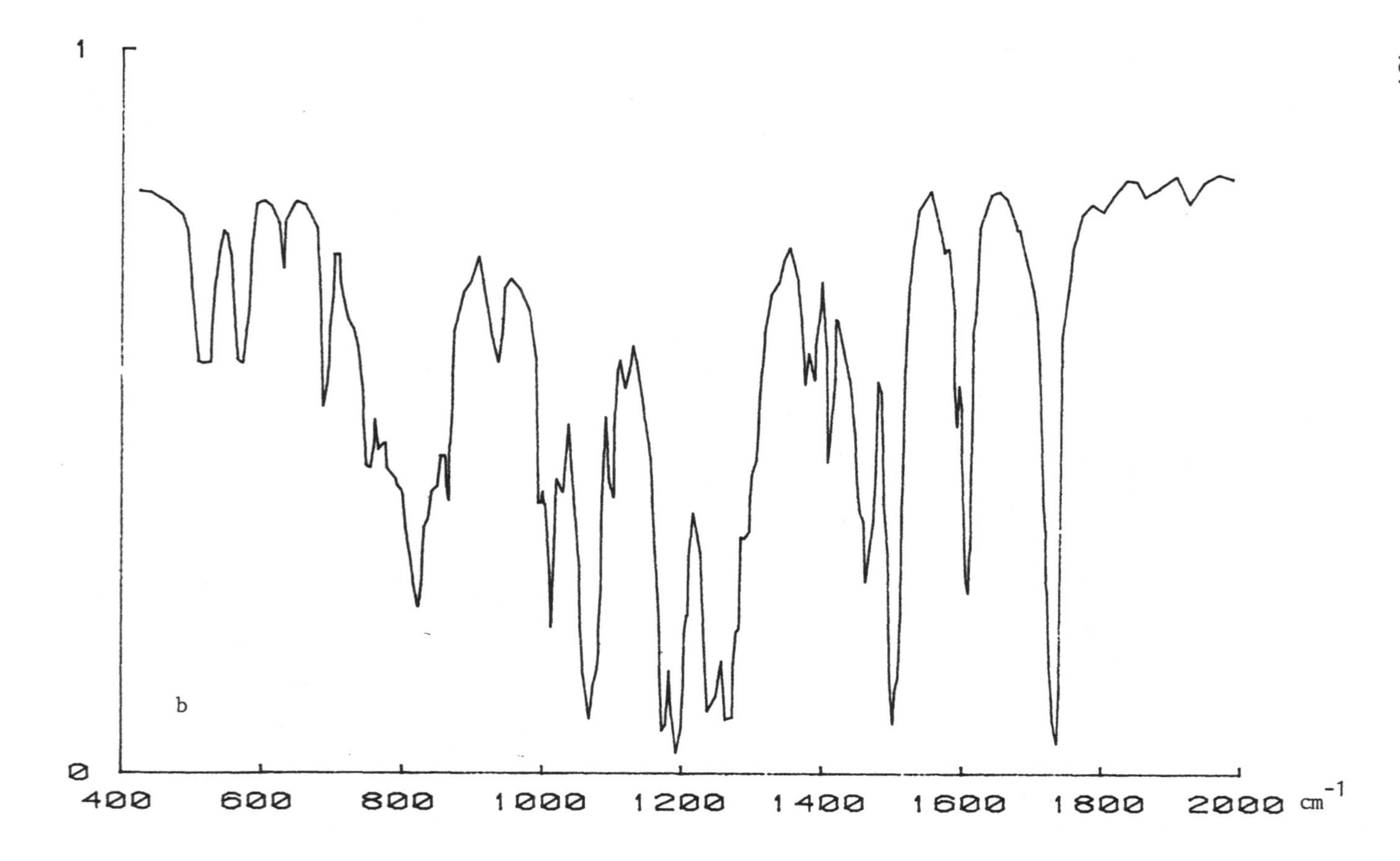
1
0
b
400
600
800
1000
1200
1400
1600
1800
2000 cm^{-1}

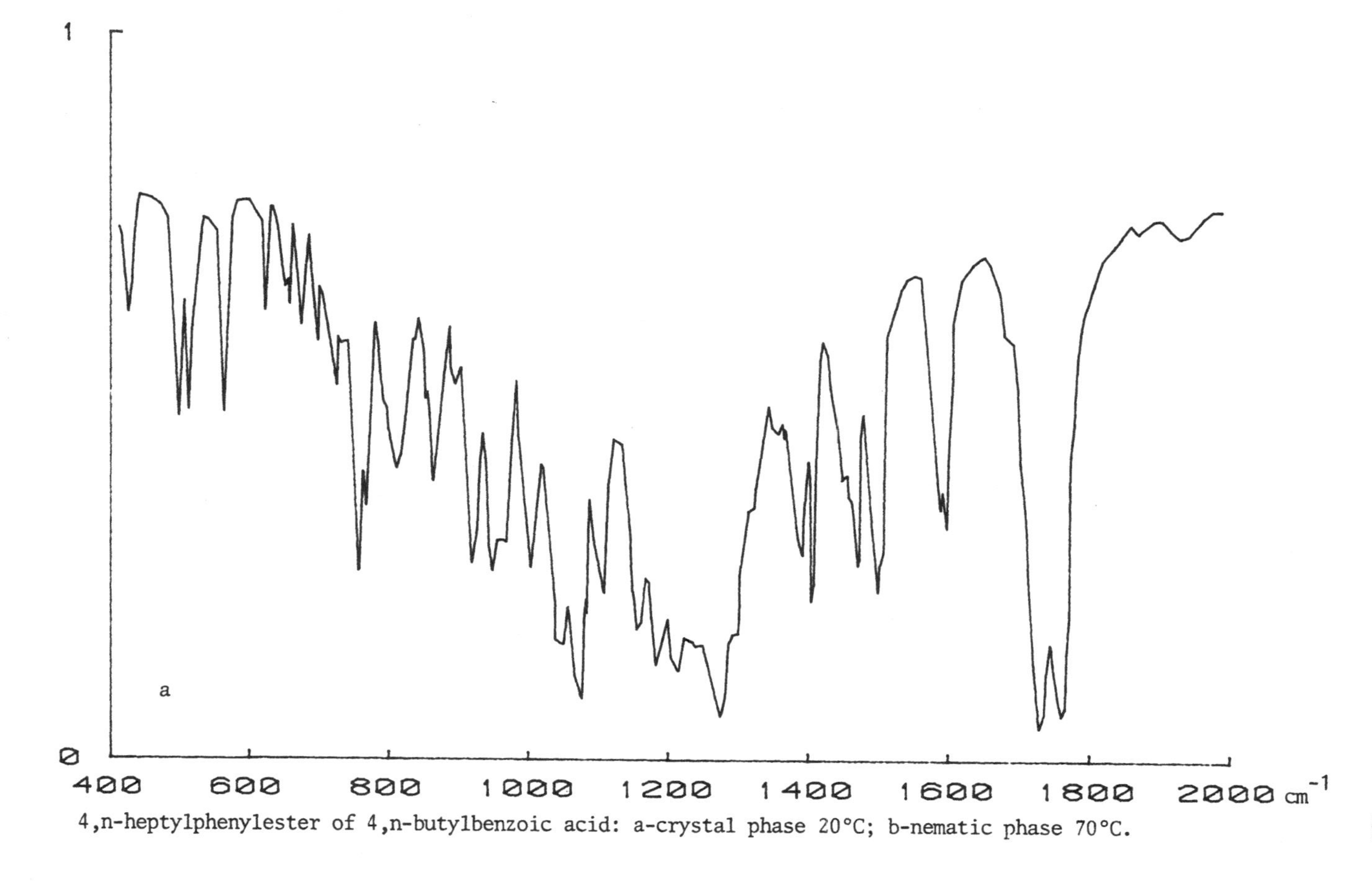

4,n-heptylphenylester of 4,n-butylbenzoic acid: a-crystal phase 20°C; b-nematic phase 70°C.

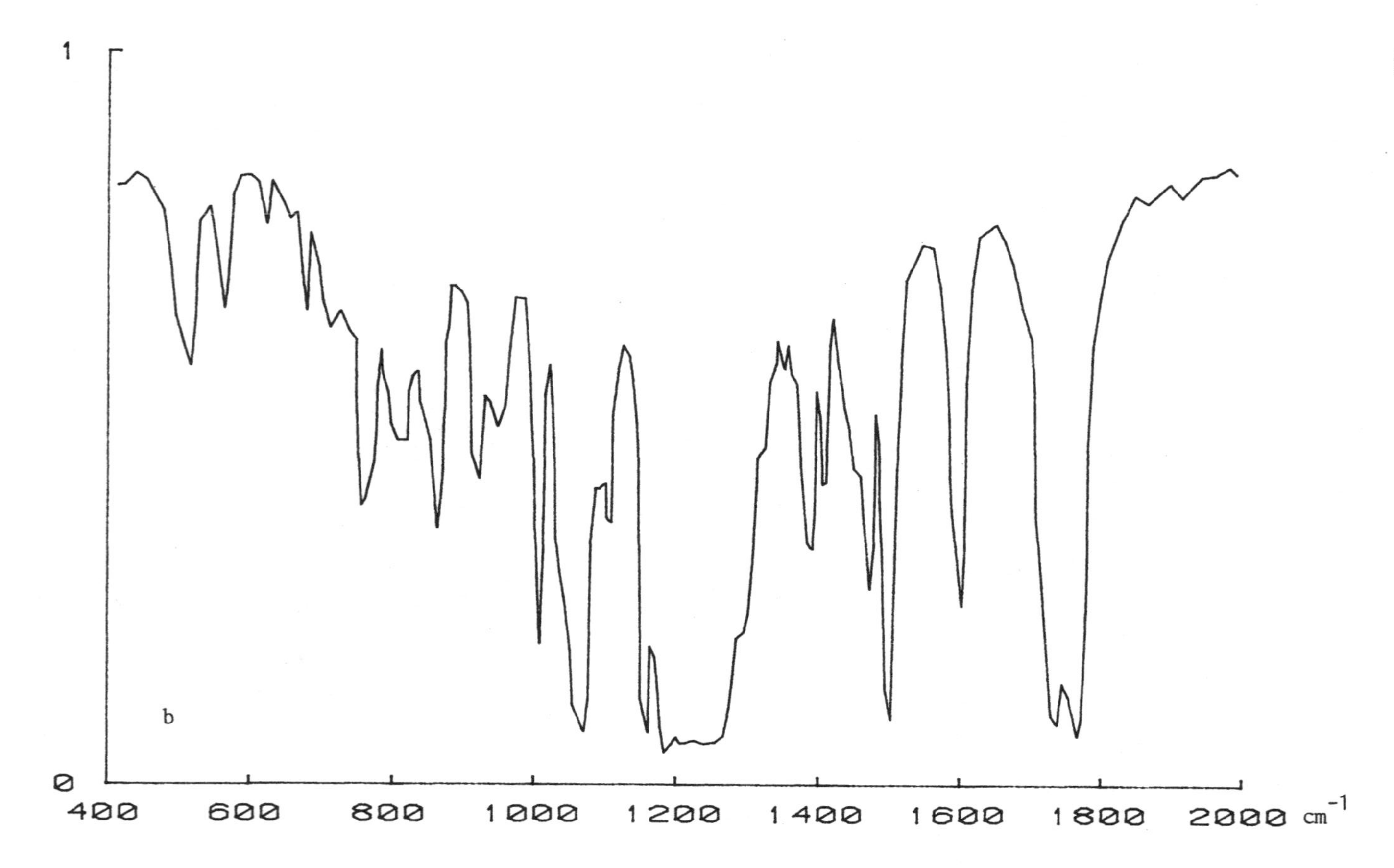
1
0
400
600
800
1000
1200
1400
1600
1800
2000 cm-1
b

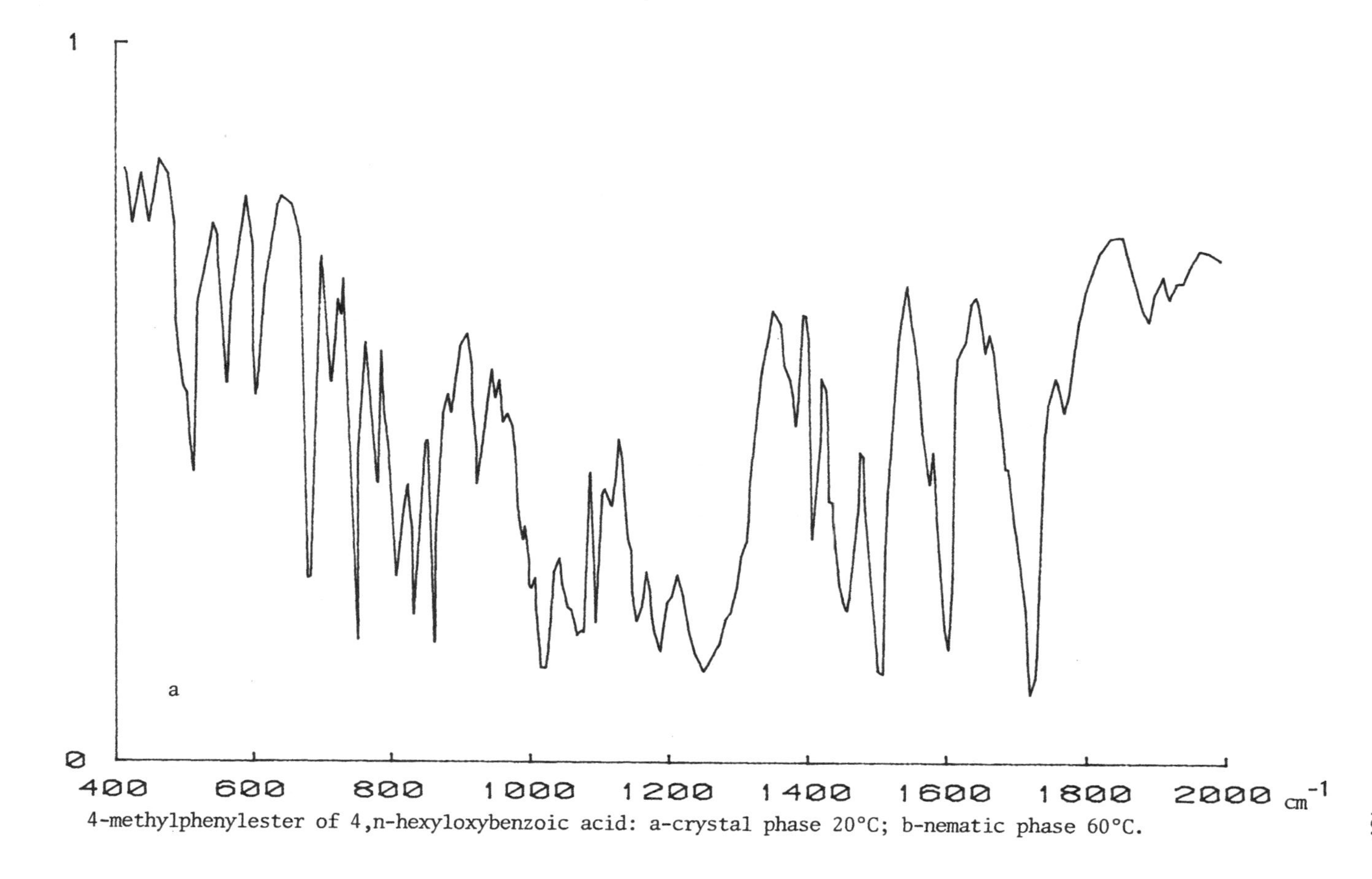

4-methylphenylester of 4,n-hexyloxybenzoic acid: a-crystal phase 20°C; b-nematic phase 60°C.

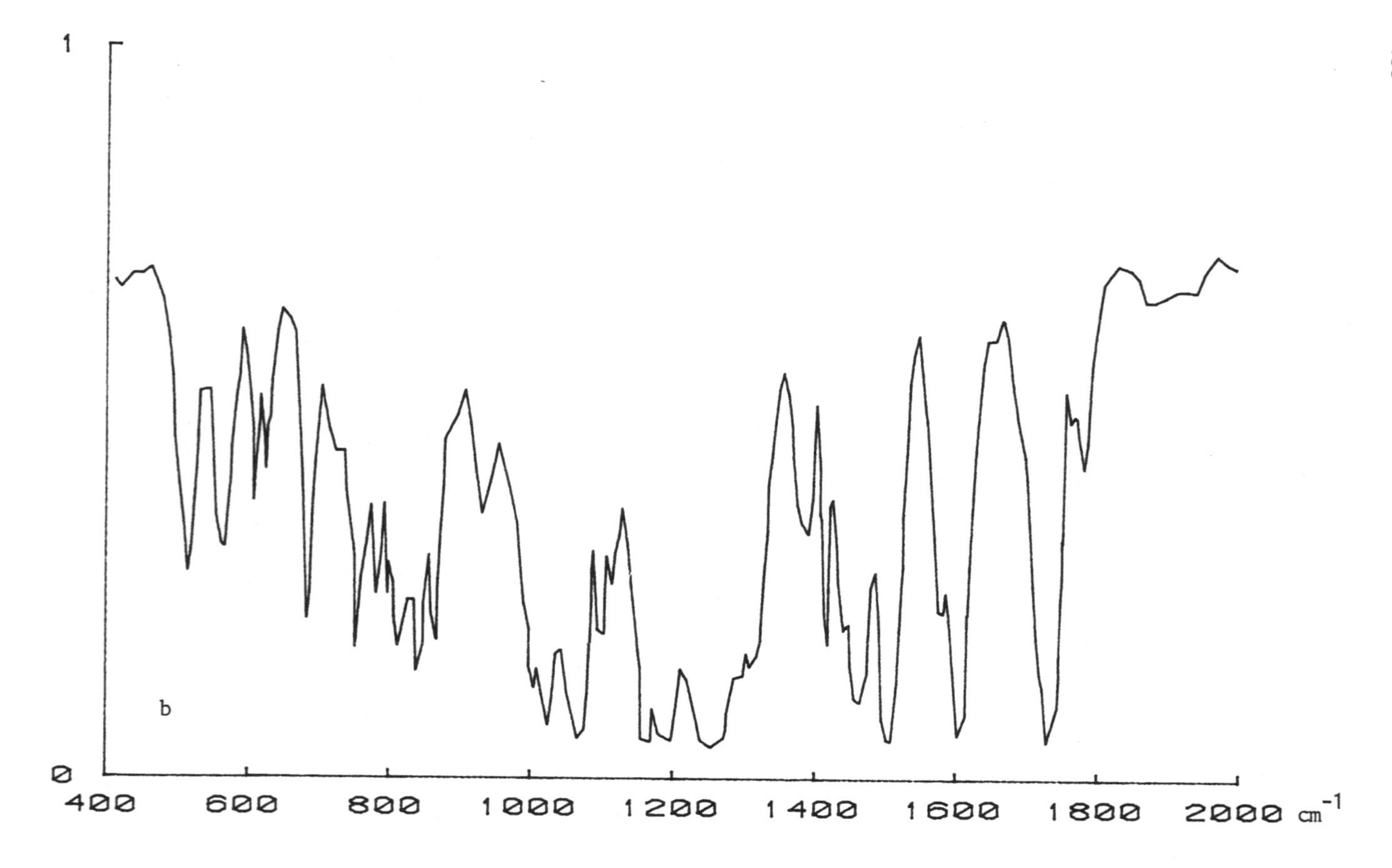
1
0
b
400
600
800
1000
1200
1400
1600
1800
2000 cm^{-1}

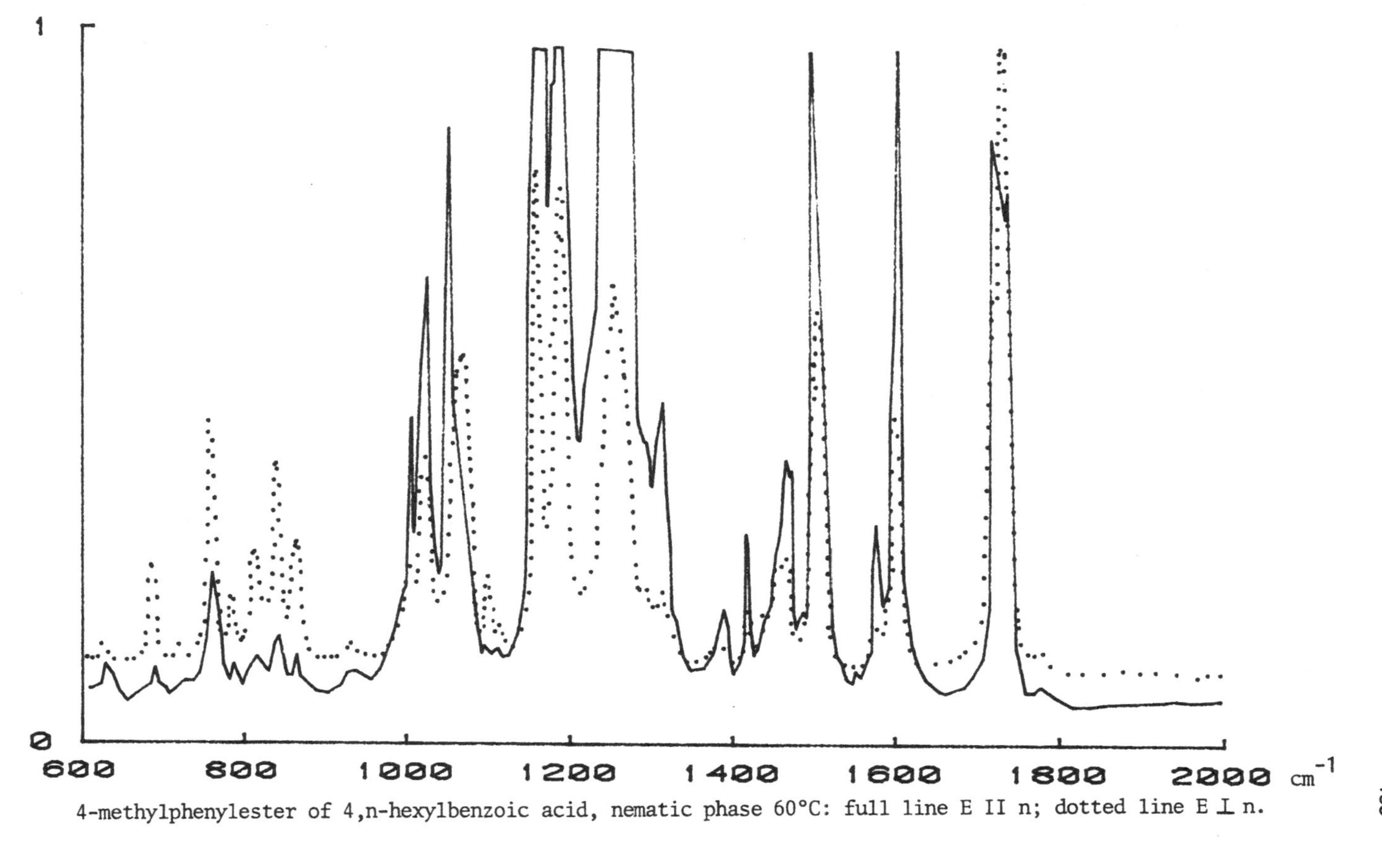

4-methylphenylester of 4,n-hexylbenzoic acid, nematic phase 60°C: full line E II n; dotted line E ⊥ n.

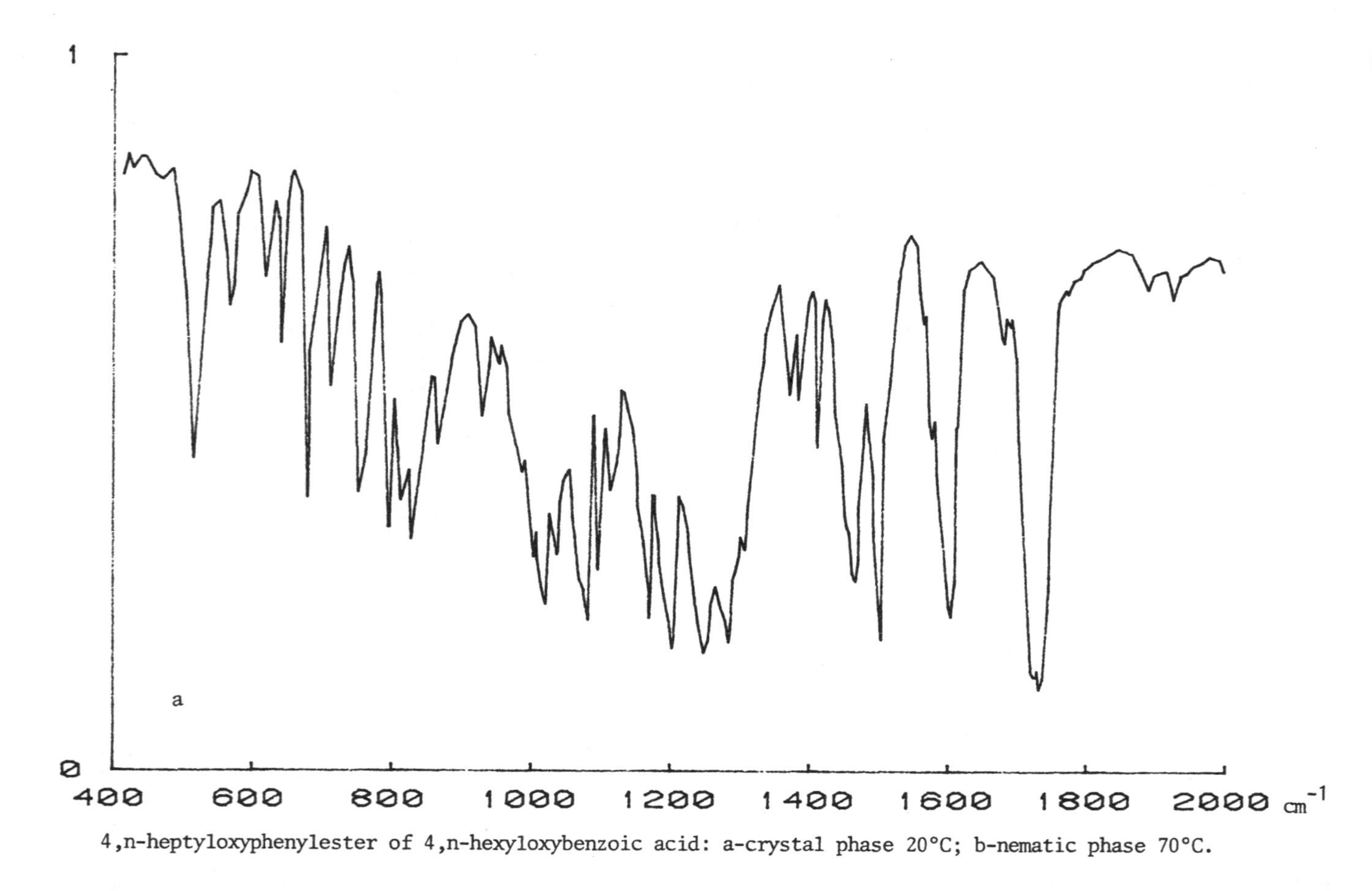

4,n-heptyloxyphenylester of 4,n-hexyloxybenzoic acid: a-crystal phase 20°C; b-nematic phase 70°C.

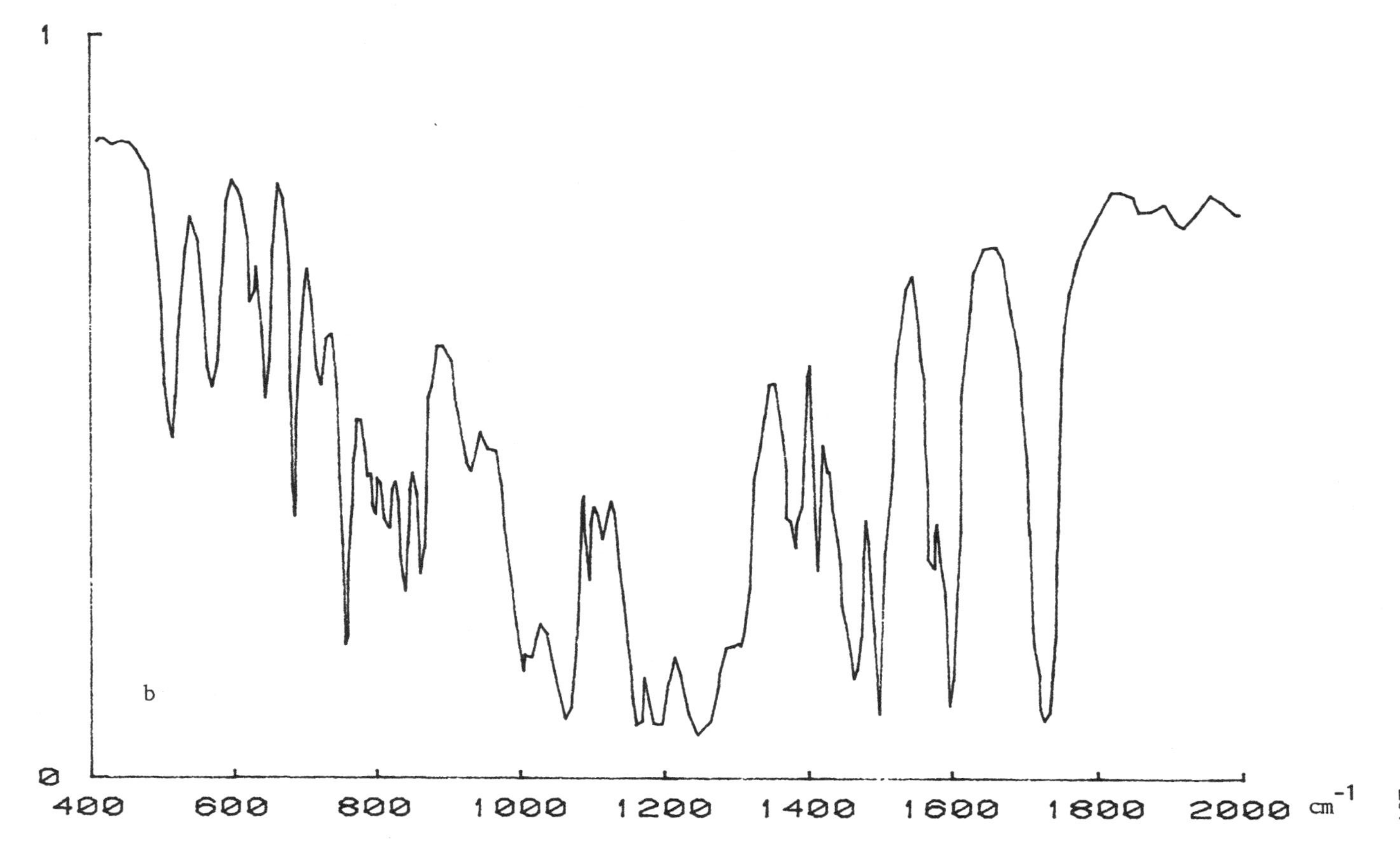
1
0
400
600
800
1000
1200
1400
1600
1800
2000
cm-1
b

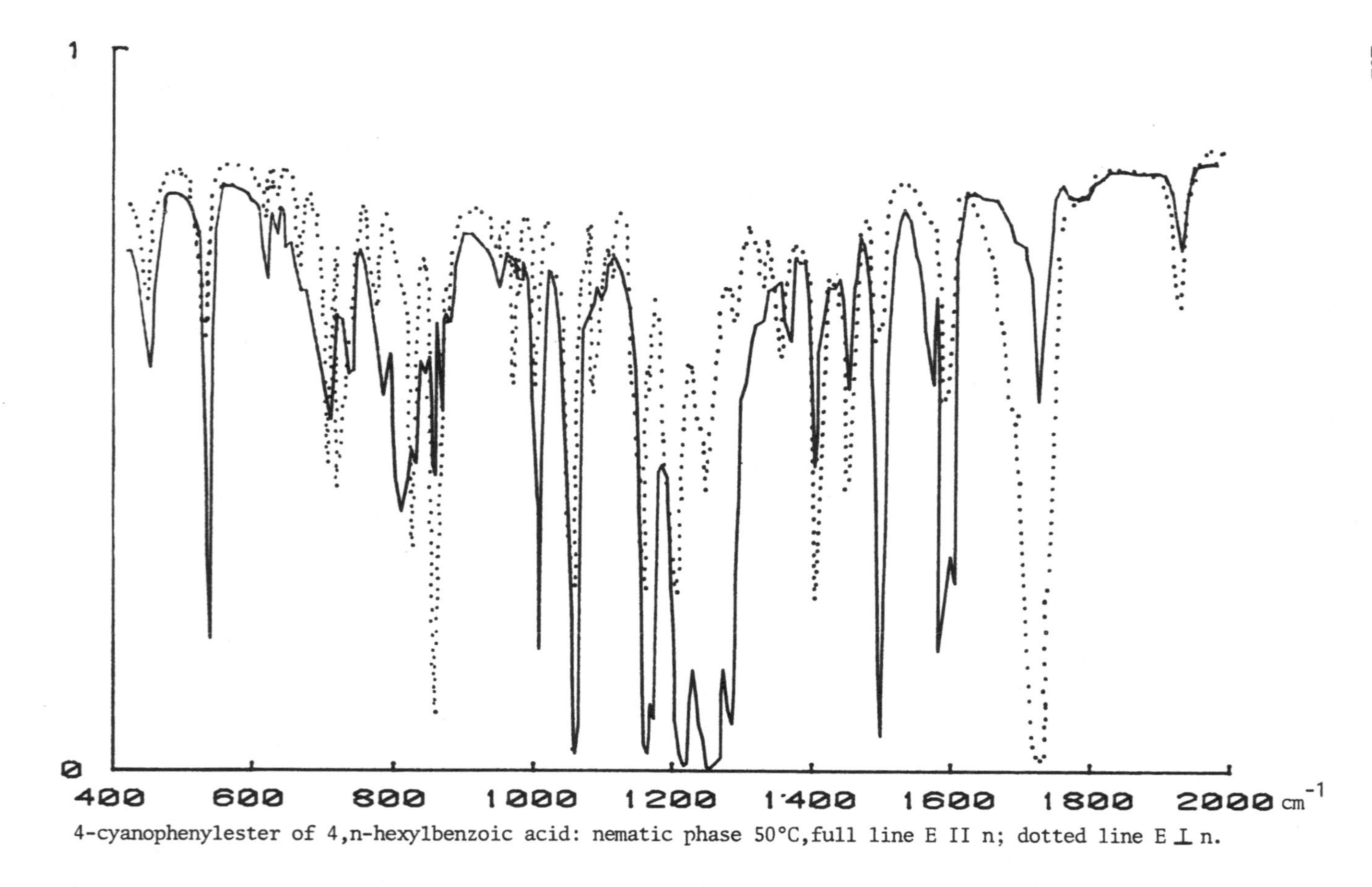

4-cyanophenylester of 4,n-hexylbenzoic acid: nematic phase 50°C, full line E II n; dotted line E ⊥ n.

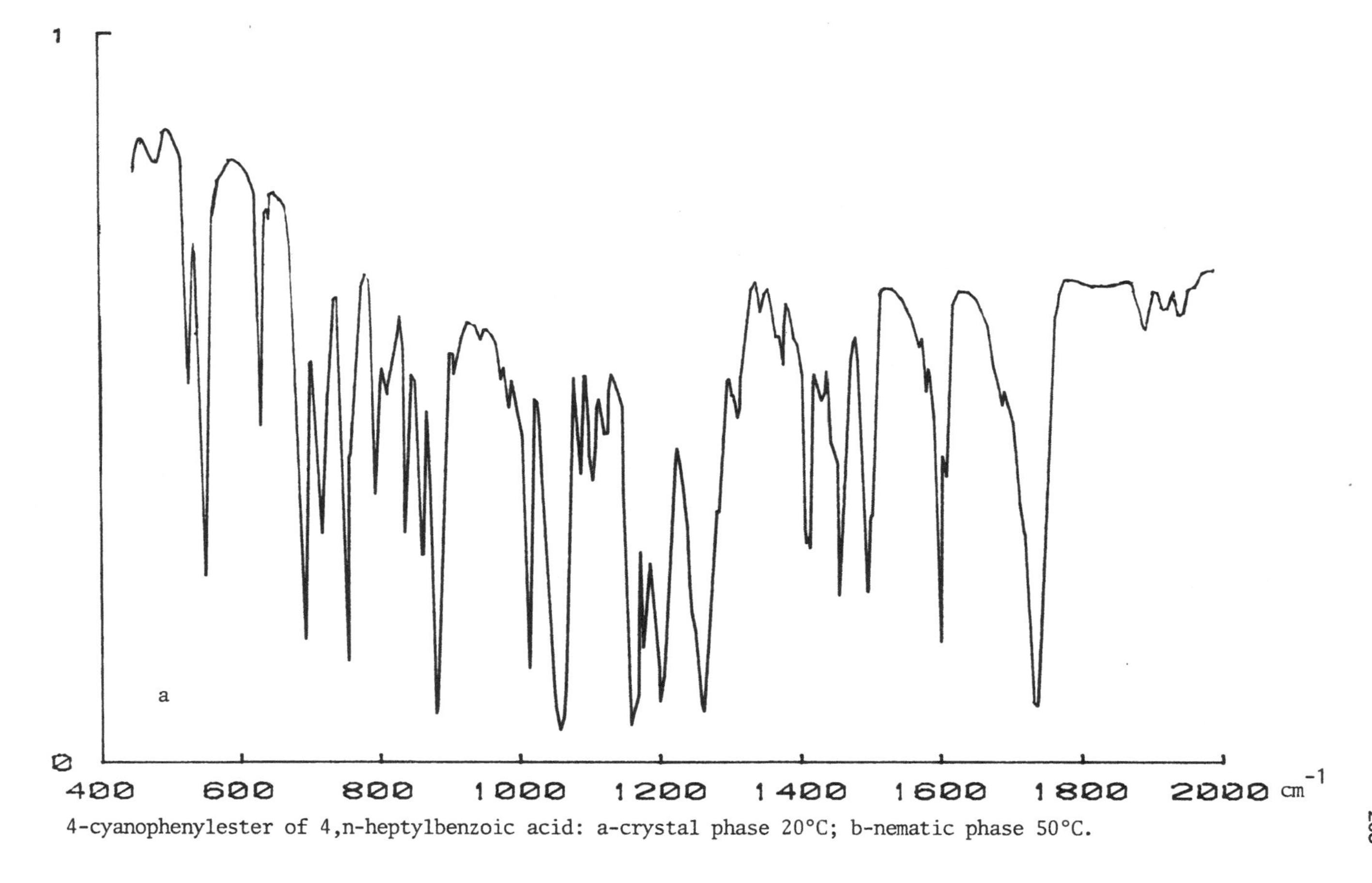

4-cyanophenylester of 4,n-heptylbenzoic acid: a-crystal phase 20°C; b-nematic phase 50°C.

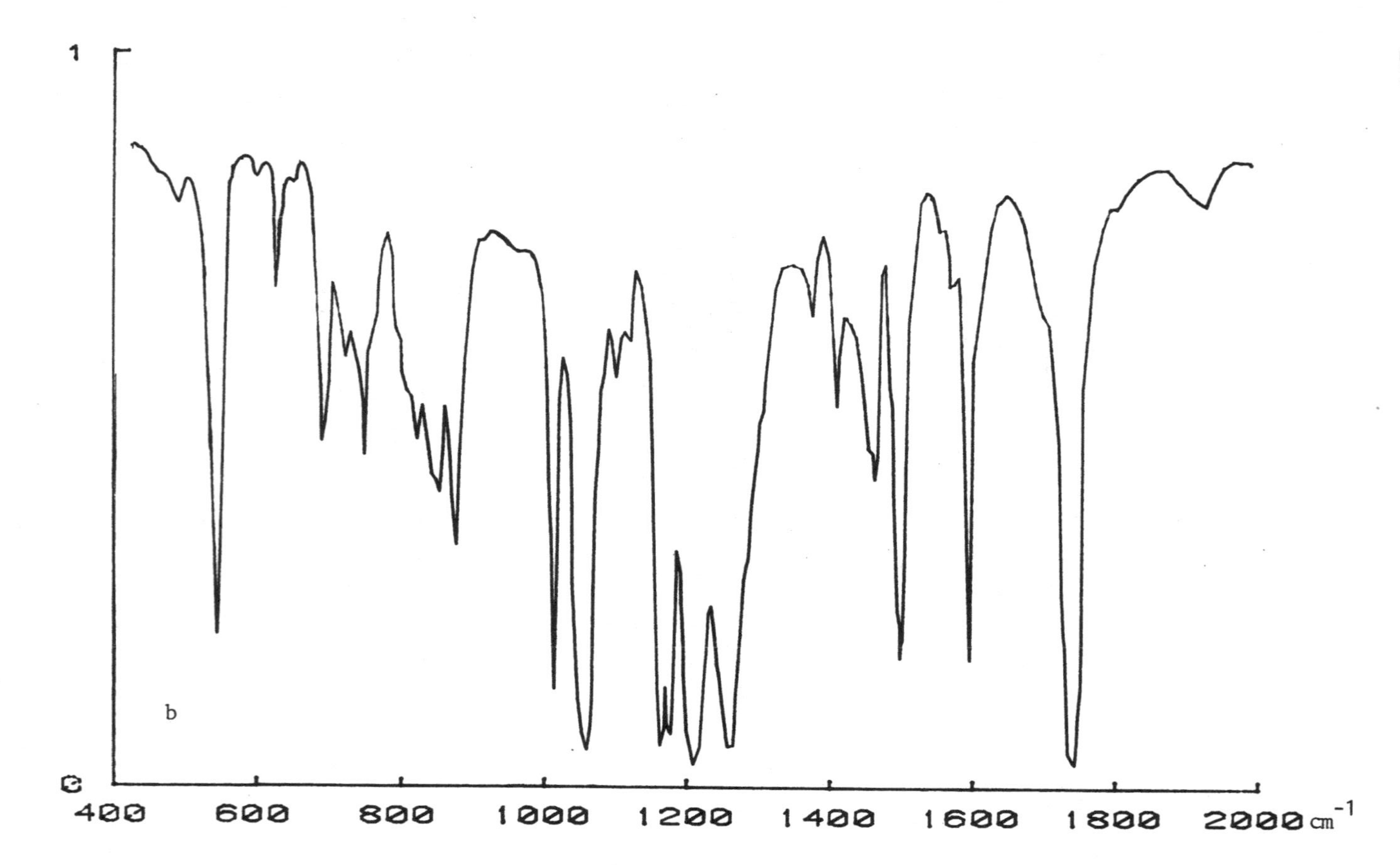
1
0
b
400
600
800
1000
1200
1400
1600
1800
2000 cm-1

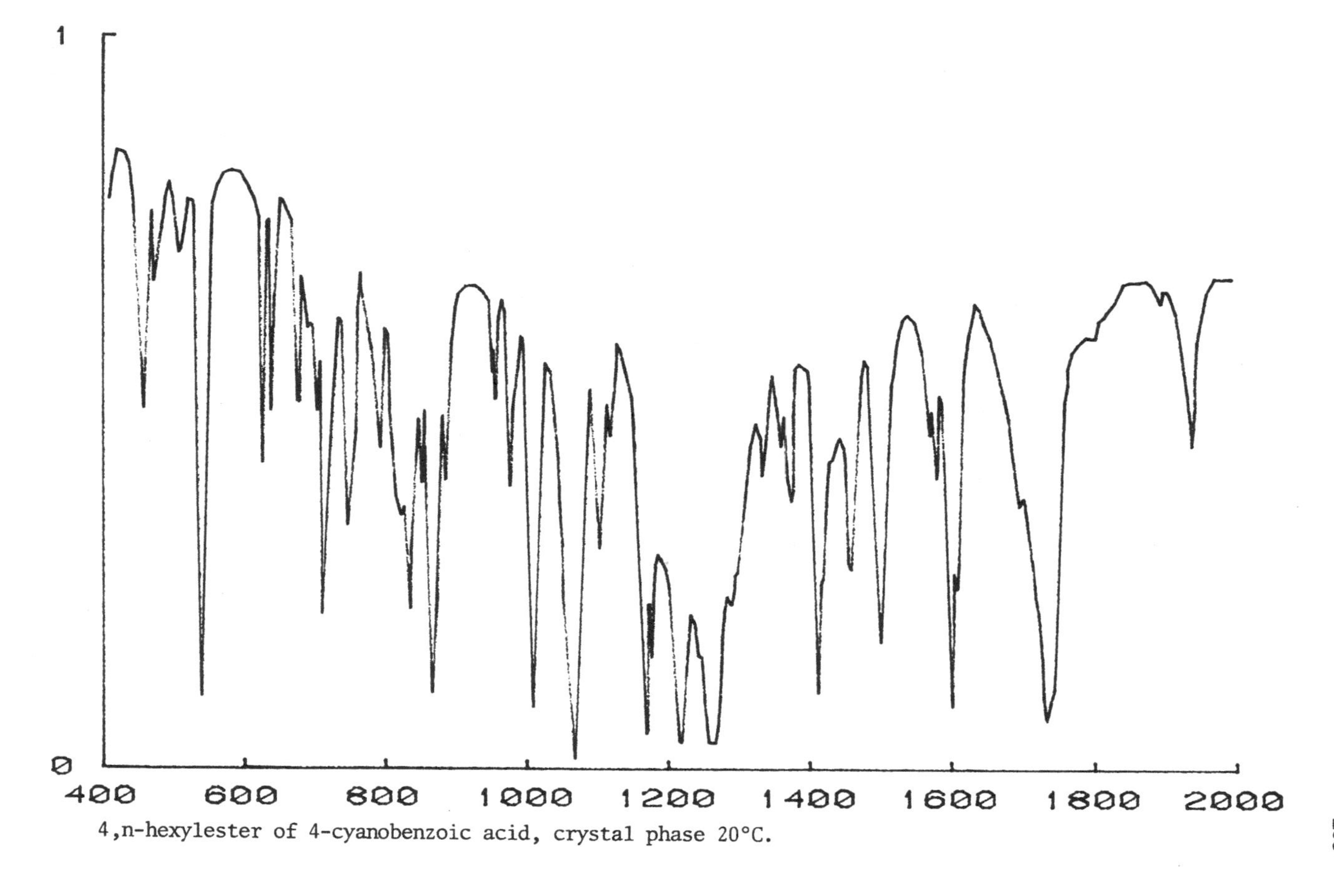

4,n-hexylester of 4-cyanobenzoic acid, crystal phase 20°C.

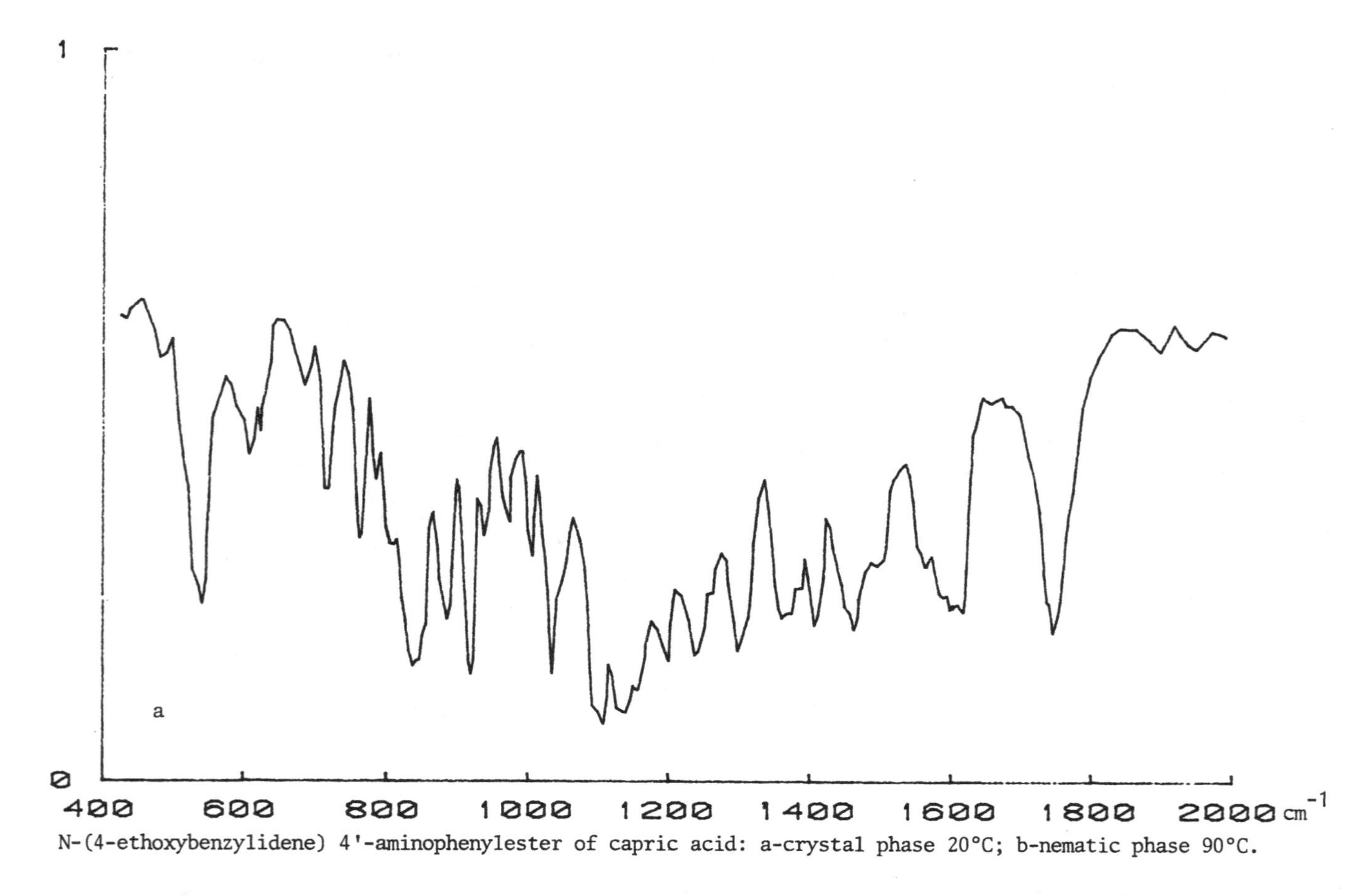

N-(4-ethoxybenzylidene) 4'-aminophenylester of capric acid: a-crystal phase 20°C; b-nematic phase 90°C.

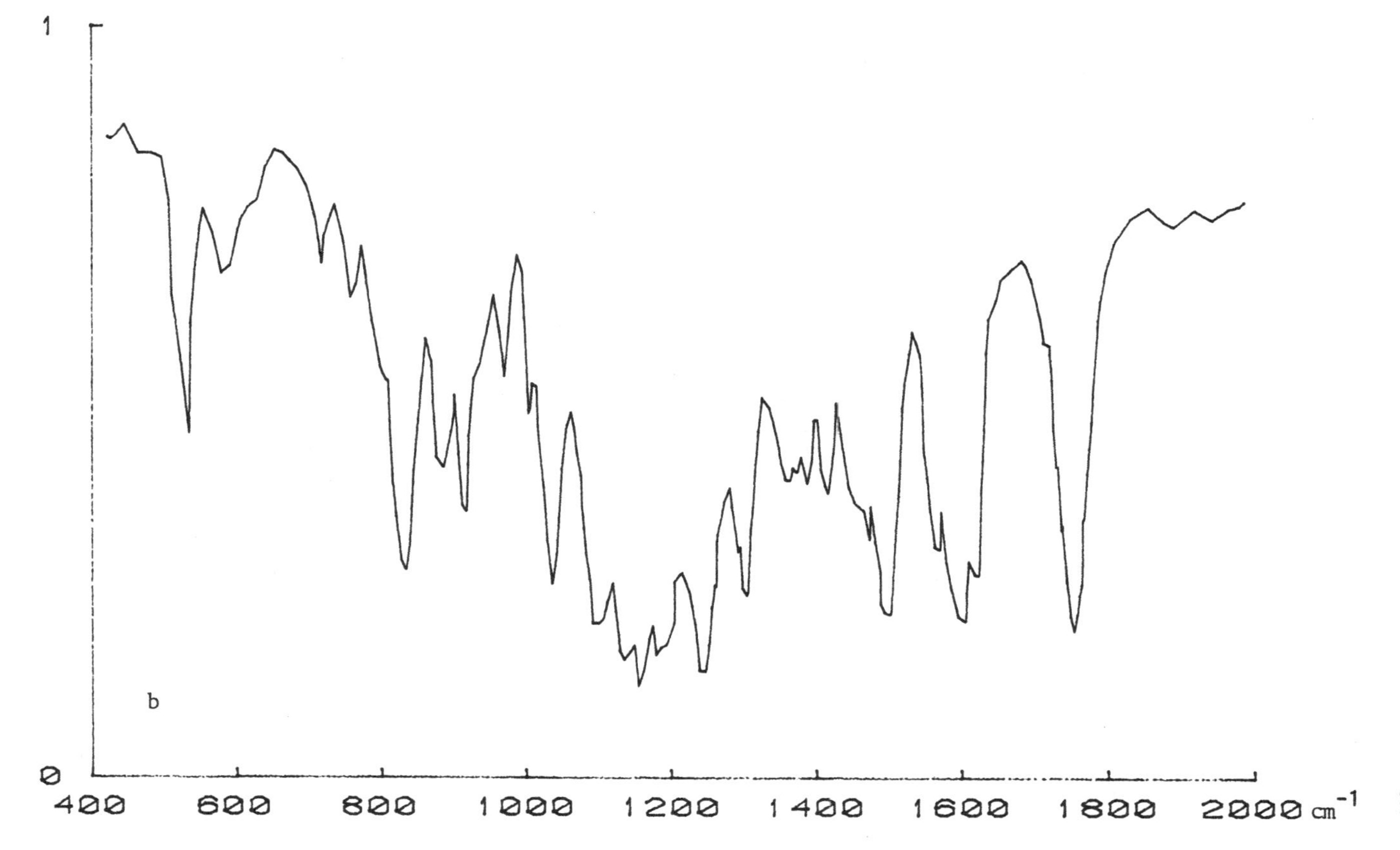
1
0
b
400
600
800
1000
1200
1400
1600
1800
2000 cm-1

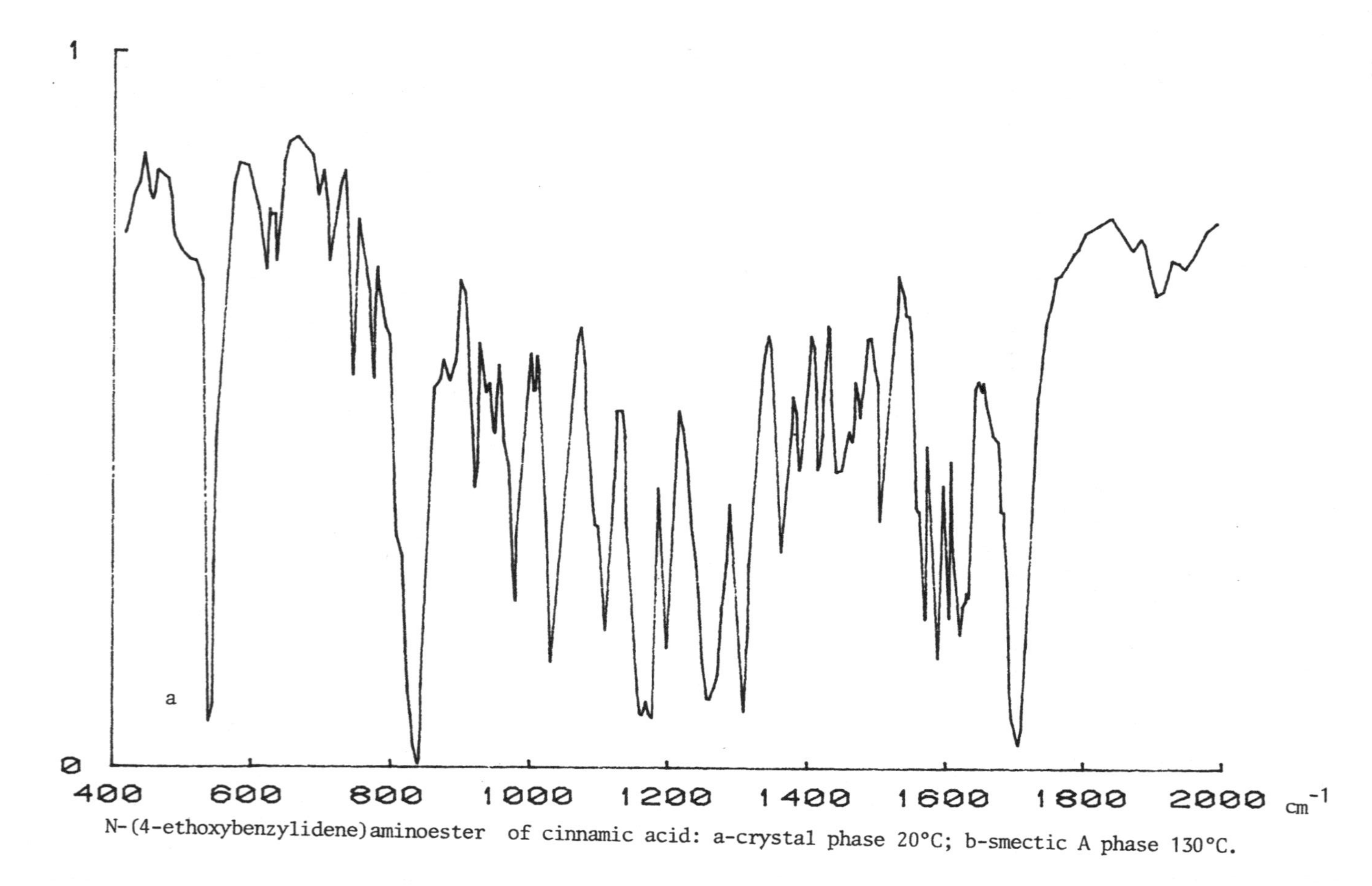

N-(4-ethoxybenzylidene)aminoester of cinnamic acid: a-crystal phase 20°C; b-smectic A phase 130°C.

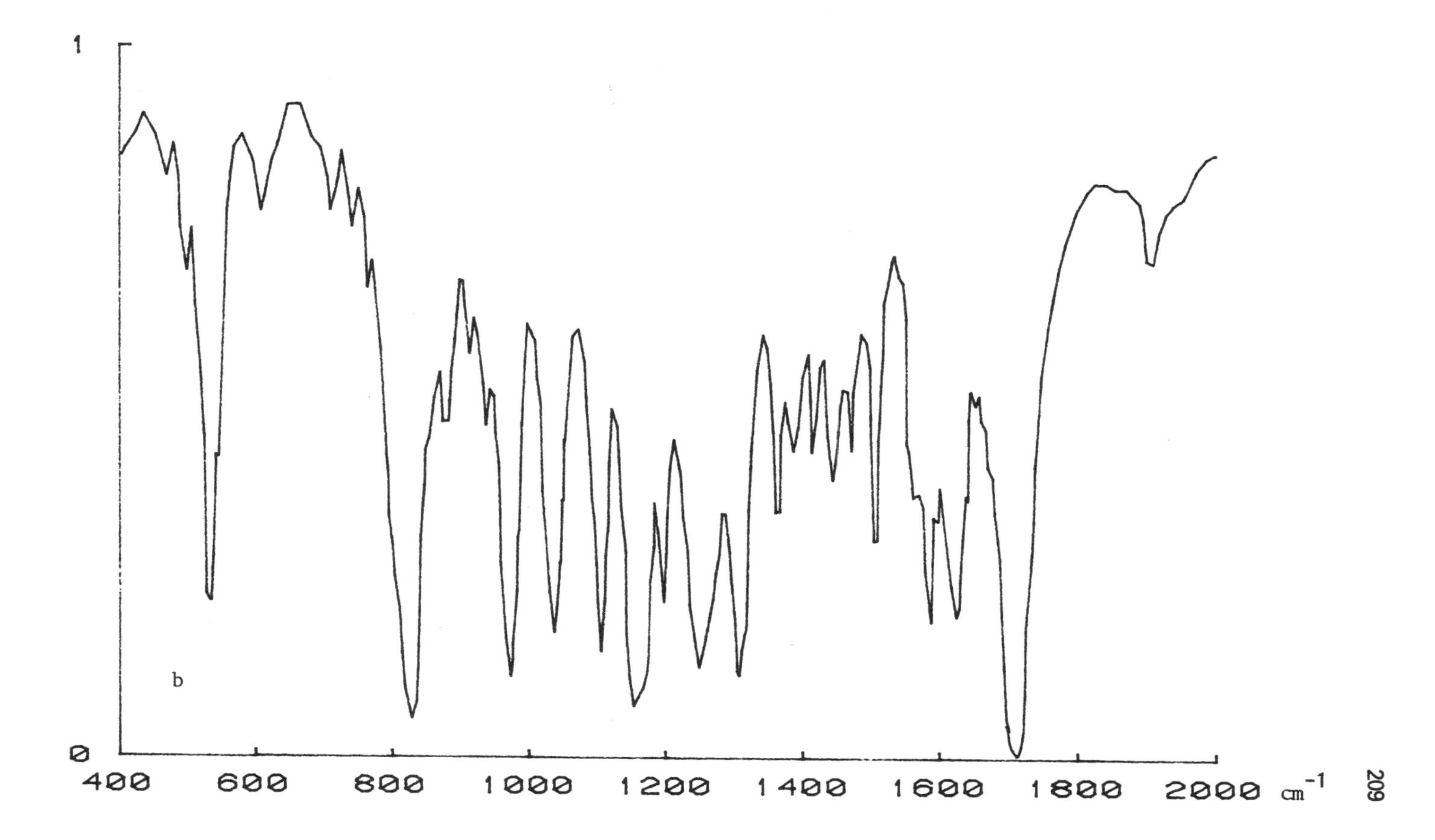
1
0
400
600
800
1000
1200
1400
1600
1800
2000
cm^{-1}
b

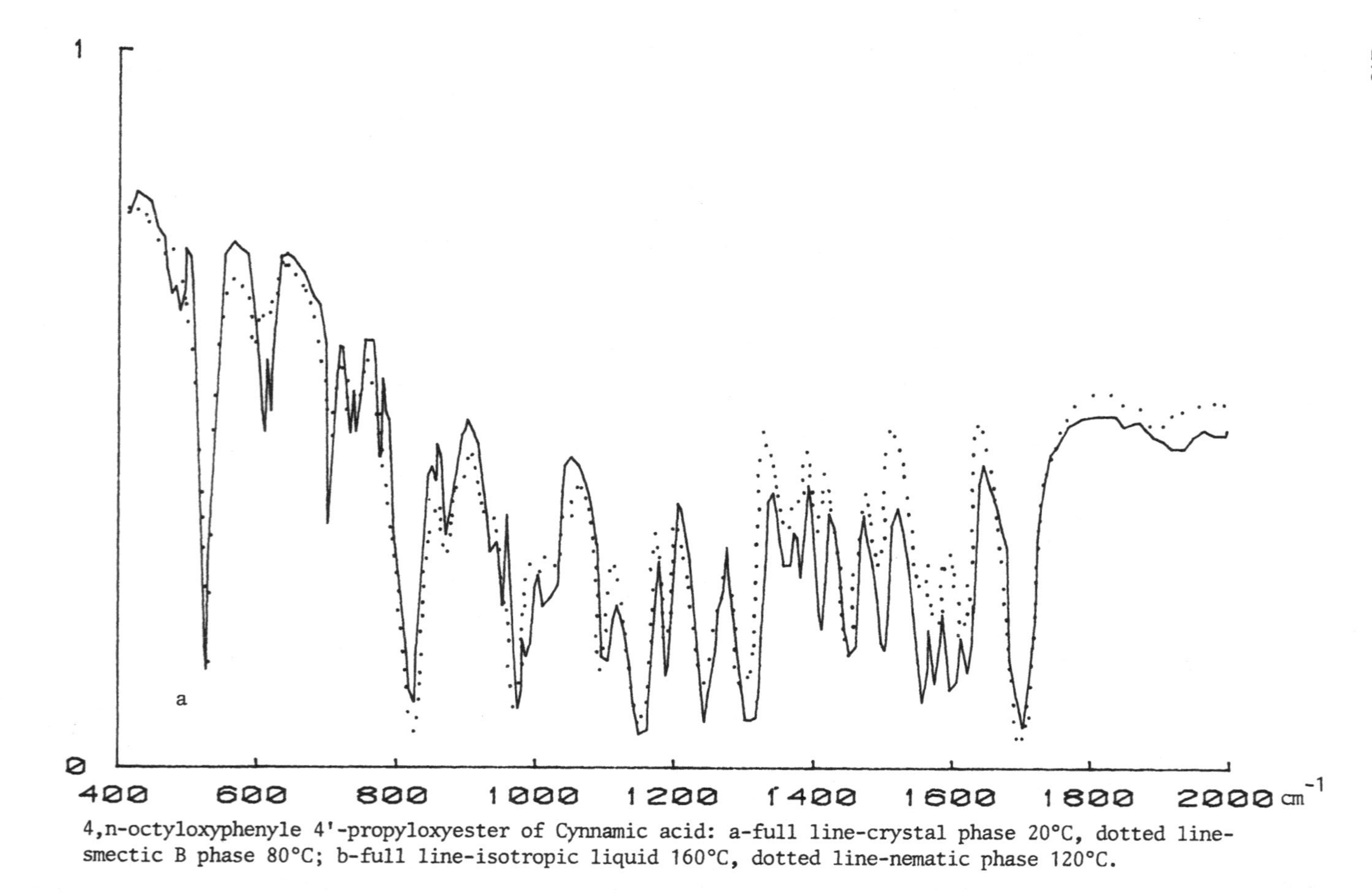

4,n-octyloxyphenyle 4'-propyloxyester of Cynnamic acid: a-full line-crystal phase 20°C, dotted line-smectic B phase 80°C; b-full line-isotropic liquid 160°C, dotted line-nematic phase 120°C.

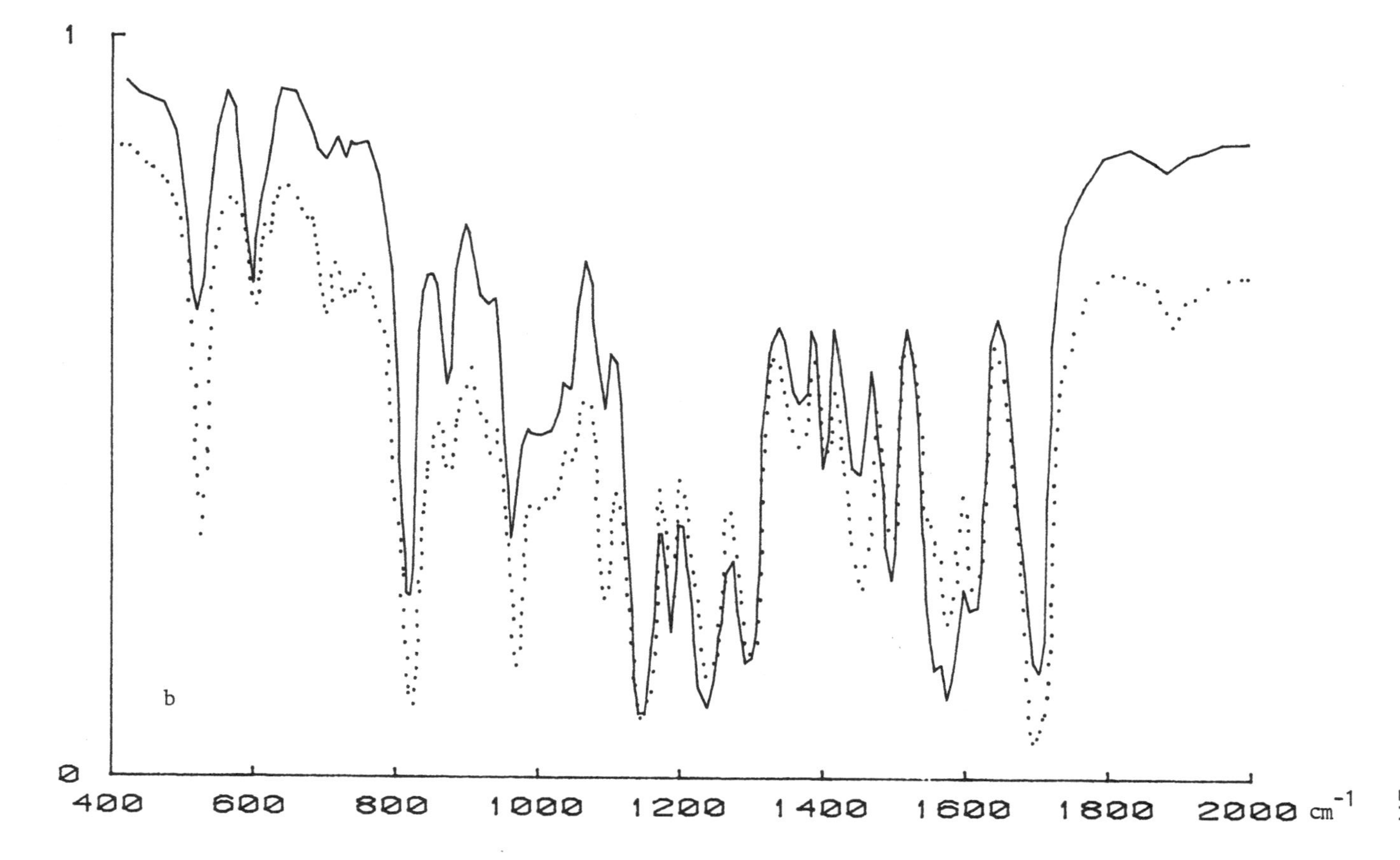
1
0
b
400
600
800
1000
1200
1400
1600
1800
2000 cm^{-1}

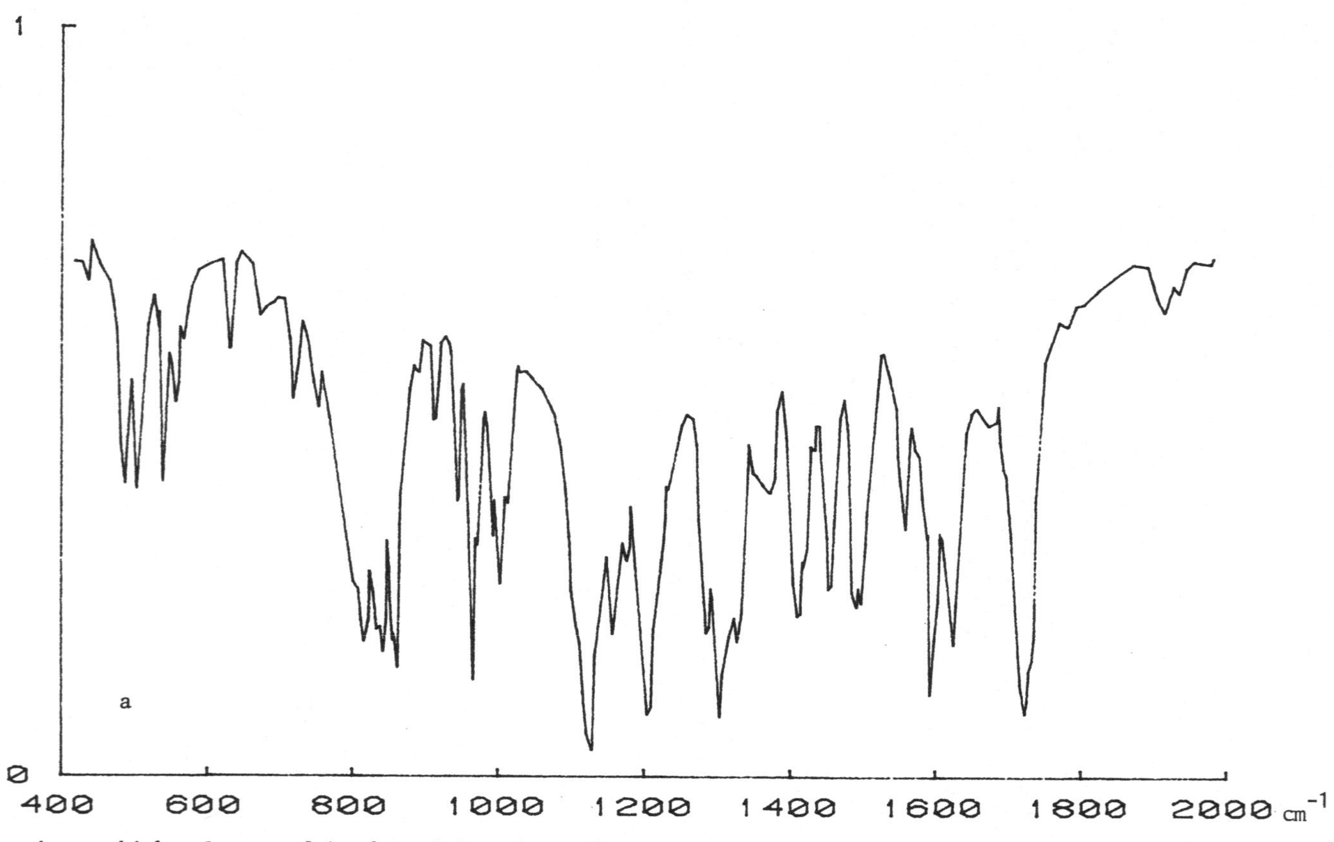

4-cyanobiphenylester of 4,n-heptylcinnamic acid: a-crystal phase 20°C; b-nematic phase 70°C.

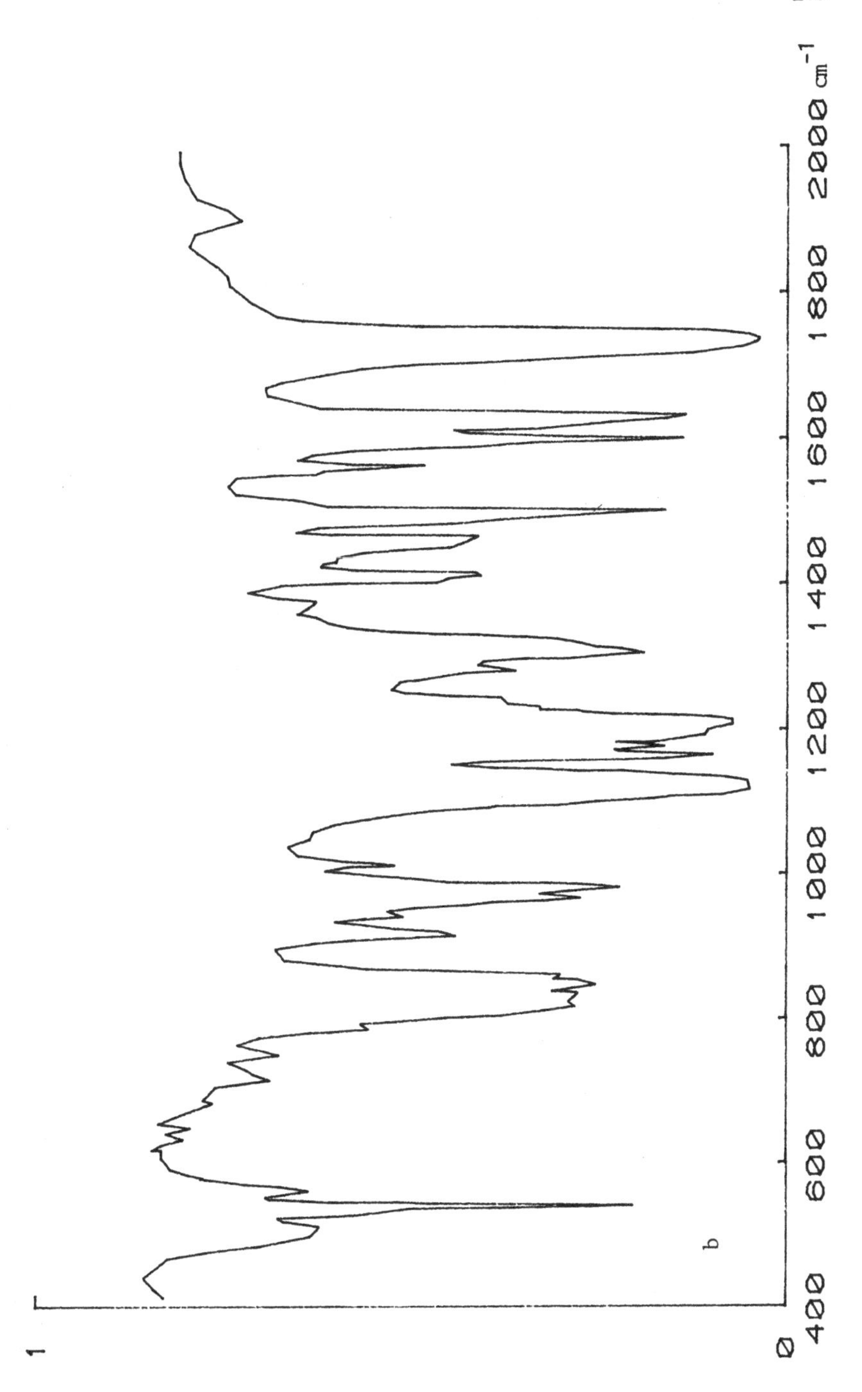

1
0
b
400
600
800
1000
1200
1400
1600
1800
2000 cm-1

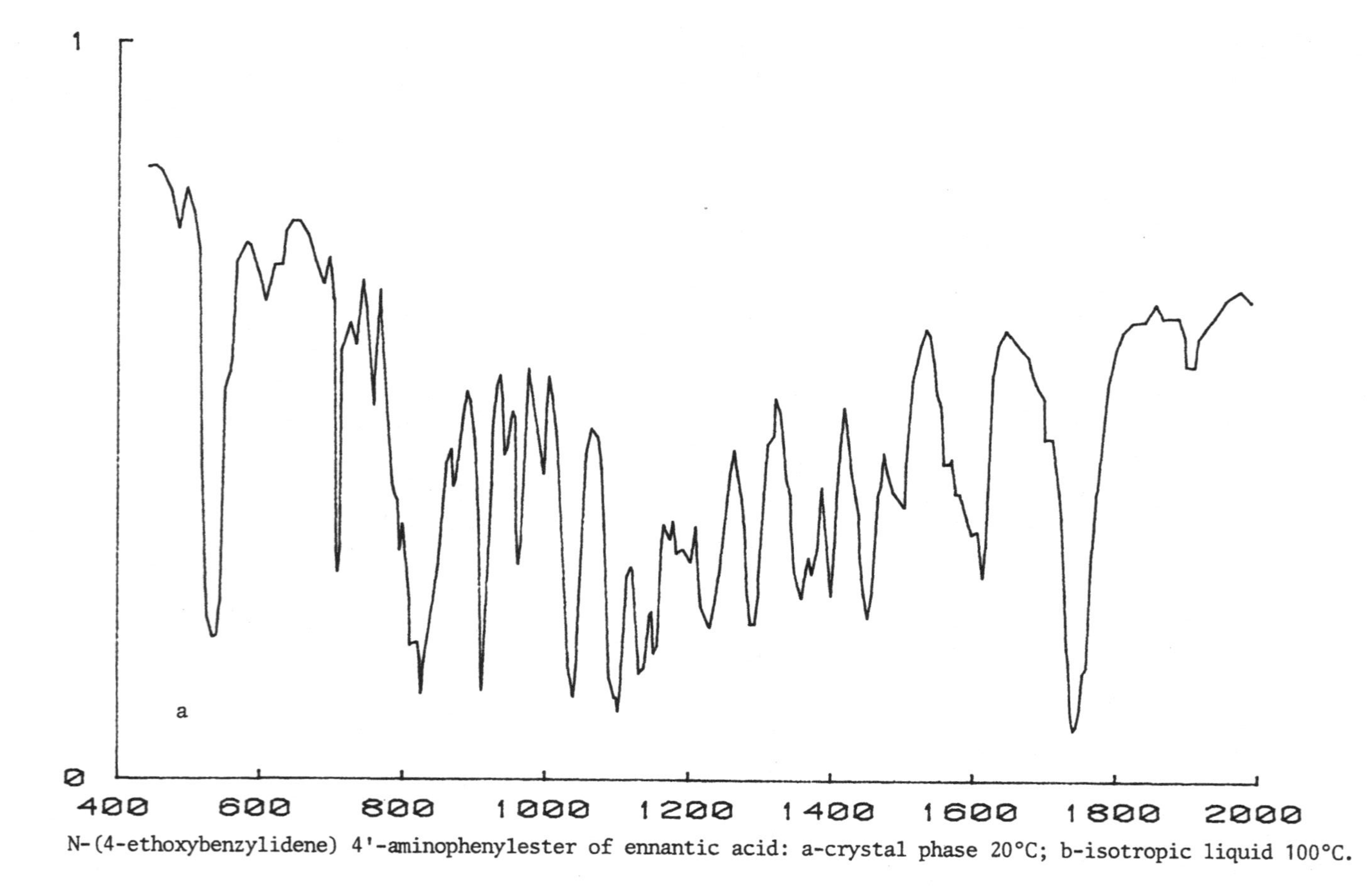

N-(4-ethoxybenzylidene) 4'-aminophenylester of ennantic acid: a-crystal phase 20°C; b-isotropic liquid 100°C.

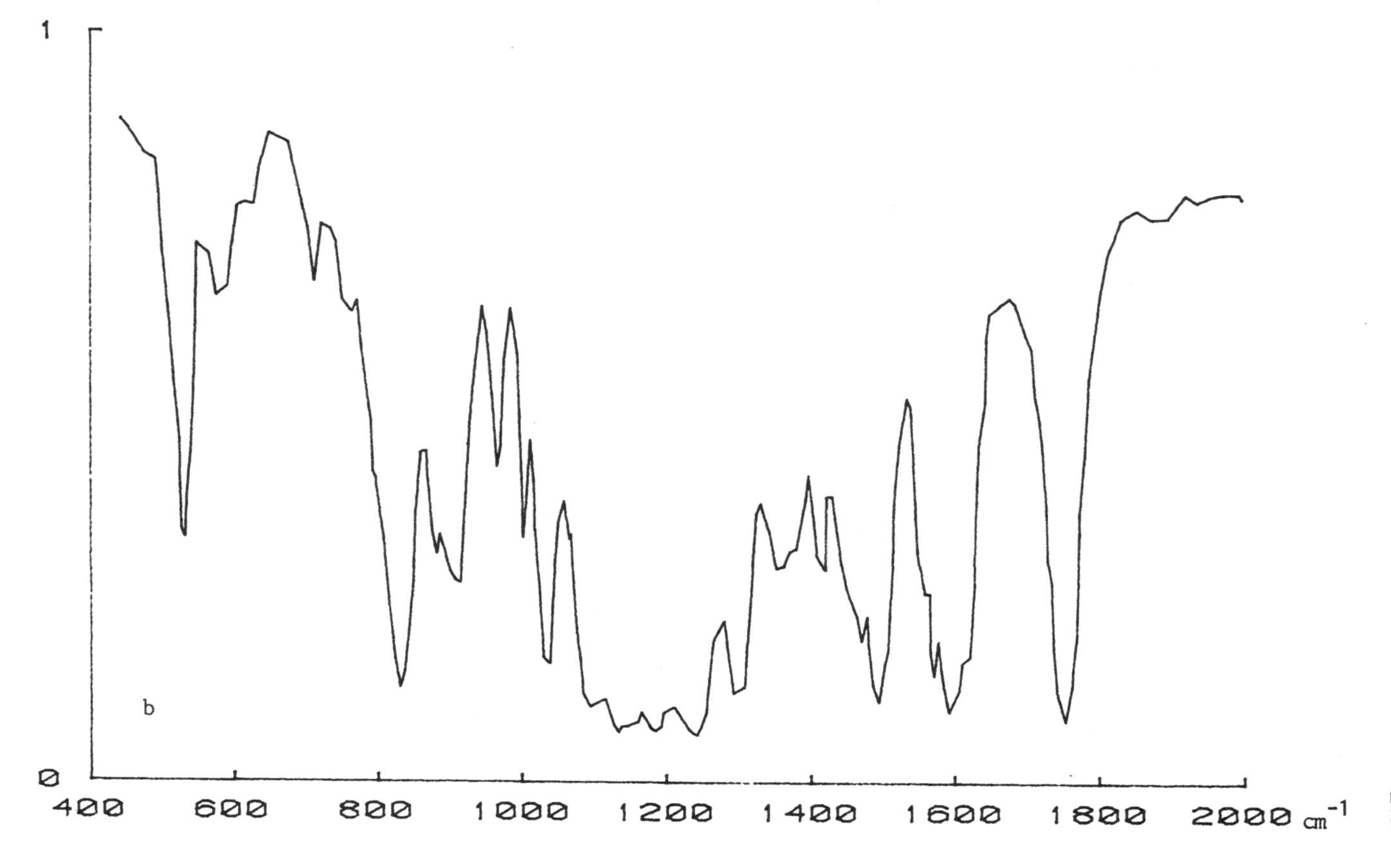

1
0
400
600
800
1000
1200
1400
1600
1800
2000
cm^{-1}
b

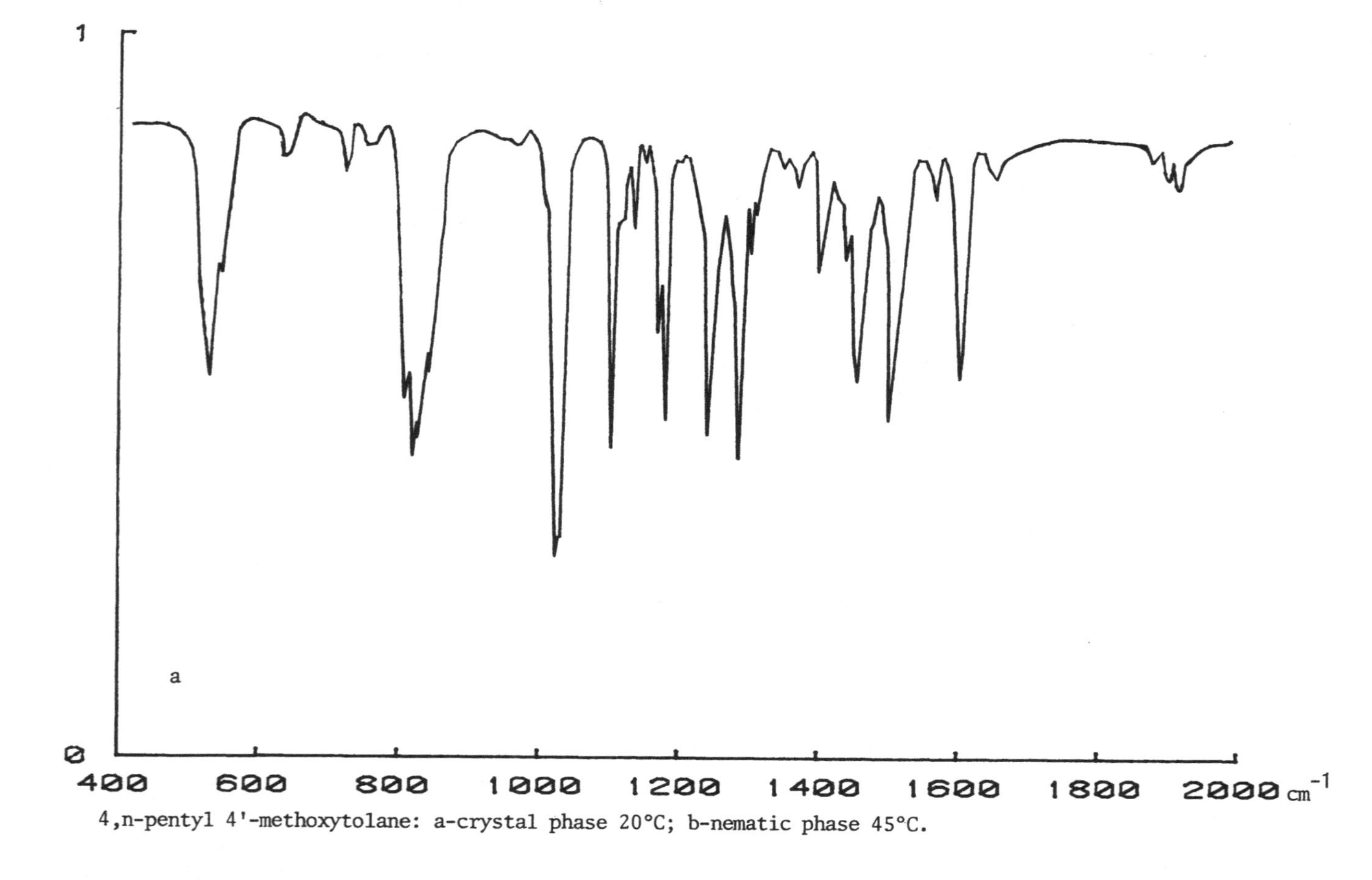

4,n-pentyl 4'-methoxytolane: a-crystal phase 20°C; b-nematic phase 45°C.

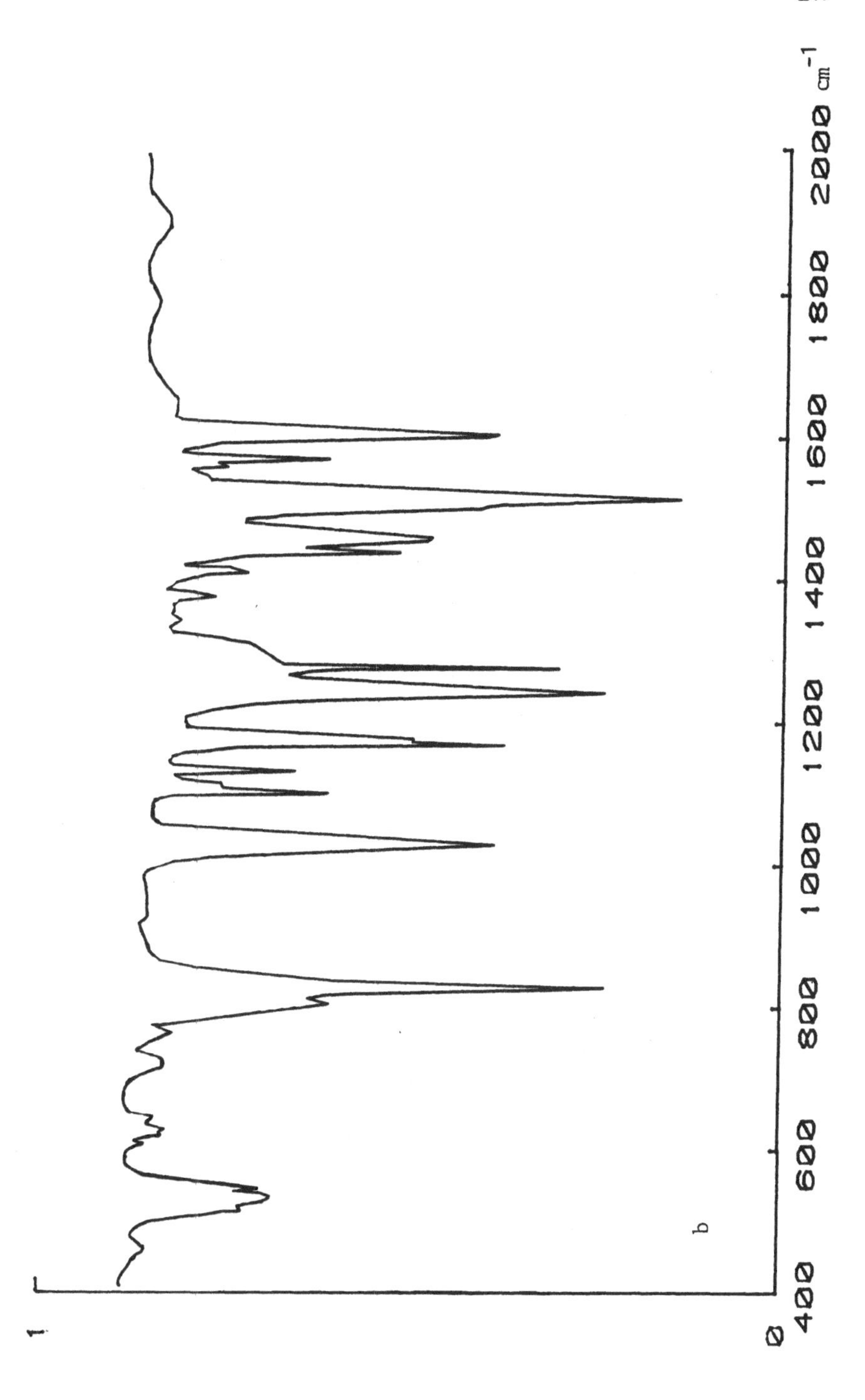

400
600
800
1000
1200
1400
1600
1800
2000
cm-1
0
1
b

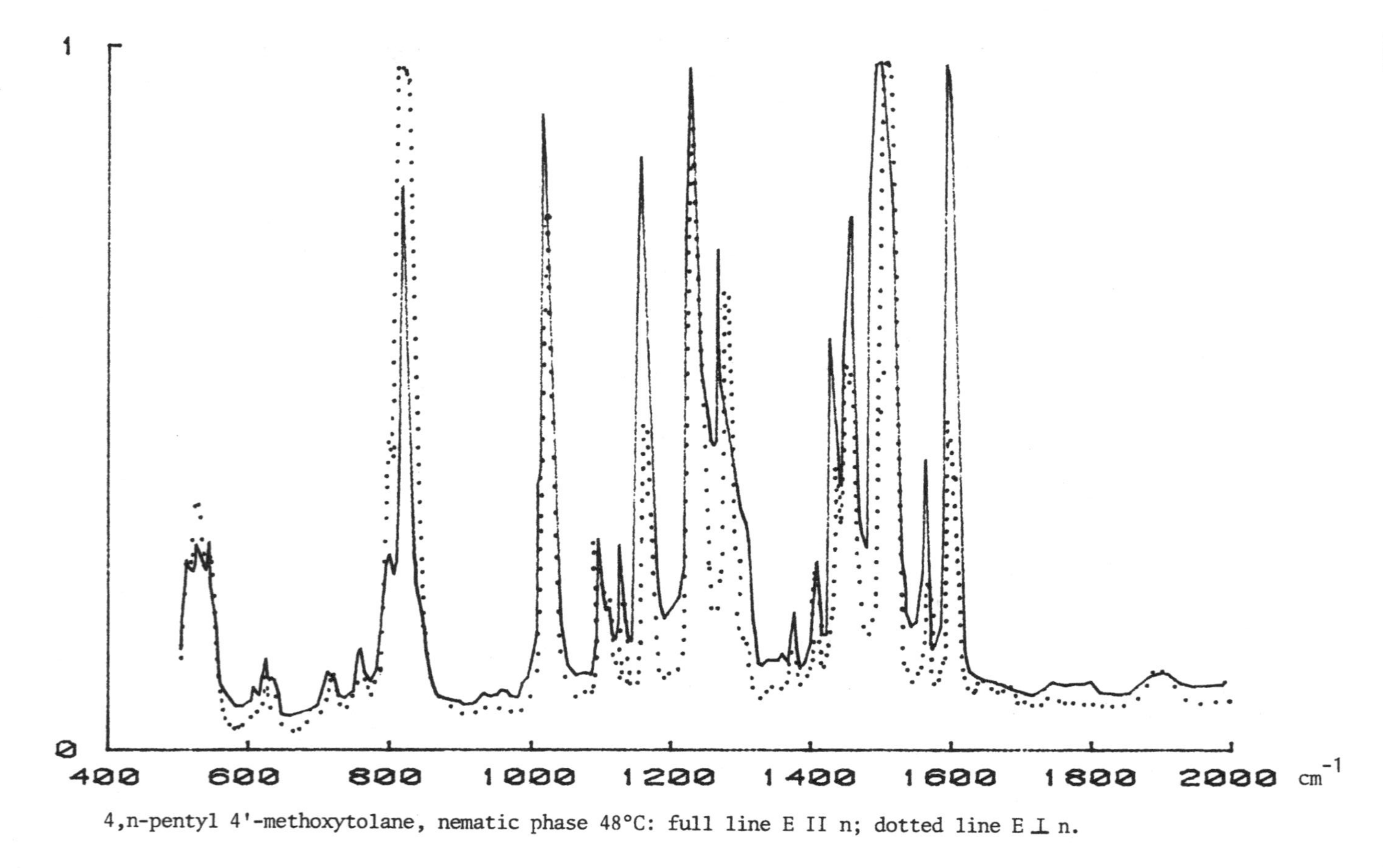

4,n-pentyl 4'-methoxytolane, nematic phase 48°C: full line E II n; dotted line E ⊥ n.

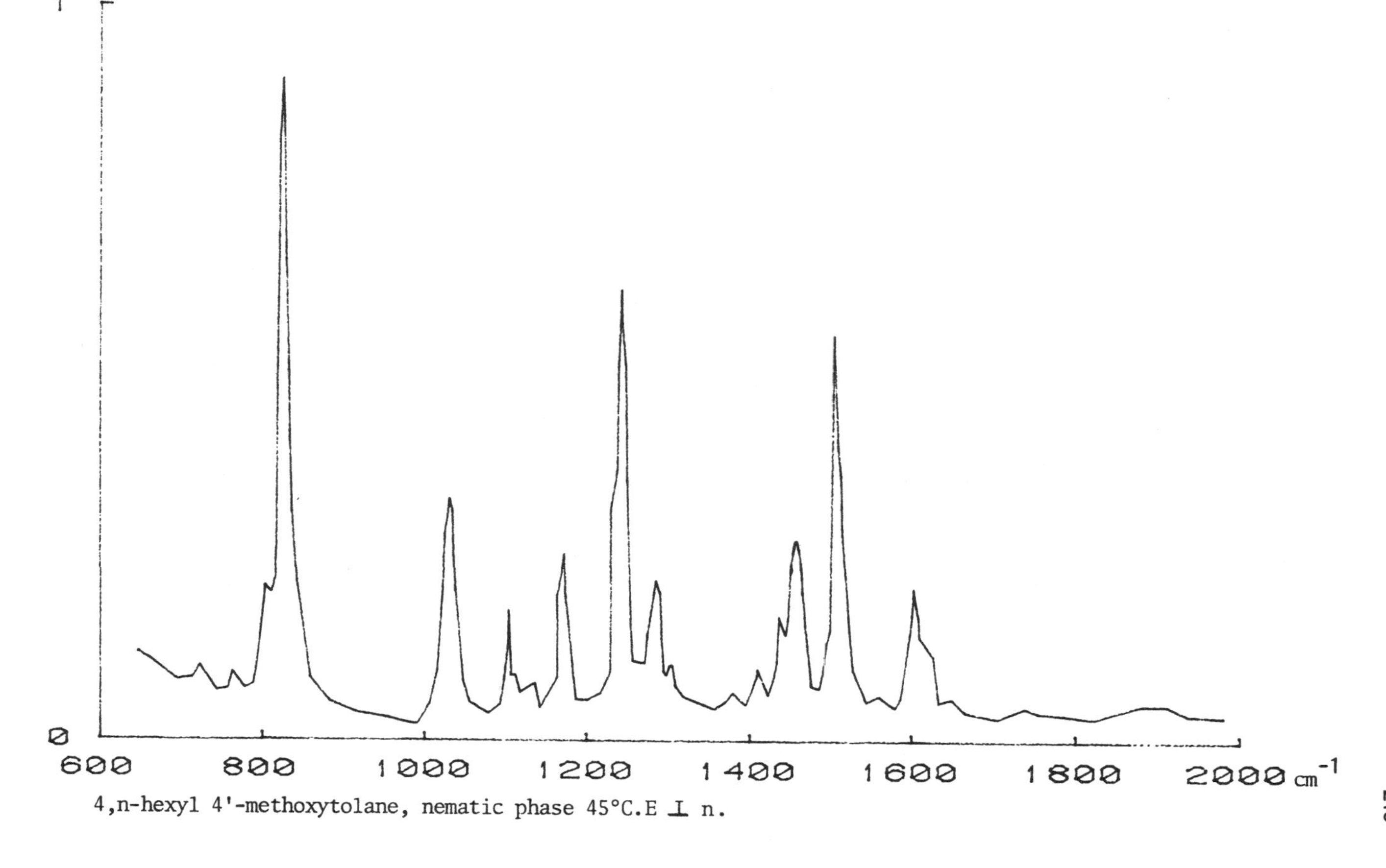

4,n-hexyl 4'-methoxytolane, nematic phase 45°C.E ⊥ n.

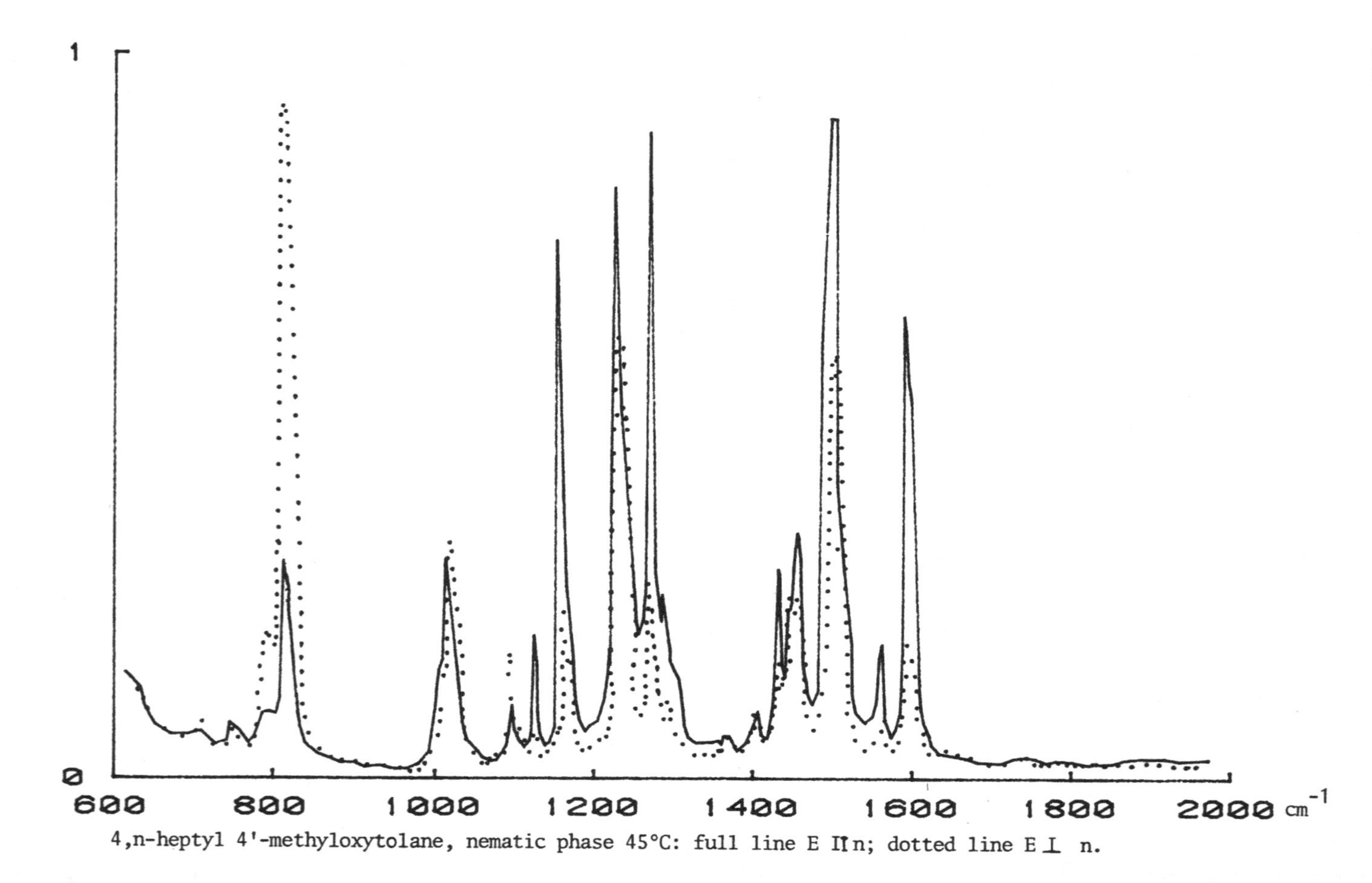

4,n-heptyl 4'-methyloxytolane, nematic phase 45°C: full line E $\parallel$ n; dotted line E $\perp$ n.

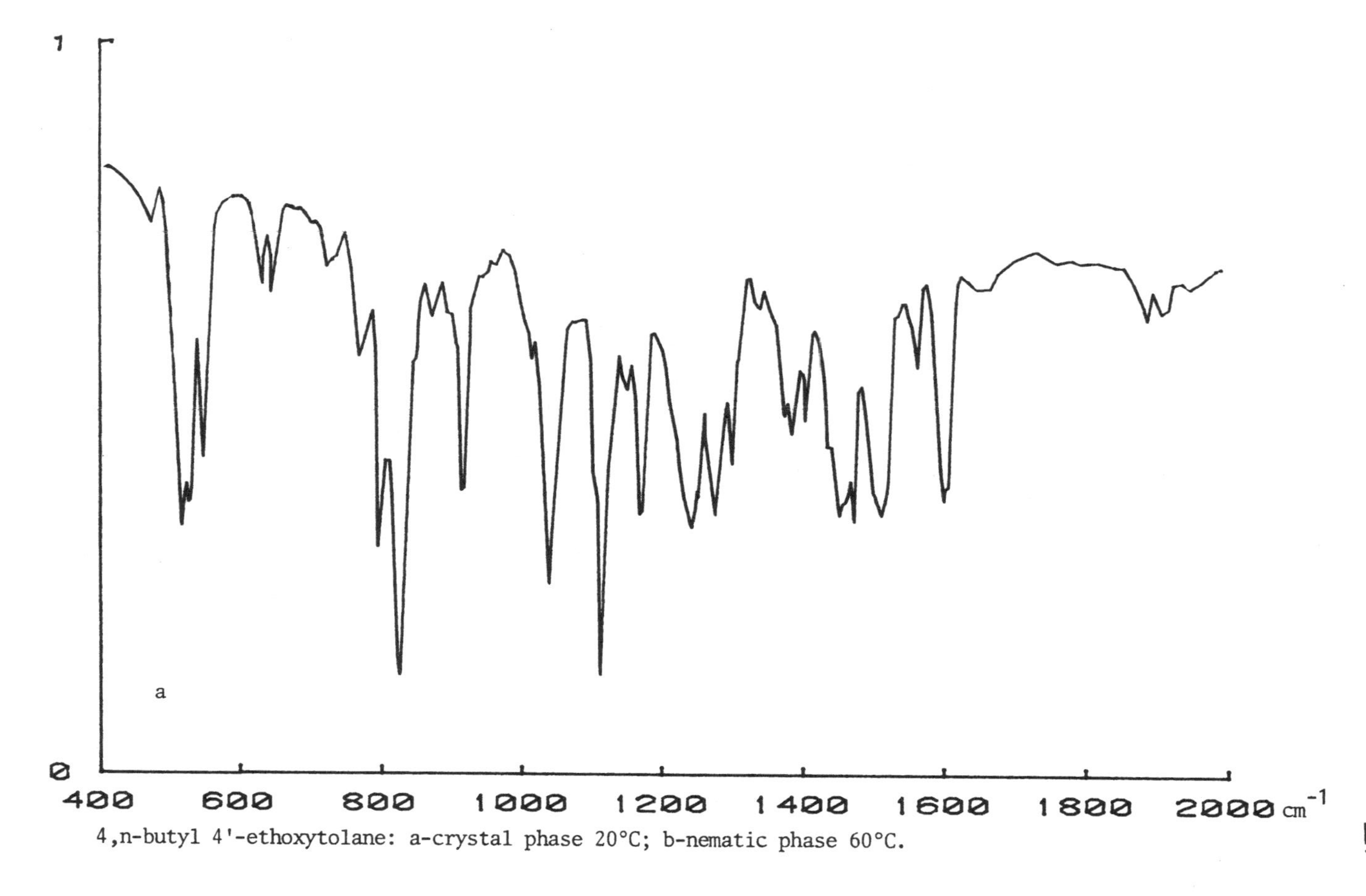

4,n-butyl 4'-ethoxytolane: a-crystal phase 20°C; b-nematic phase 60°C.

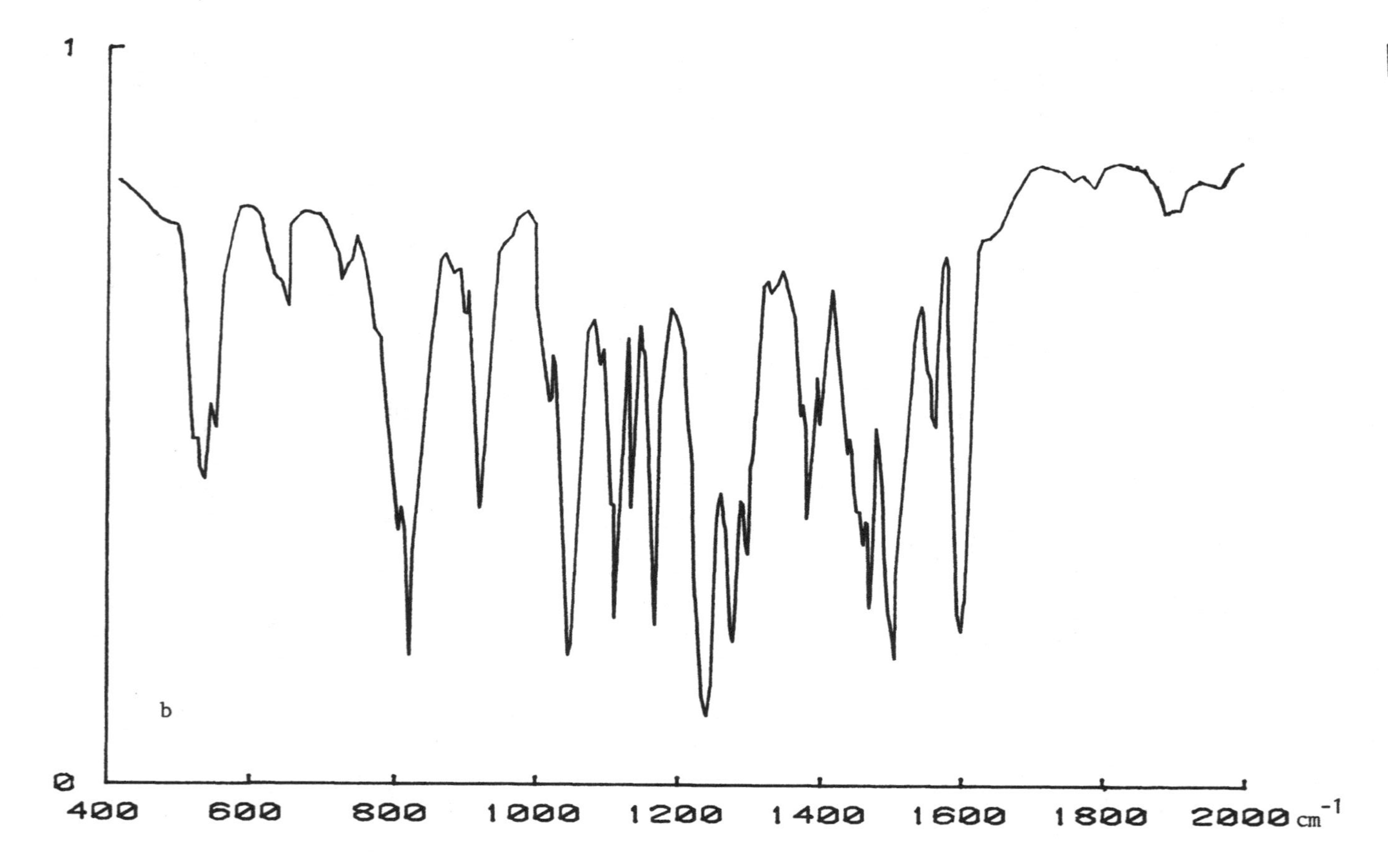

1
0
b
400
600
800
1000
1200
1400
1600
1800
2000 cm-1

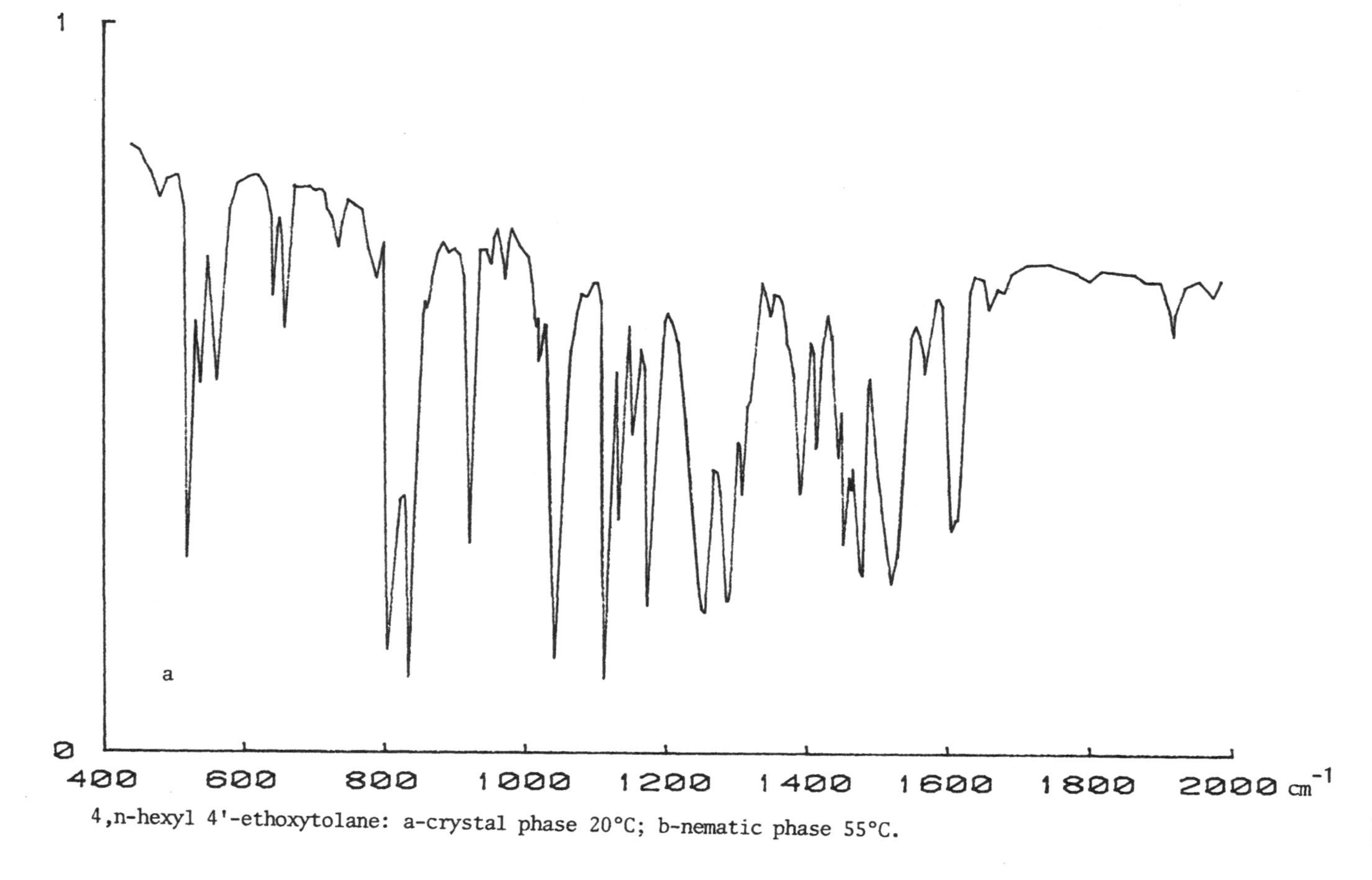

4,n-hexyl 4'-ethoxytolane: a-crystal phase 20°C; b-nematic phase 55°C.

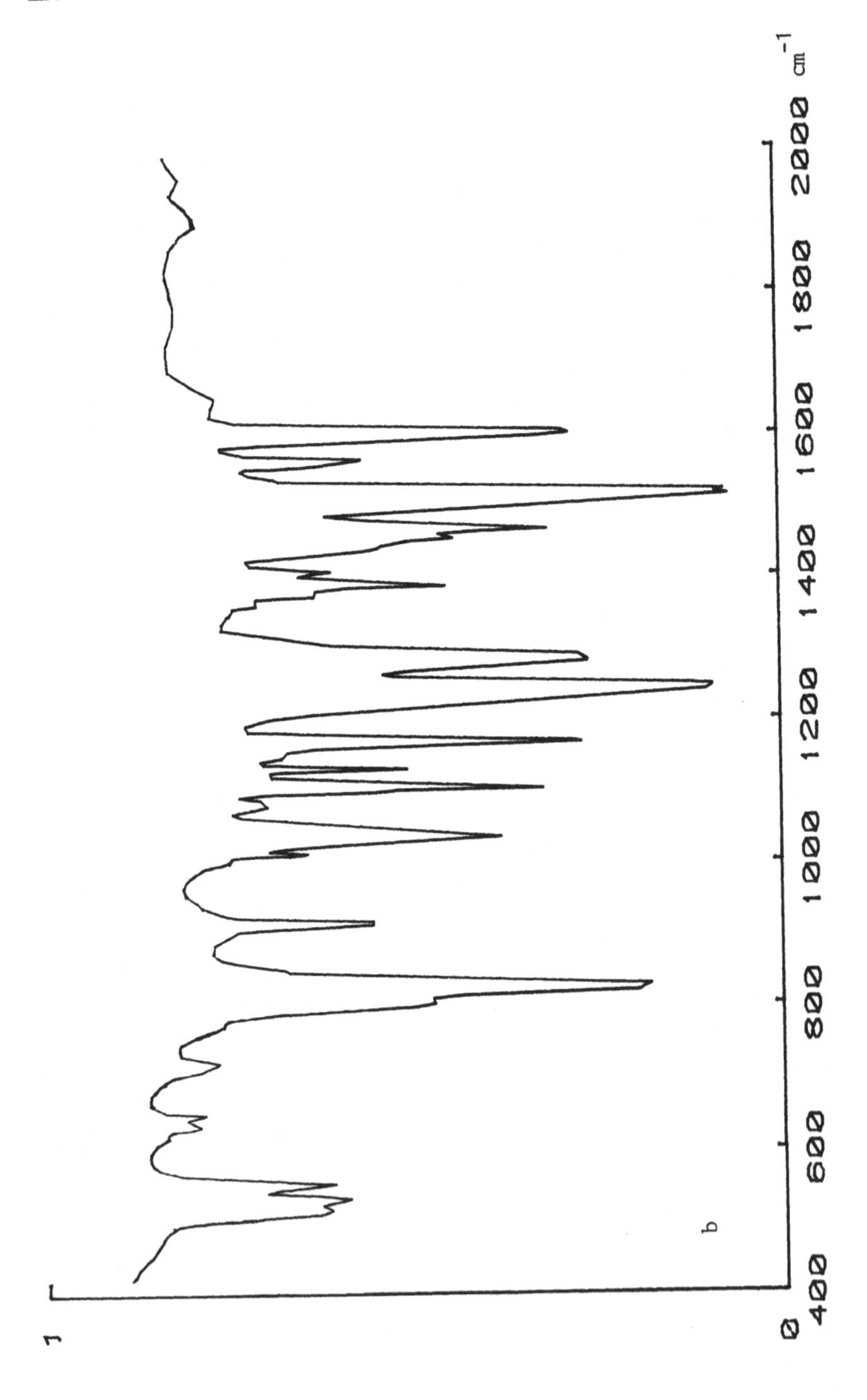
b
400
600
800
1000
1200
1400
1600
1800
2000
cm^{-1}
0
1

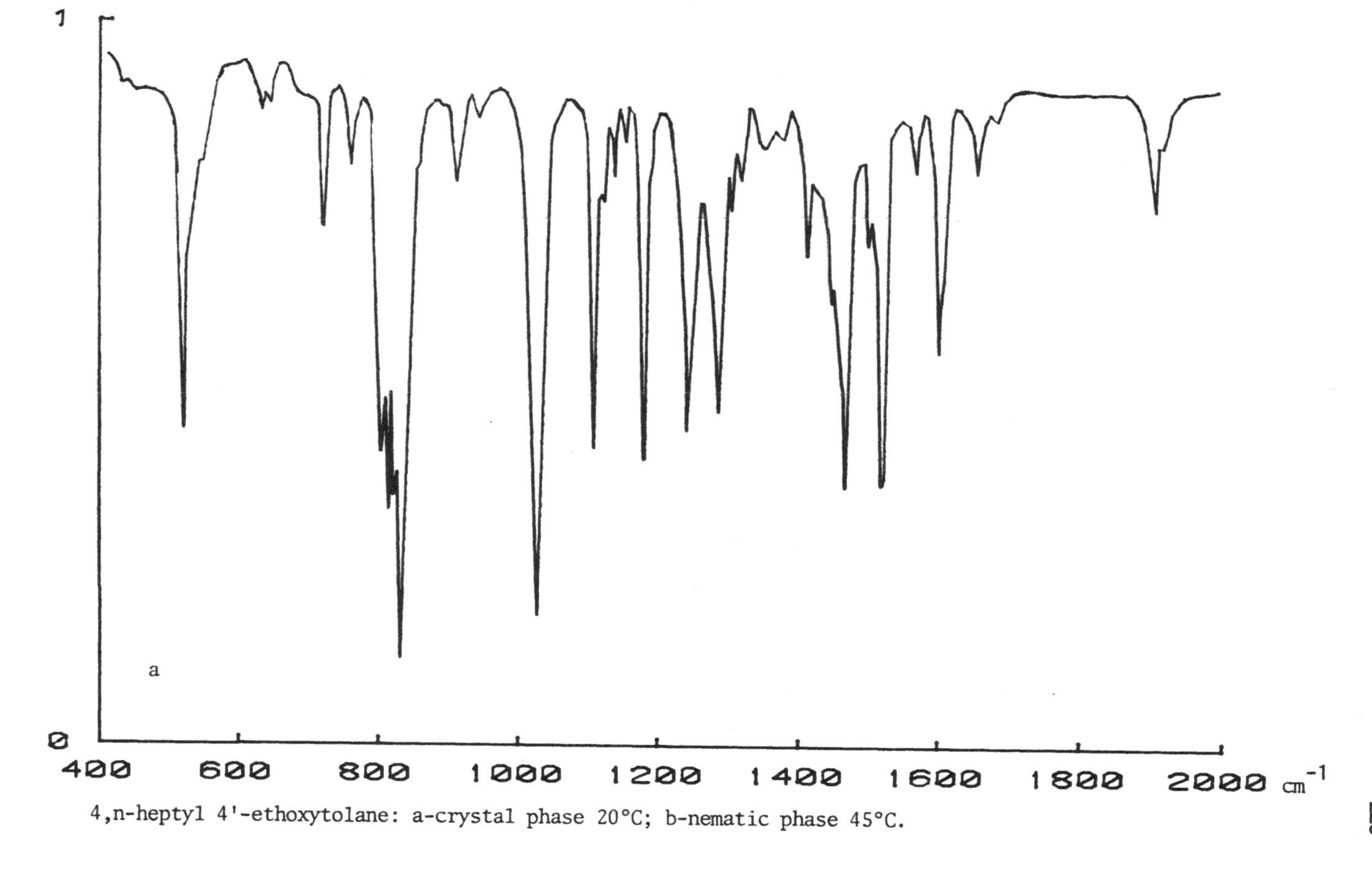

4,n-heptyl 4'-ethoxytolane: a-crystal phase 20°C; b-nematic phase 45°C.

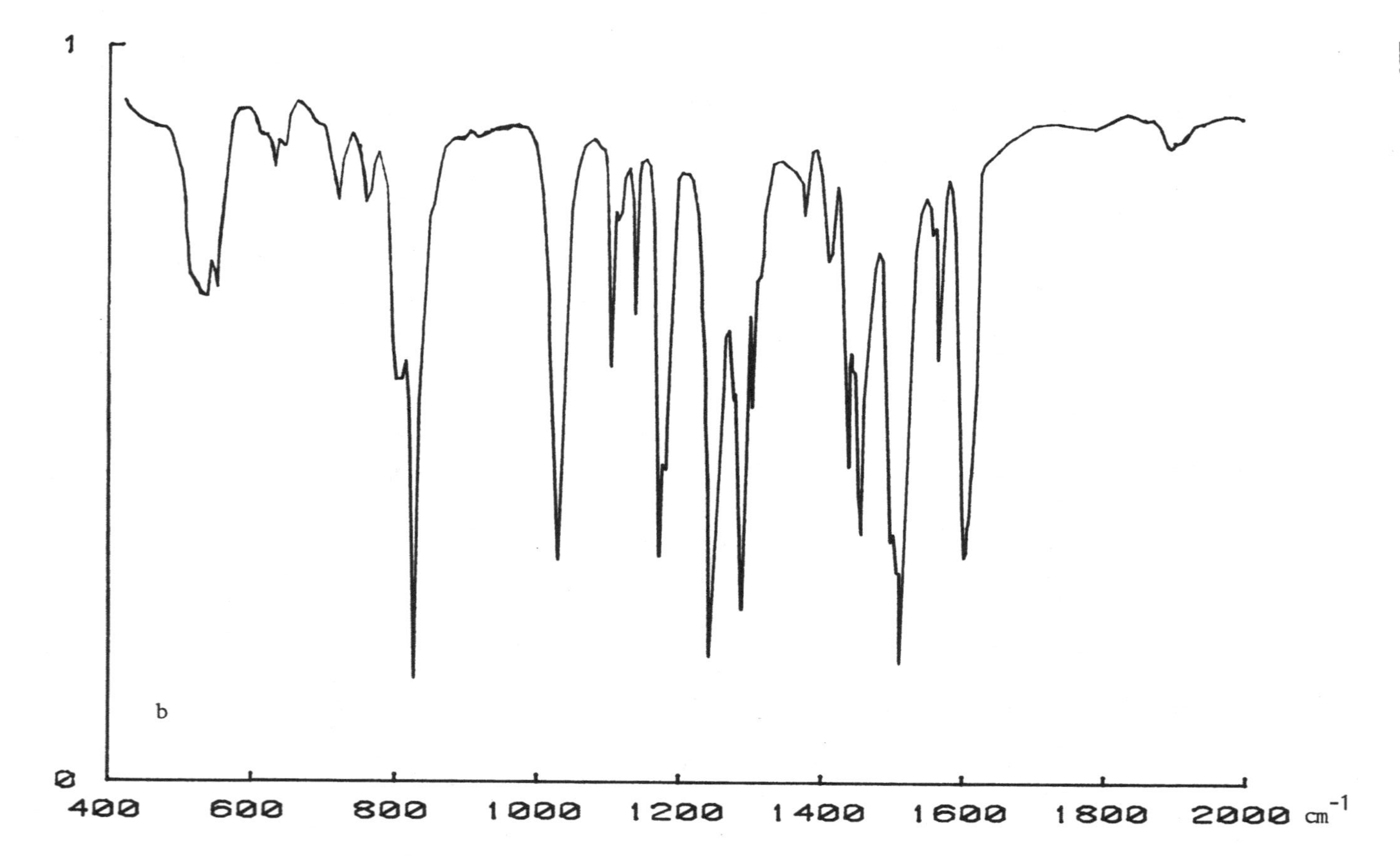

1
b
0
400
600
800
1000
1200
1400
1600
1800
2000
cm-1

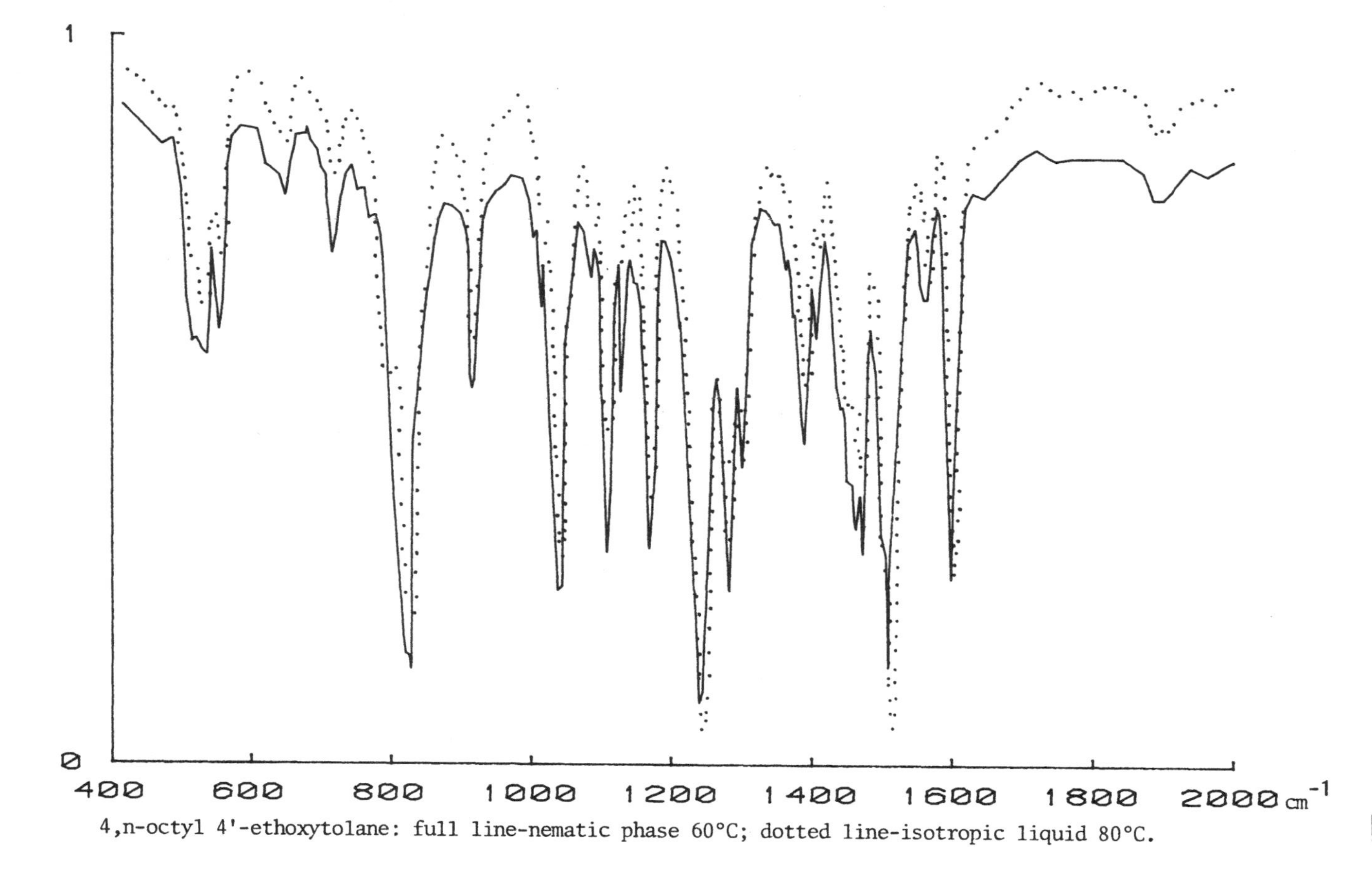

4,n-octyl 4'-ethoxytolane: full line-nematic phase 60°C; dotted line-isotropic liquid 80°C.

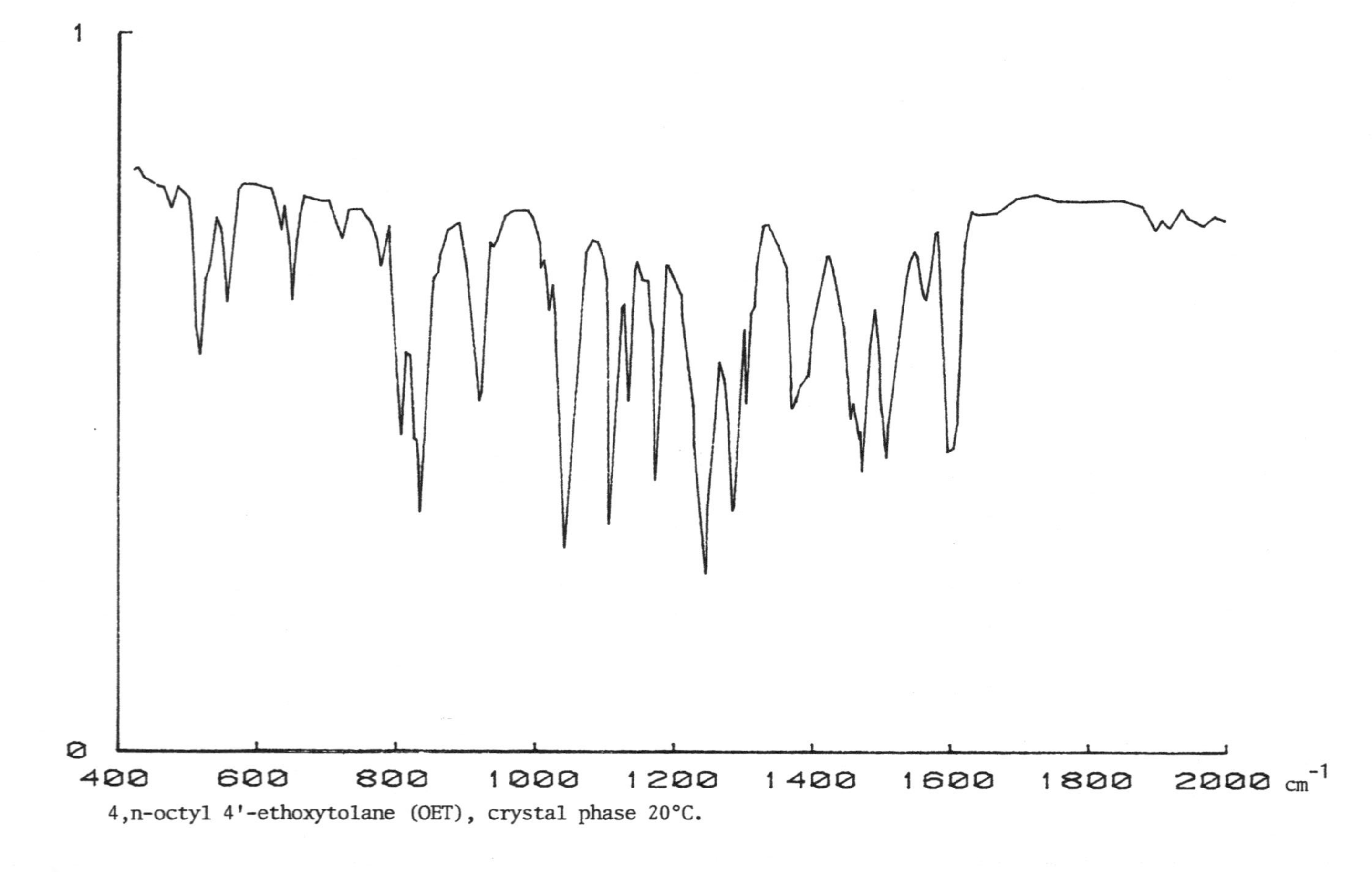

4,n-octyl 4'-ethoxytolane (OET), crystal phase 20°C.

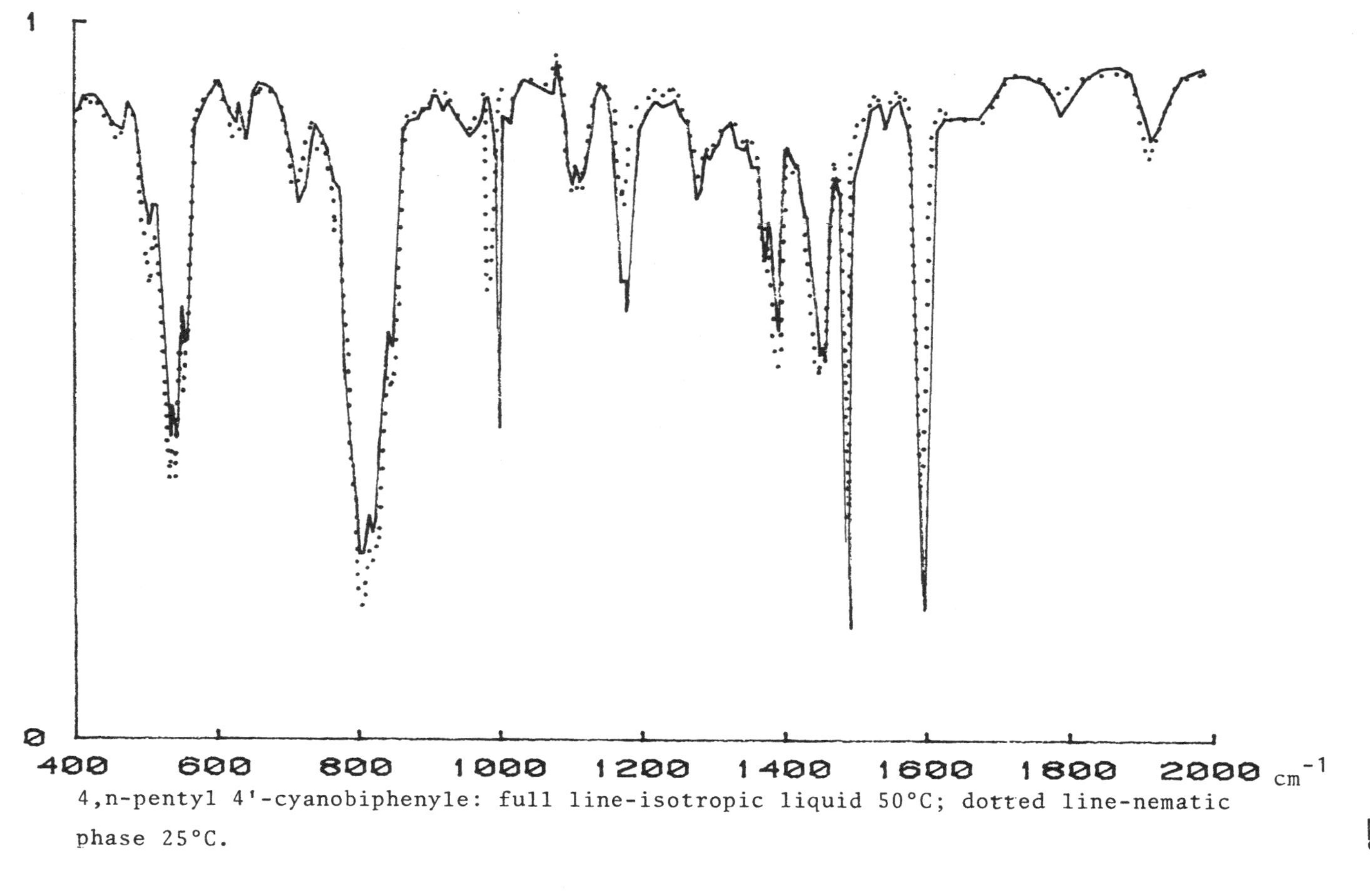

4,n-pentyl 4'-cyanobiphenyle: full line-isotropic liquid 50°C; dotted line-nematic phase 25°C.

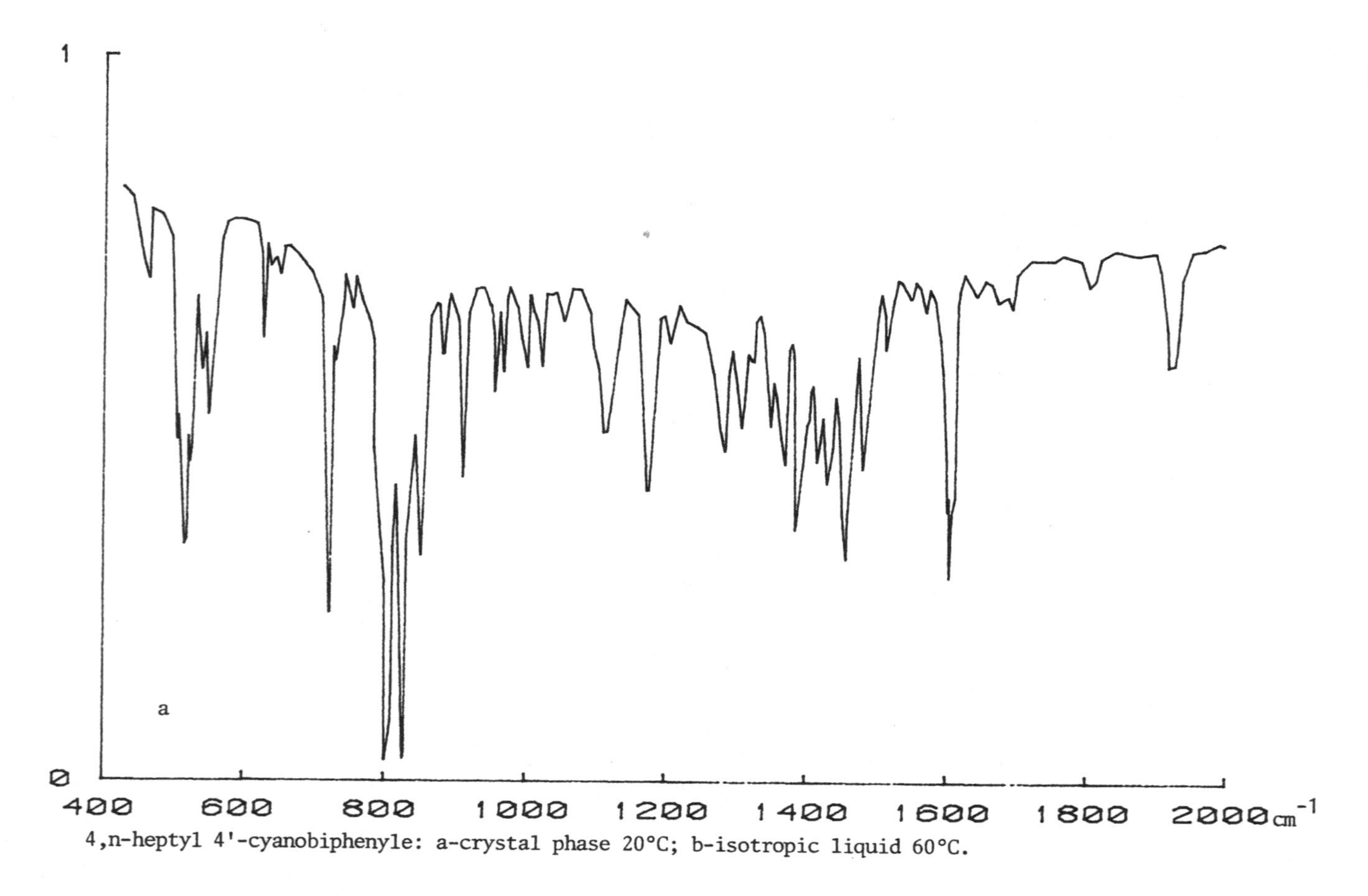

4,n-heptyl 4'-cyanobiphenyle: a-crystal phase 20°C; b-isotropic liquid 60°C.

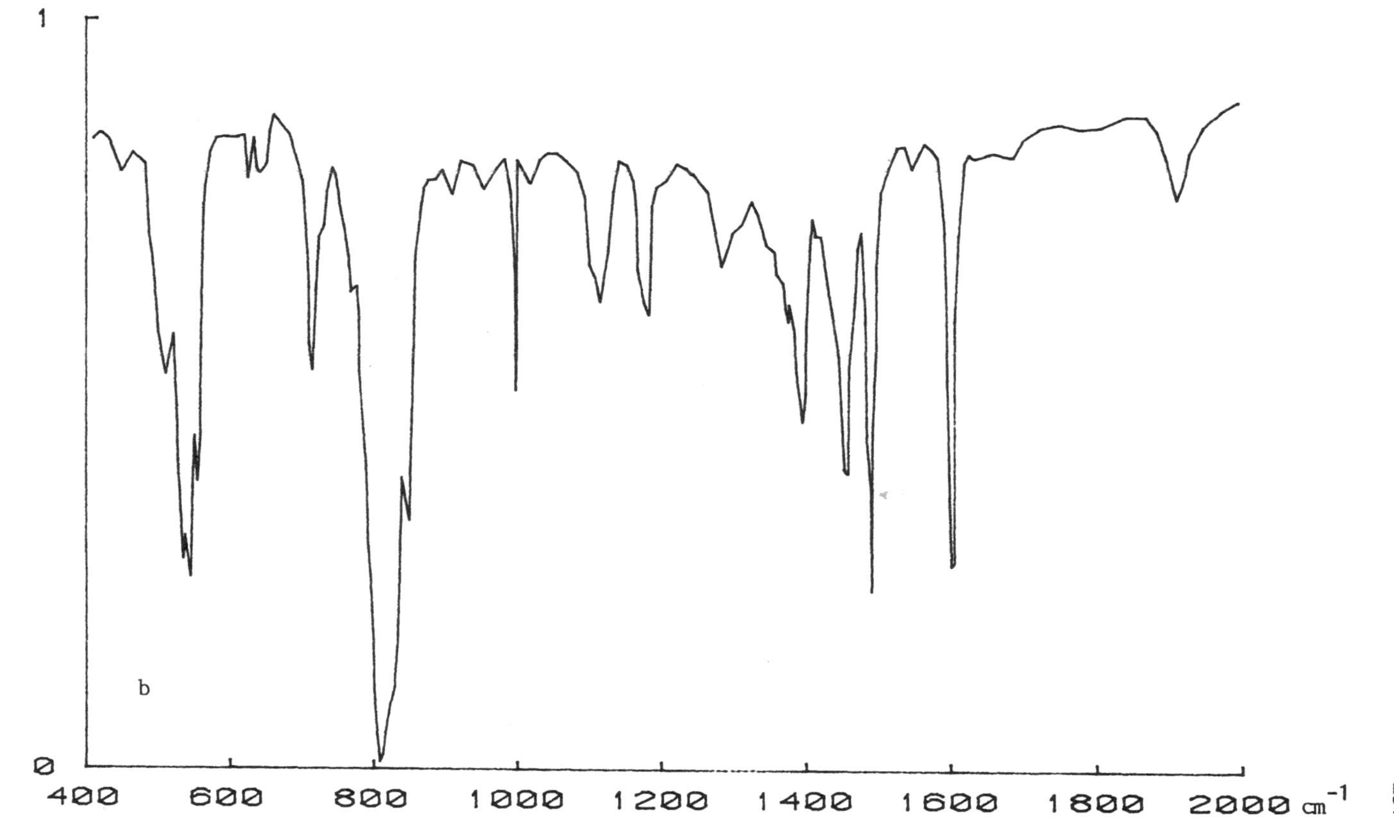
1
0
400
600
800
1000
1200
1400
1600
1800
2000 cm^{-1}
b

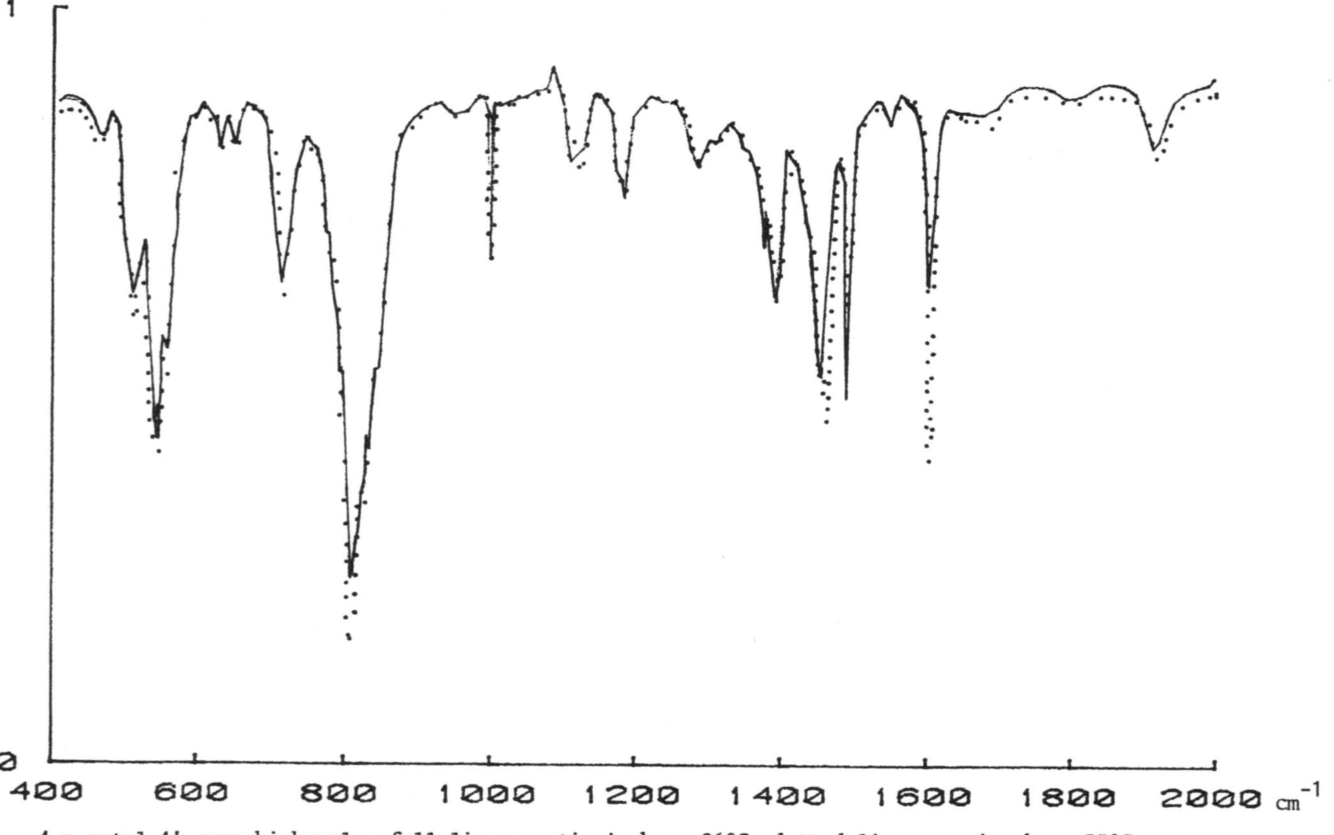

4,n-octyl 4'-cyanobiphenyle: full line-smectic A phase 26°C; dotted line-nematic phase 35°C.

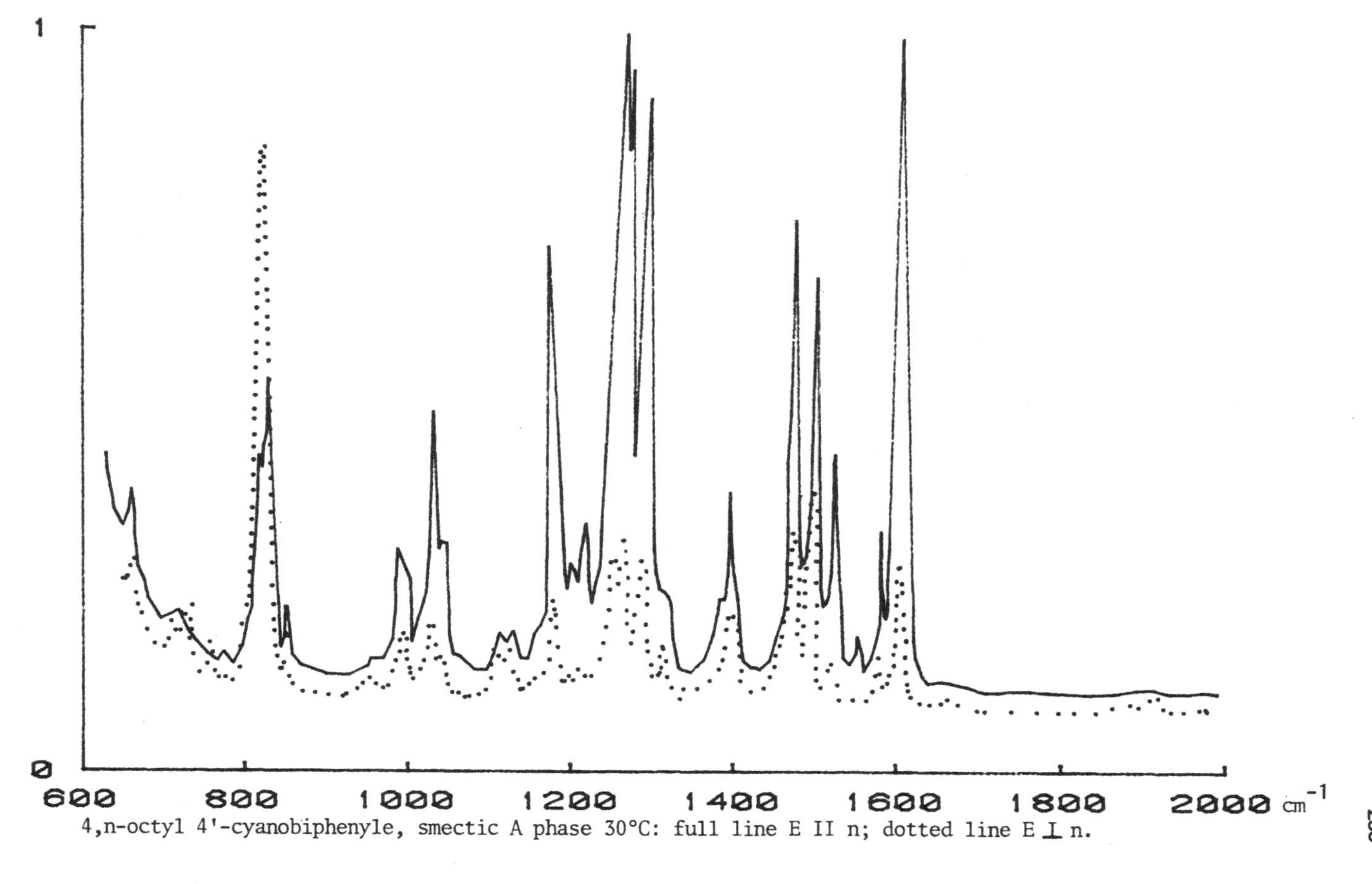

4,n-octyl 4'-cyanobiphenyle, smectic A phase 30°C: full line E II n; dotted line E ⊥ n.

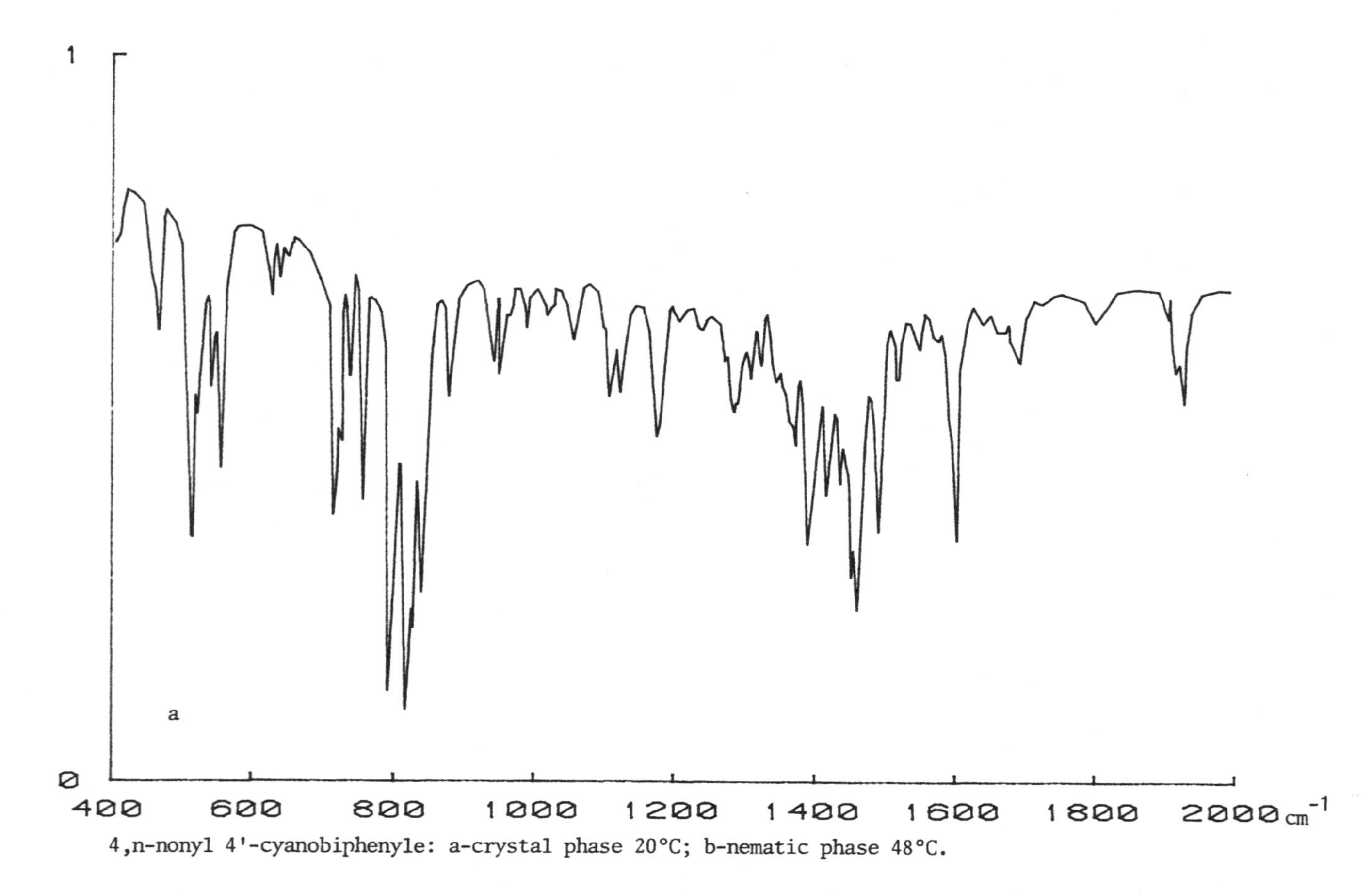

4,n-nonyl 4'-cyanobiphenyle: a-crystal phase 20°C; b-nematic phase 48°C.

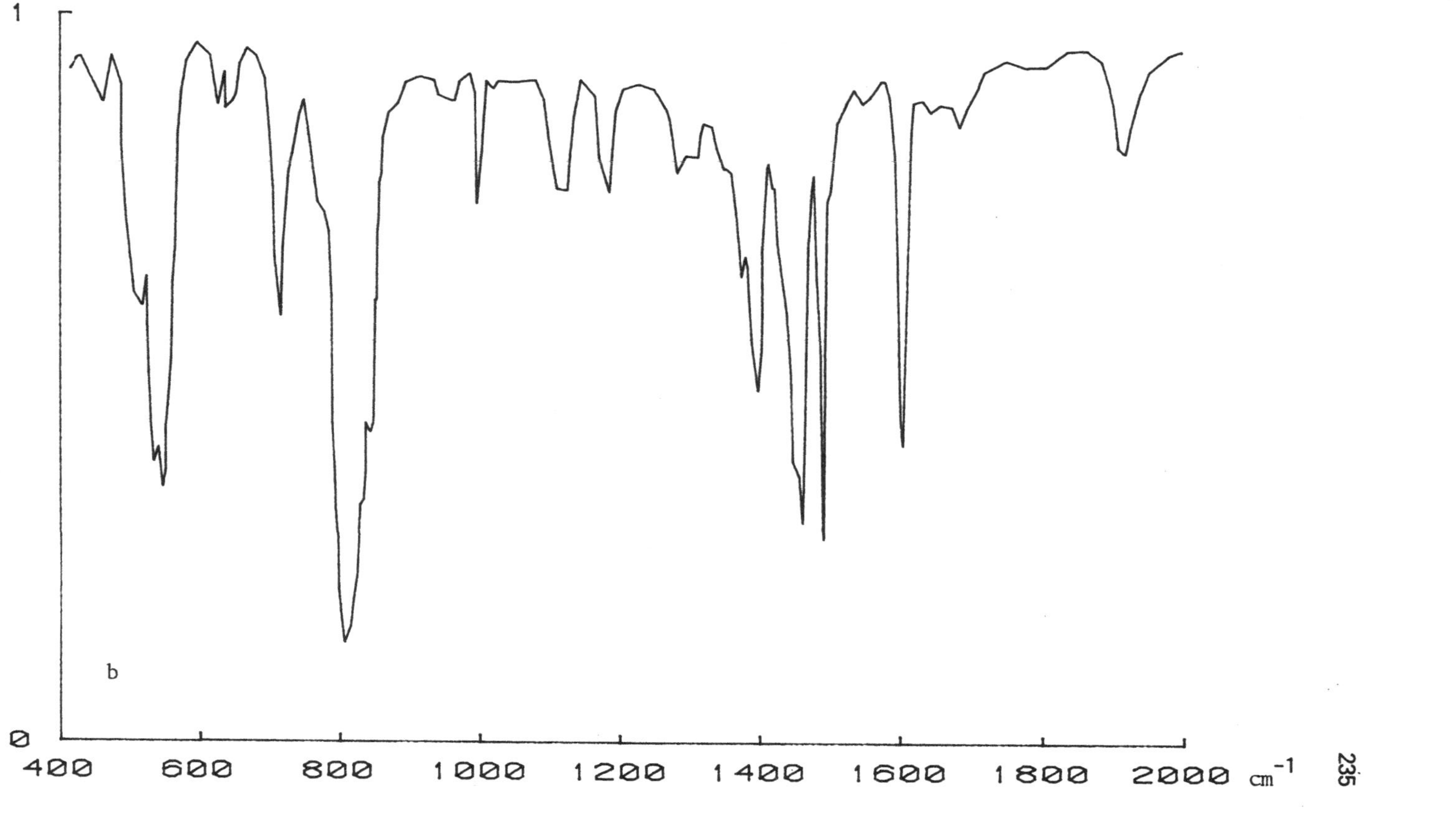

1
0
400
600
800
1000
1200
1400
1600
1800
2000
cm-1
b

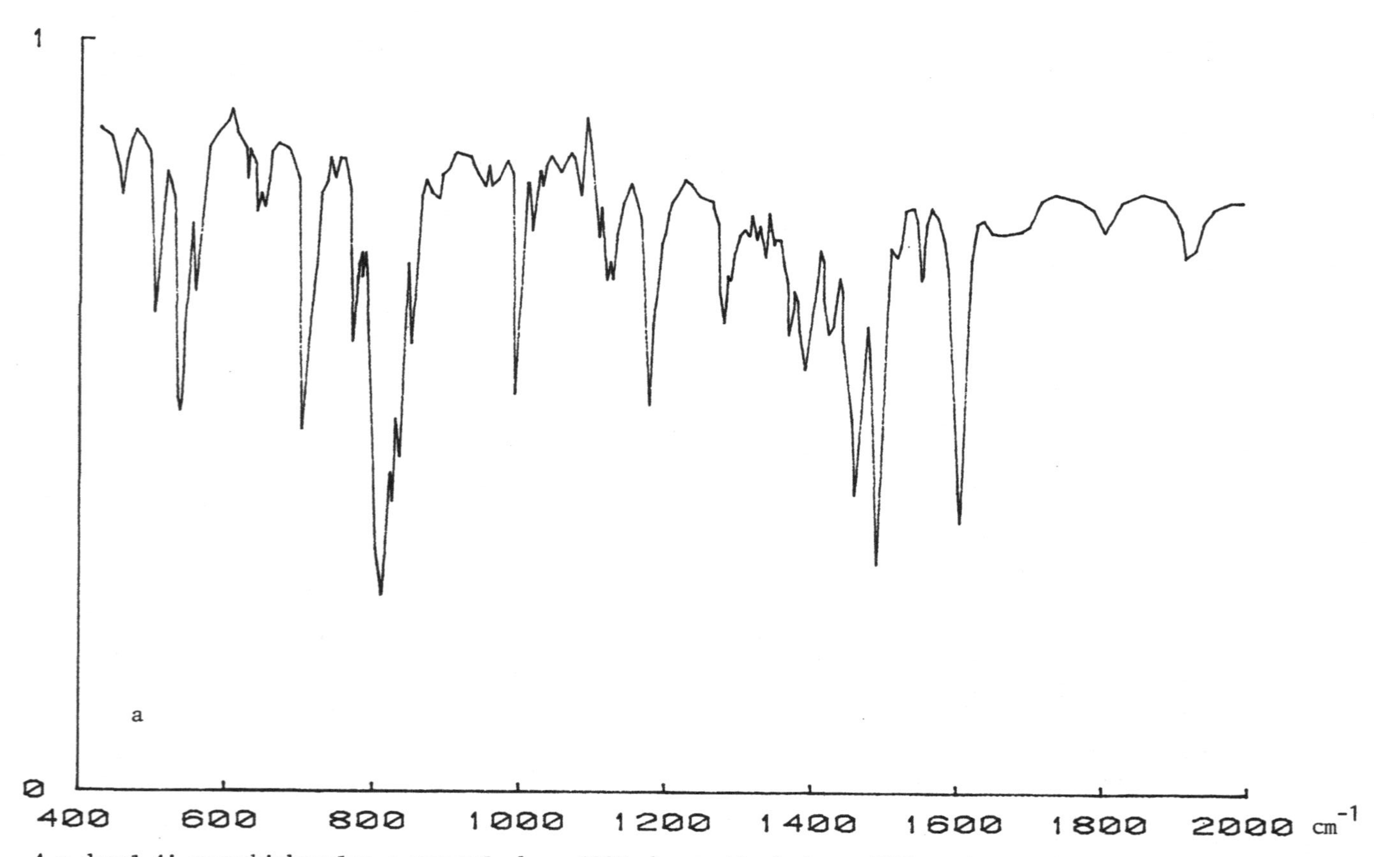

4,n-decyl 4'-cyanobiphenyle: a-crystal phase 20°C; b-smectic A phase 48°C; c-isotropic liquid 55°C.

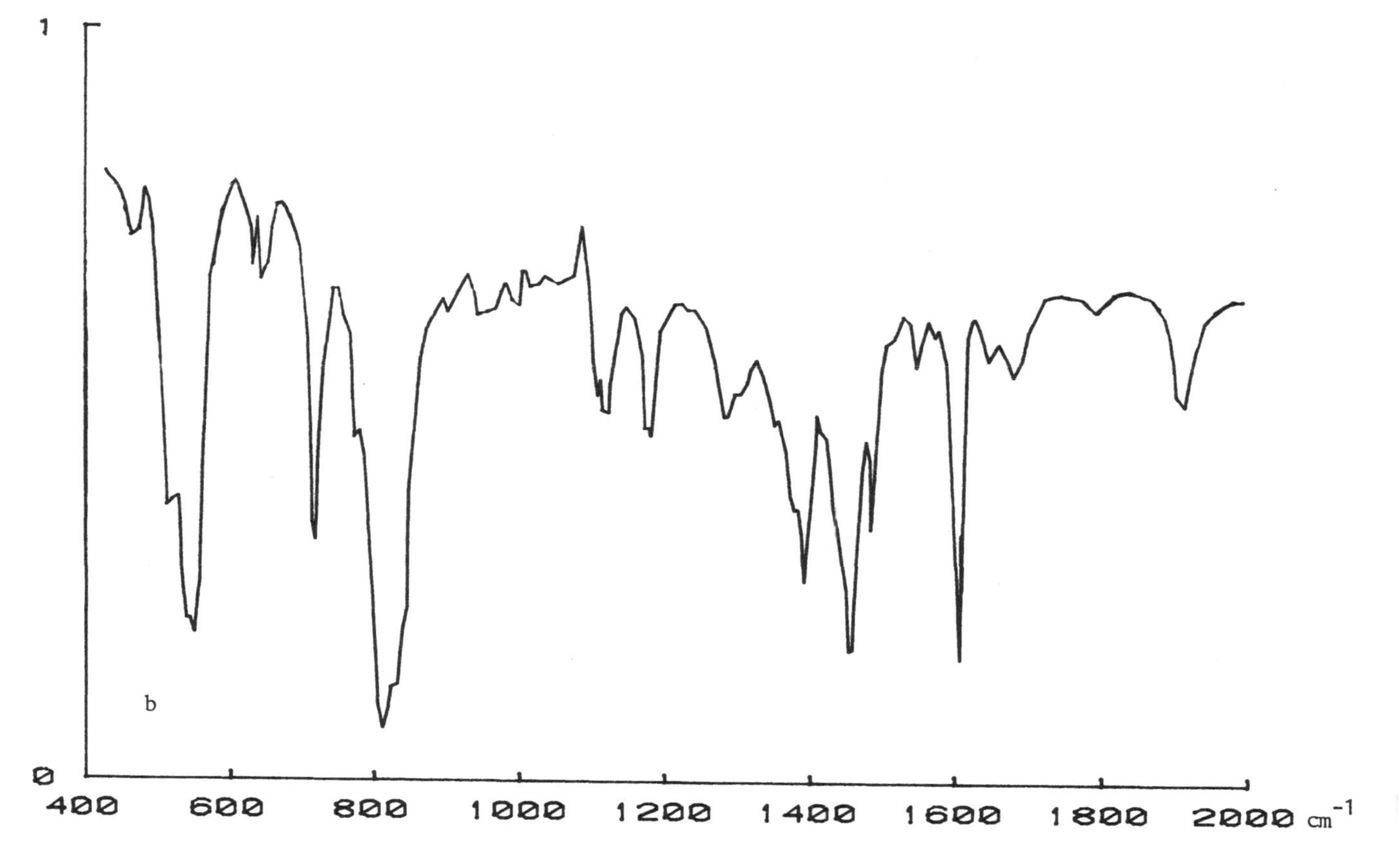

1
0
400
600
800
1000
1200
1400
1600
1800
2000
cm^{-1}
b

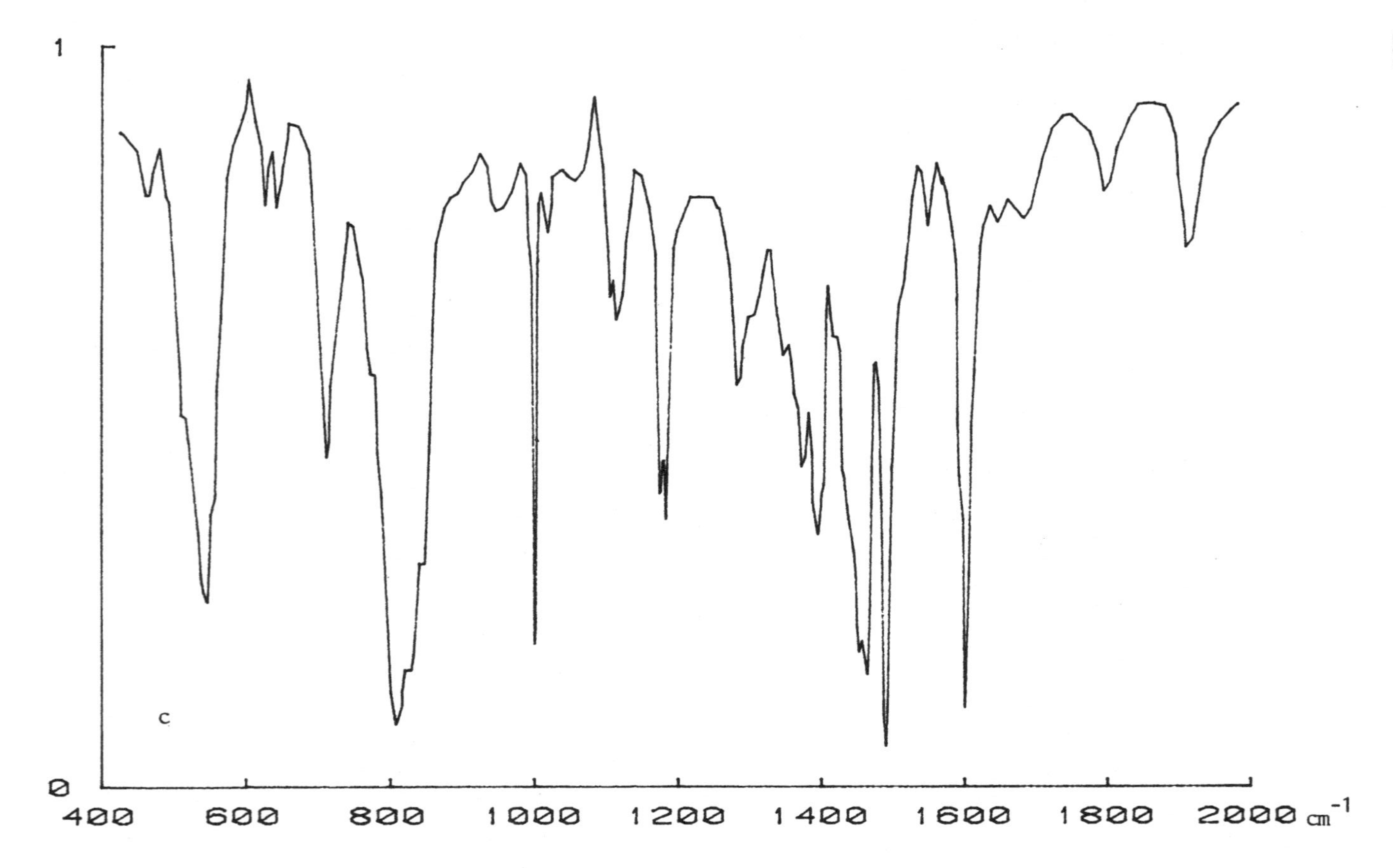
1
0
400
600
800
1000
1200
1400
1600
1800
2000 cm⁻¹
c

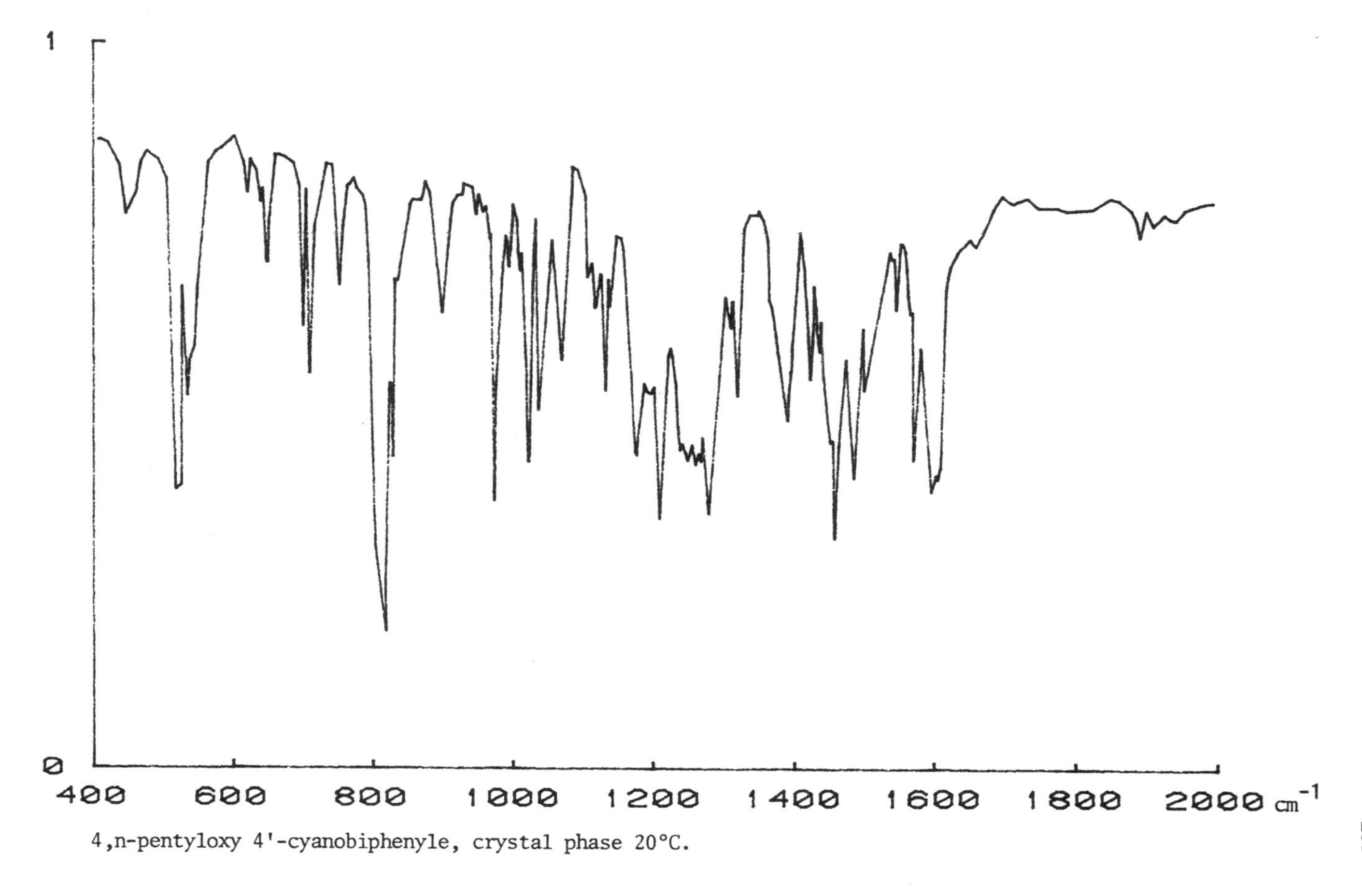

4,n-pentyloxy 4'-cyanobiphenyle, crystal phase 20°C.

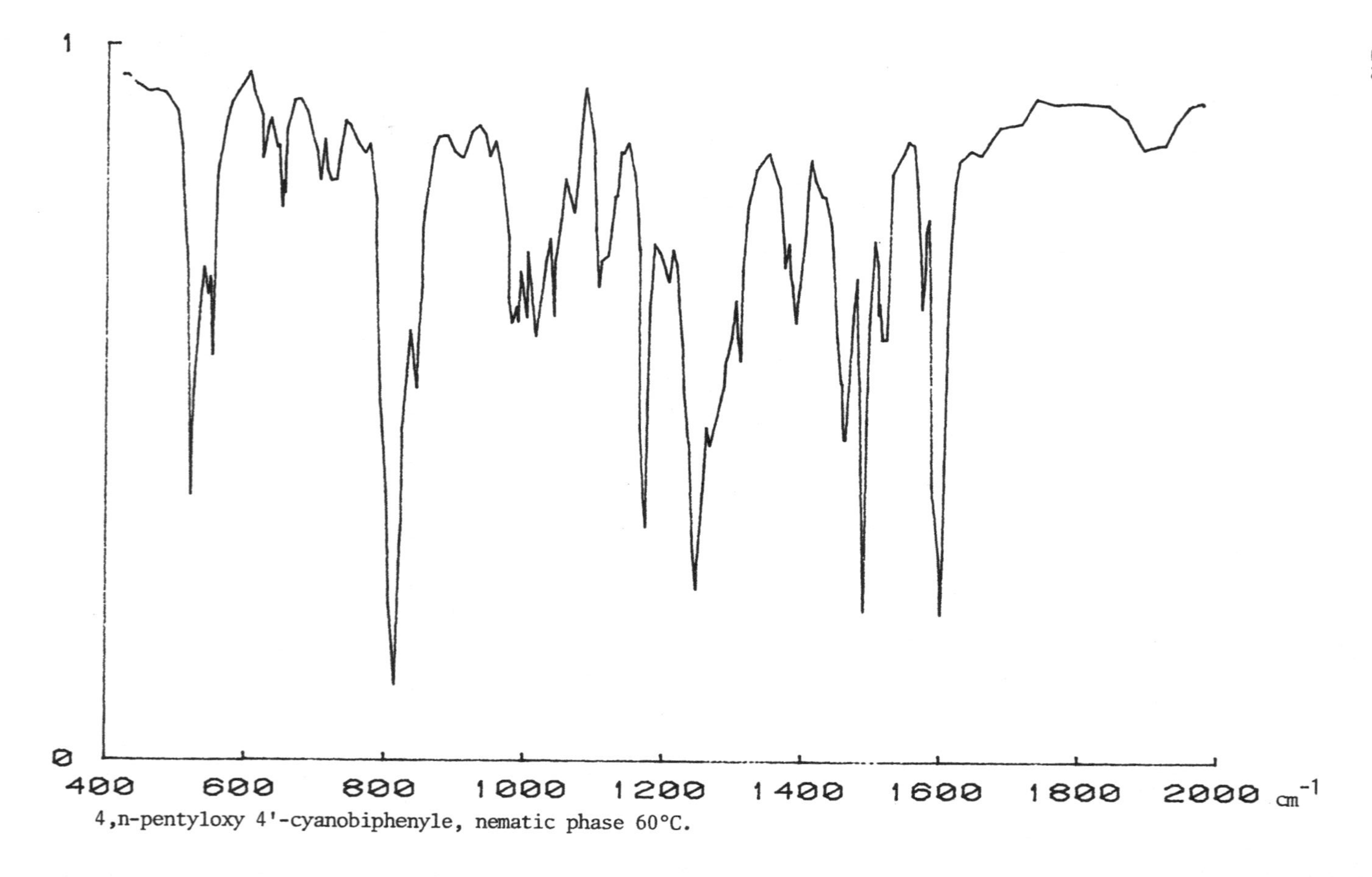

4,n-pentyloxy 4'-cyanobiphenyle, nematic phase 60°C.

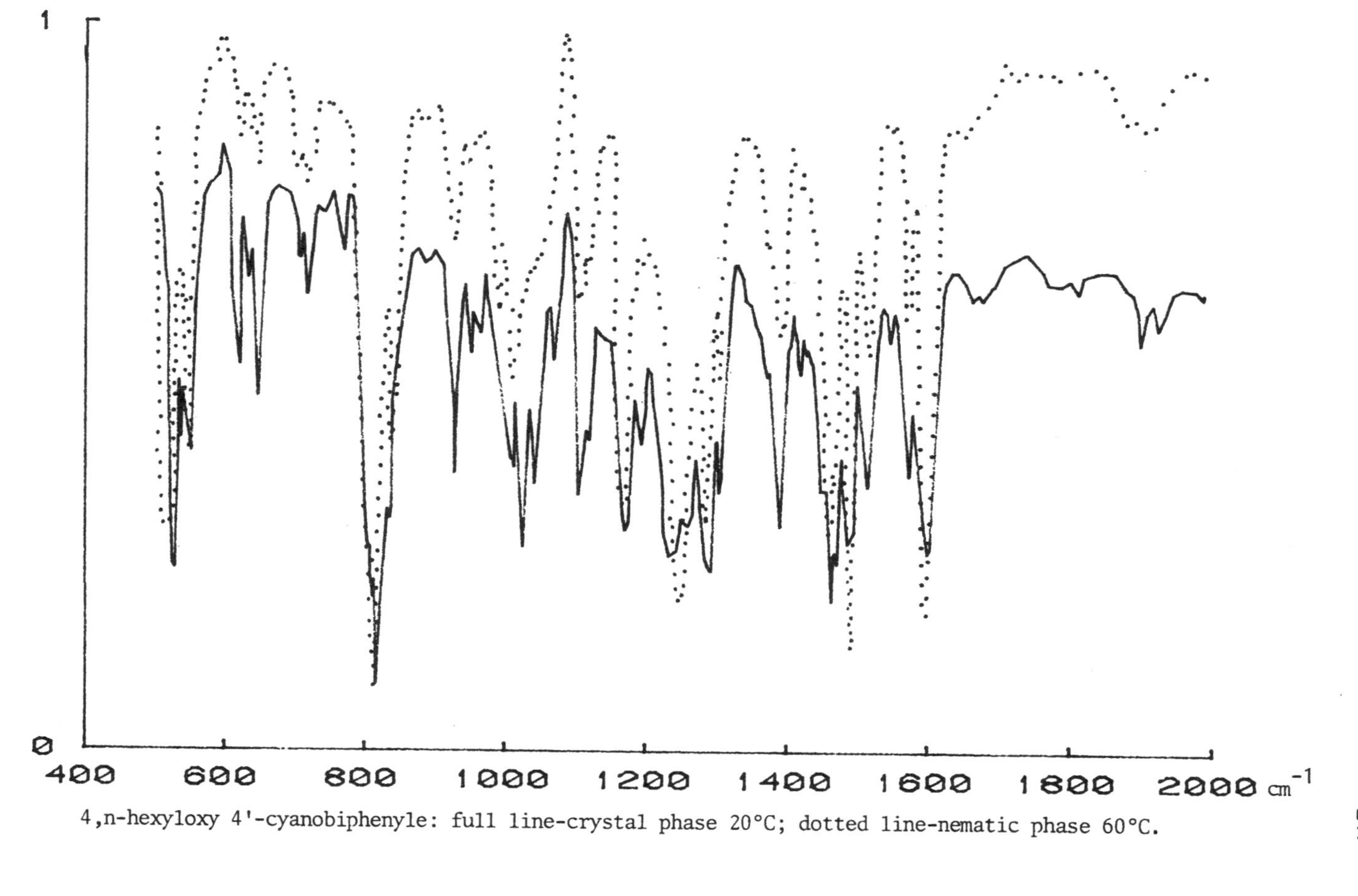

4,n-hexyloxy 4'-cyanobiphenyle: full line-crystal phase 20°C; dotted line-nematic phase 60°C.

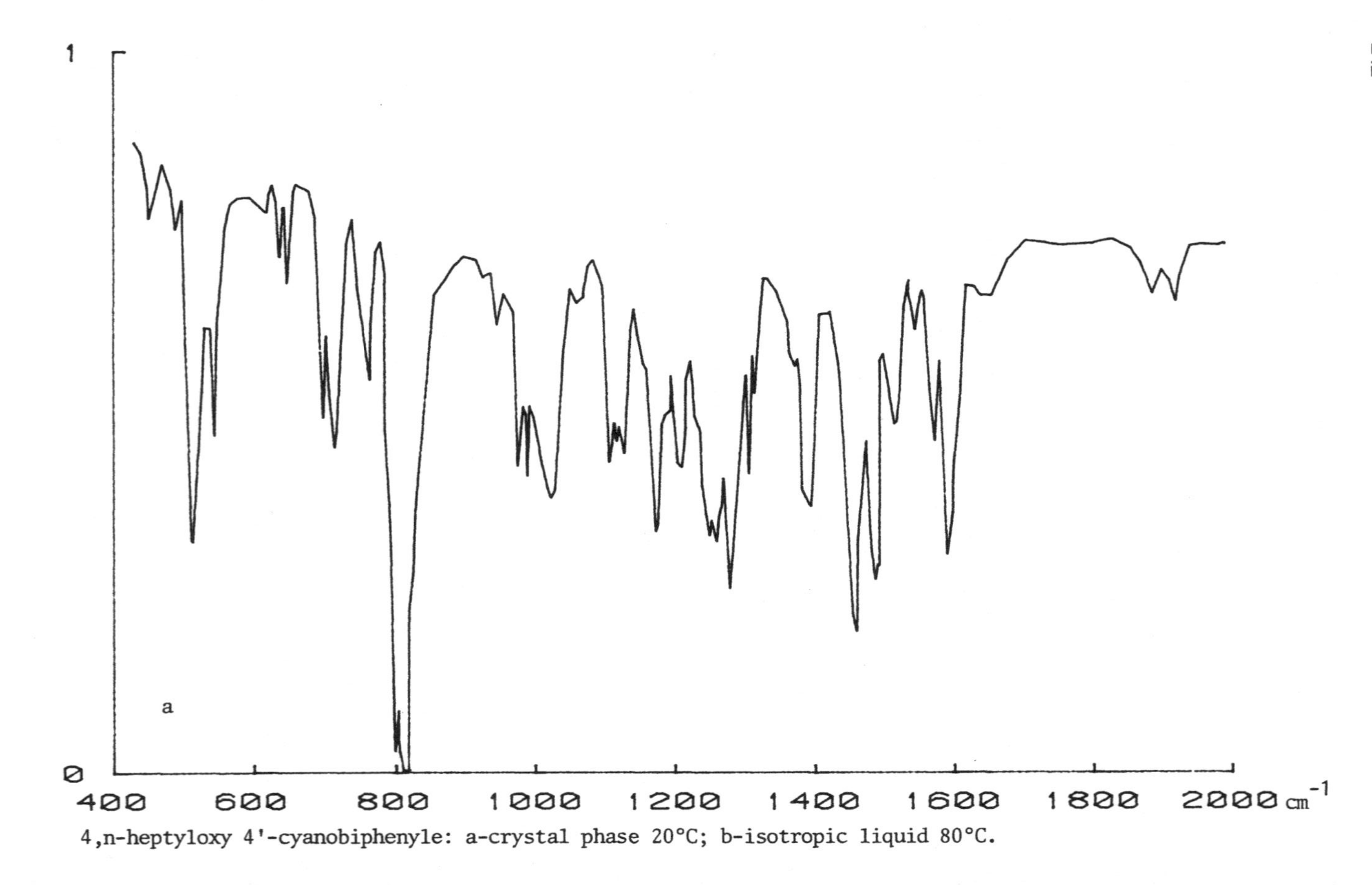

4,n-heptyloxy 4'-cyanobiphenyle: a-crystal phase 20°C; b-isotropic liquid 80°C.

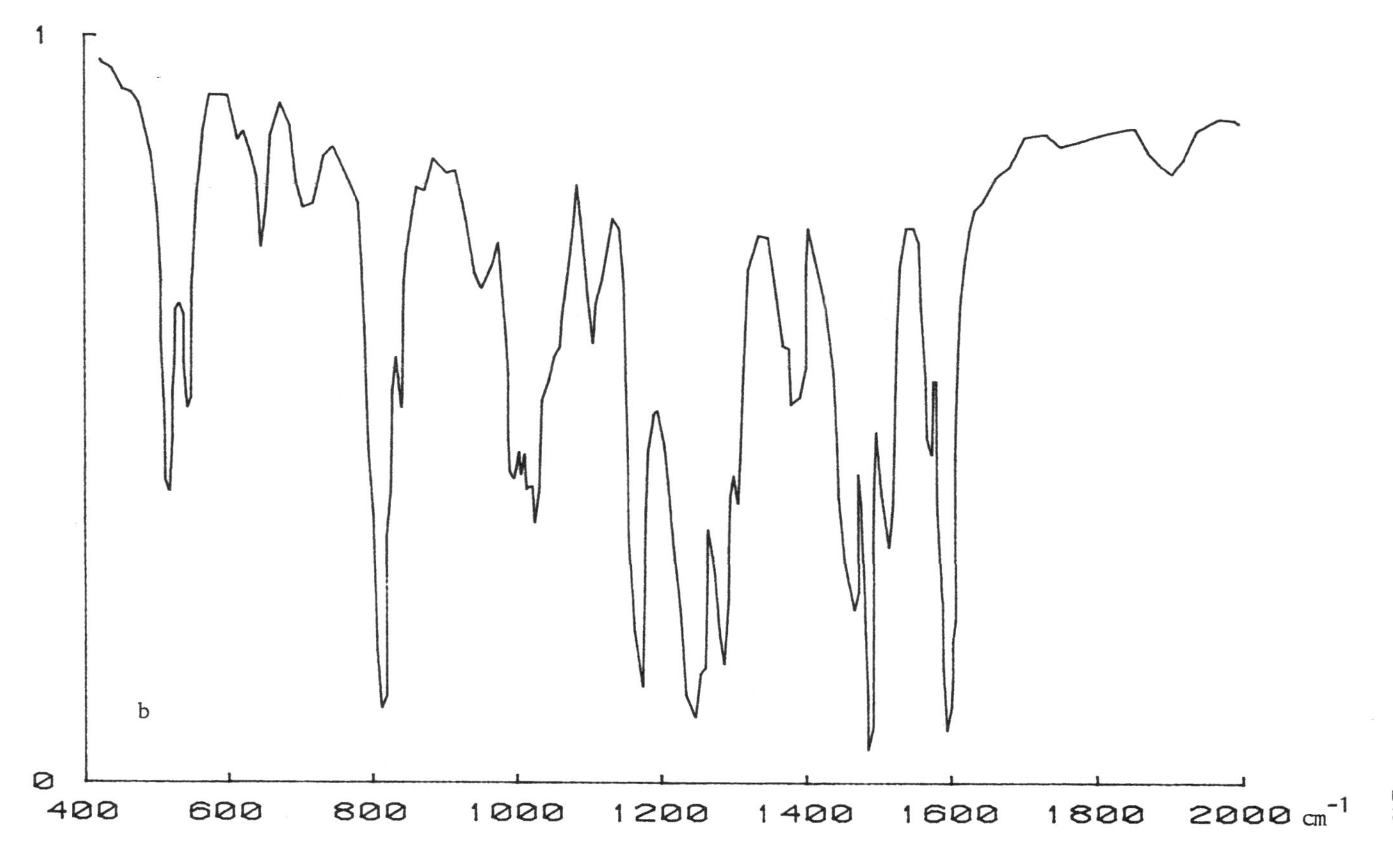

1
0
400
600
800
1000
1200
1400
1600
1800
2000 cm-1
b

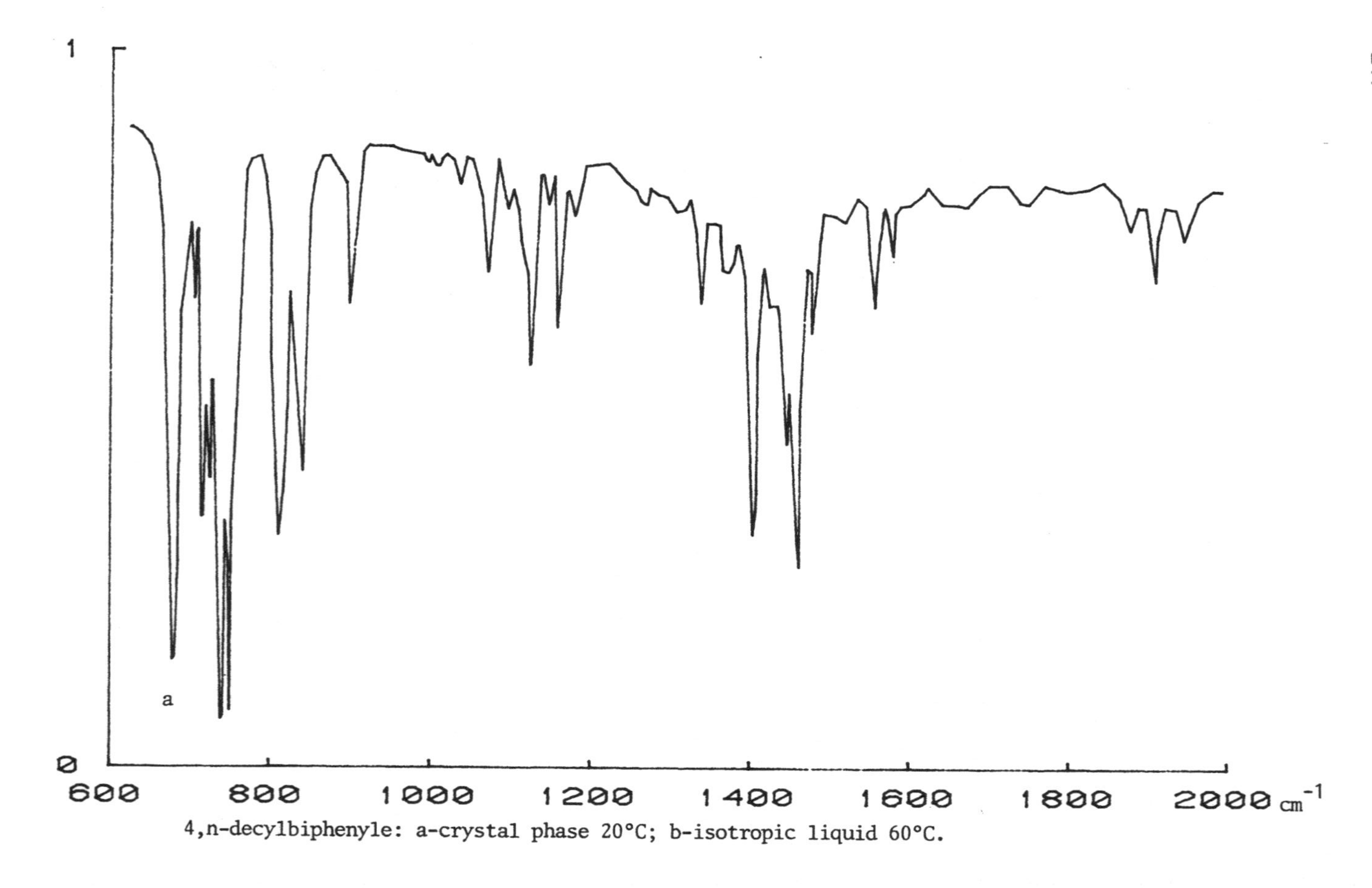

4,n-decylbiphenyle: a-crystal phase 20°C; b-isotropic liquid 60°C.

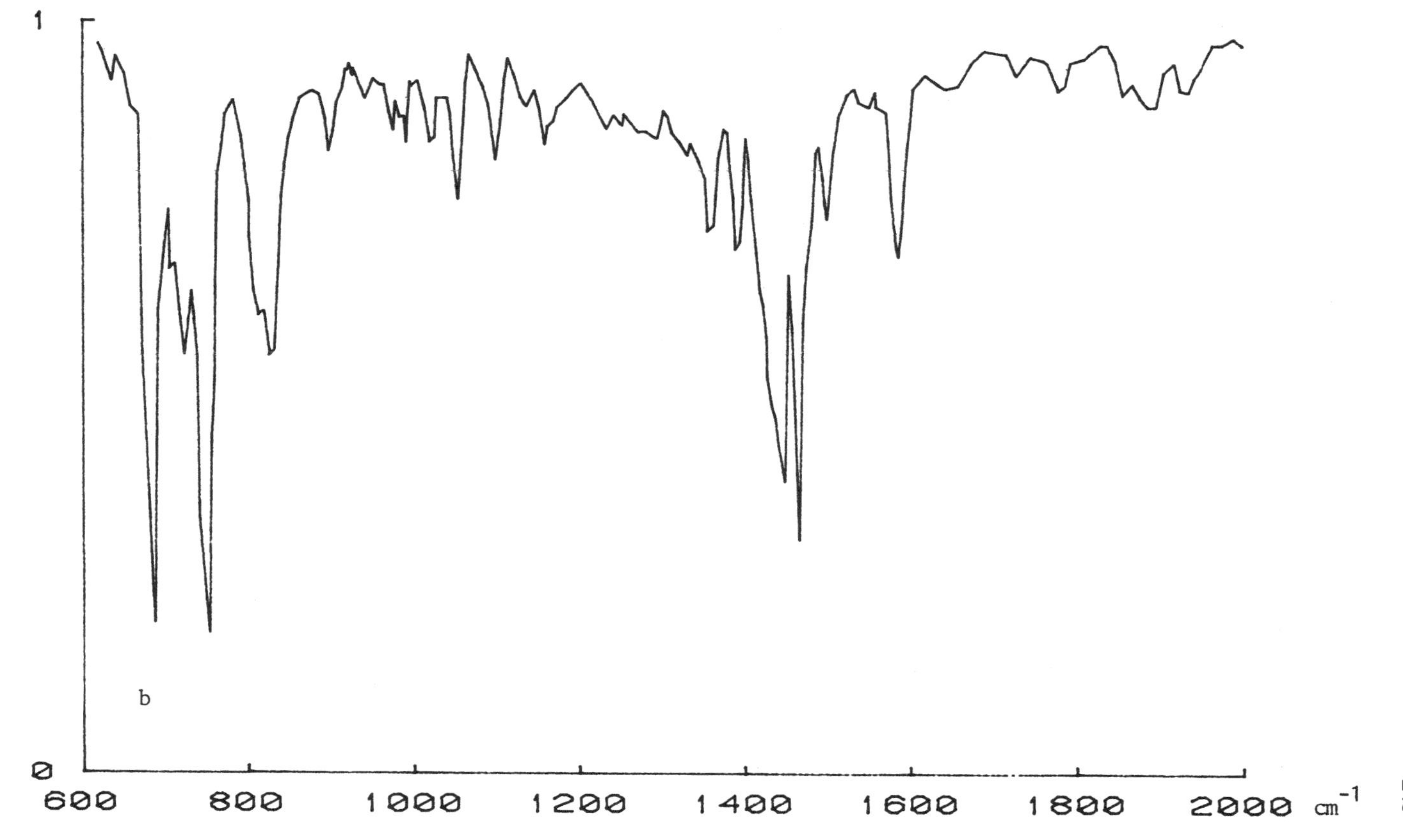
1
0
600
800
1000
1200
1400
1600
1800
2000
cm-1
b

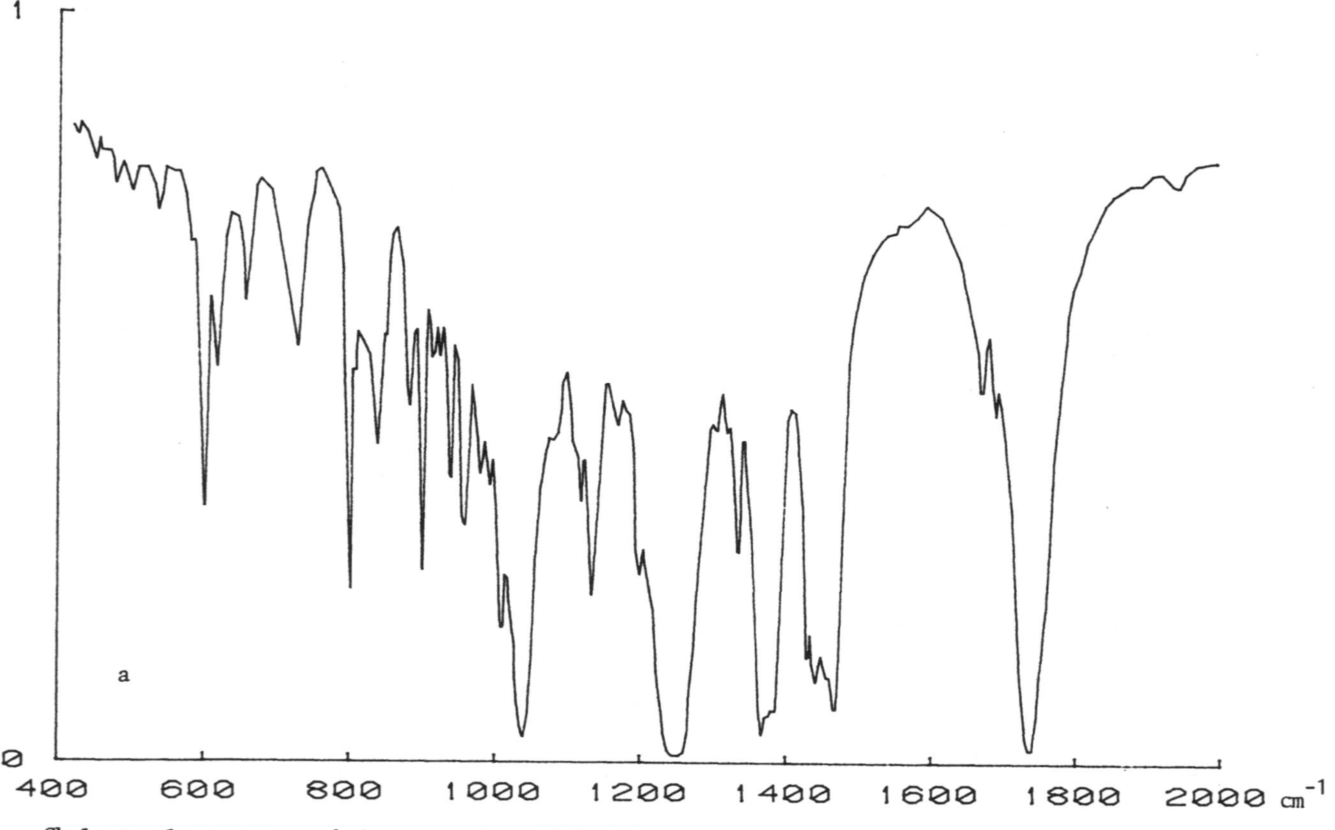

Cholesteryl acetate: a-cholesteric phase 95°C; b-isotropic liquid 120°C.

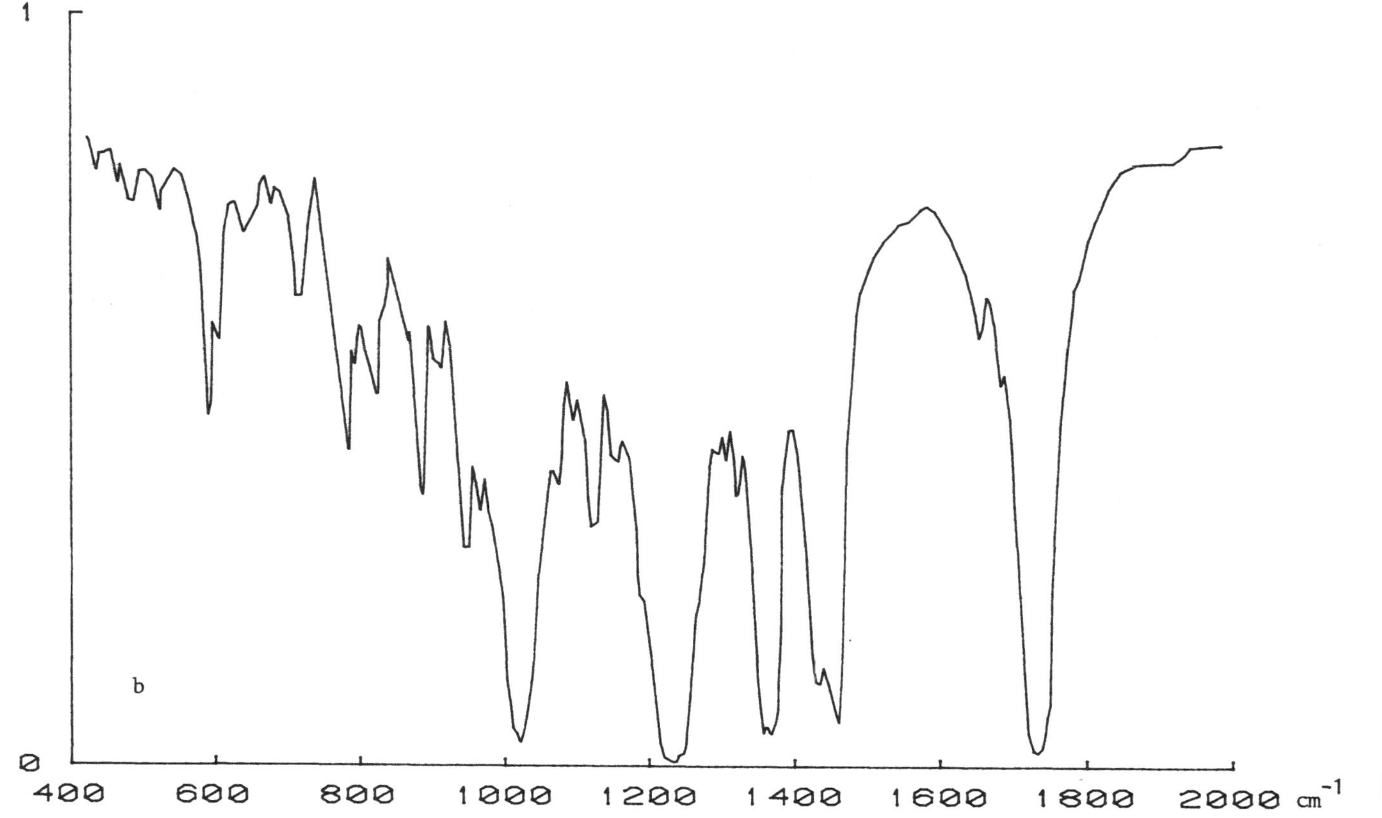

1
Ø
400
600
800
1000
1200
1400
1600
1800
2000 cm⁻¹
b

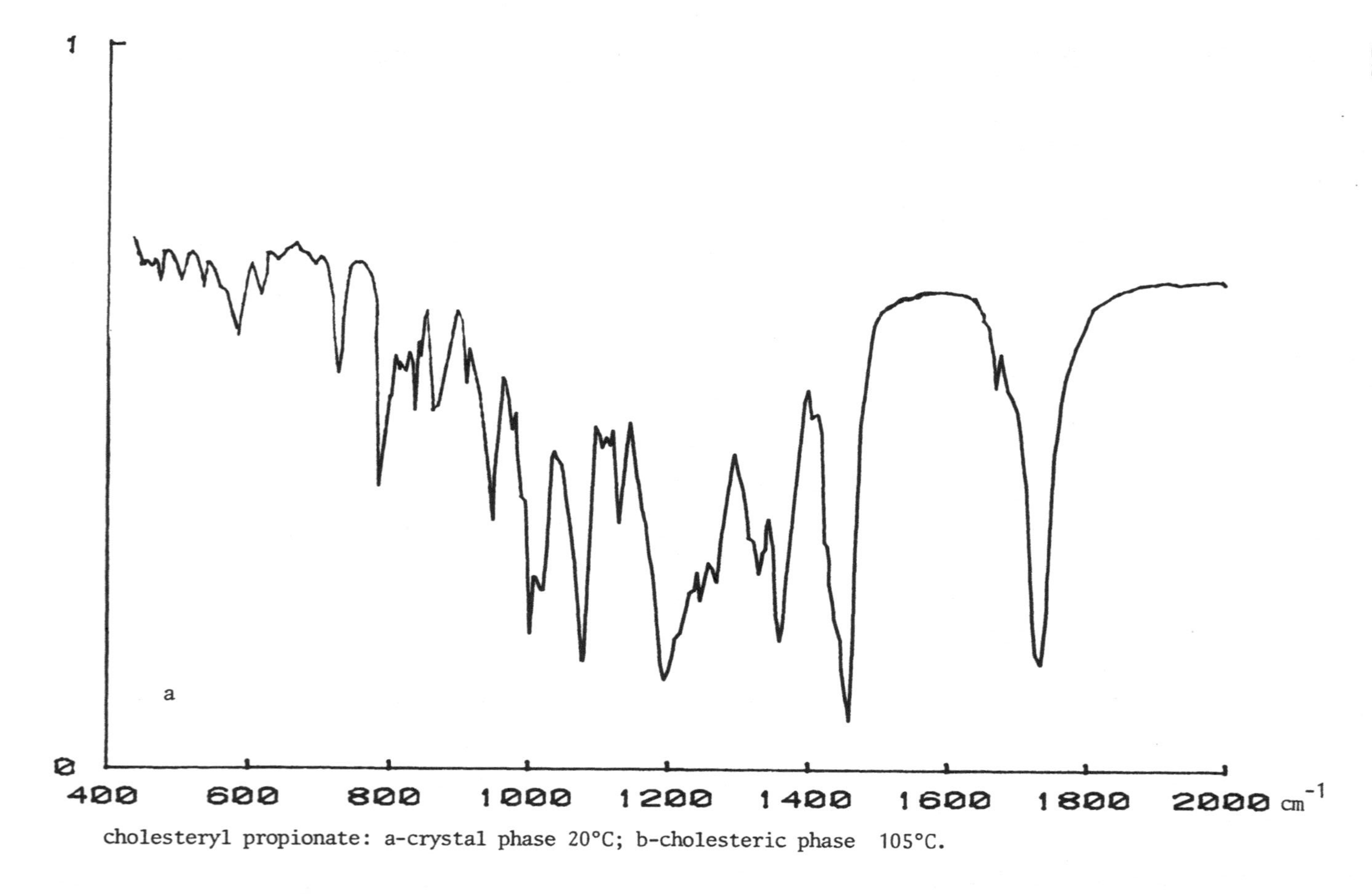

cholesteryl propionate: a-crystal phase 20°C; b-cholesteric phase 105°C.

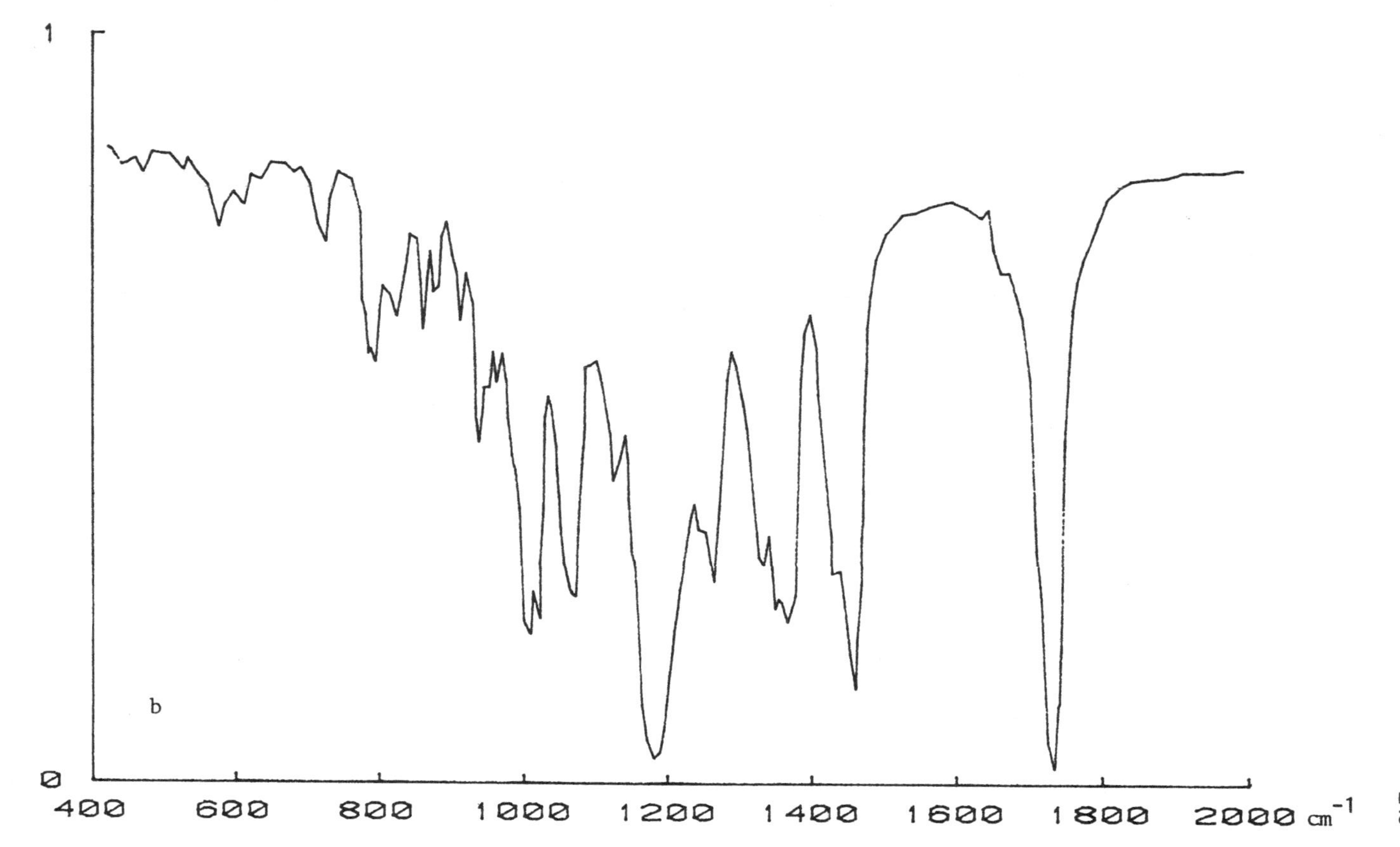
1
b
0
400
600
800
1000
1200
1400
1600
1800
2000 cm-1

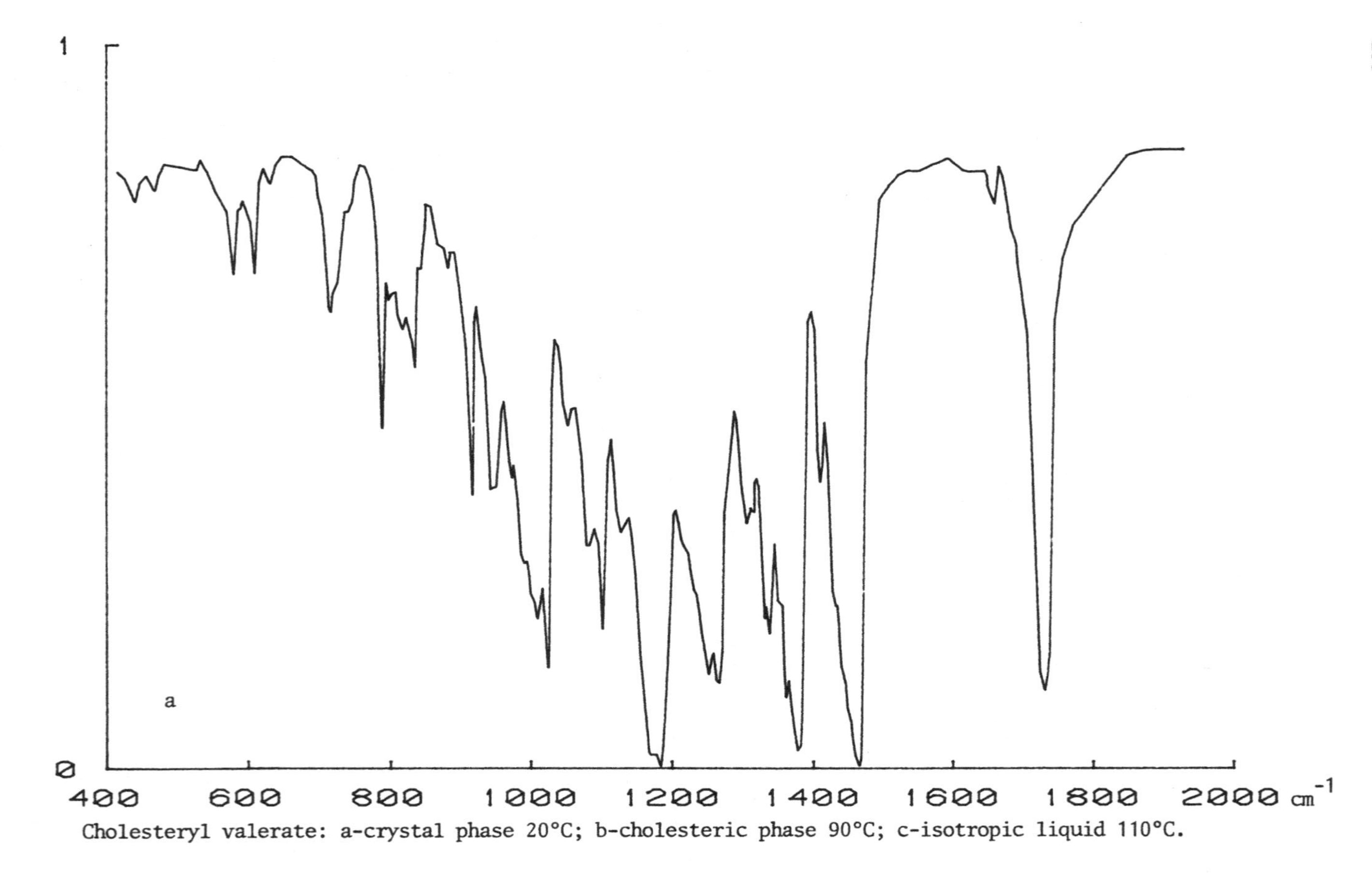

Cholesteryl valerate: a-crystal phase 20°C; b-cholesteric phase 90°C; c-isotropic liquid 110°C.

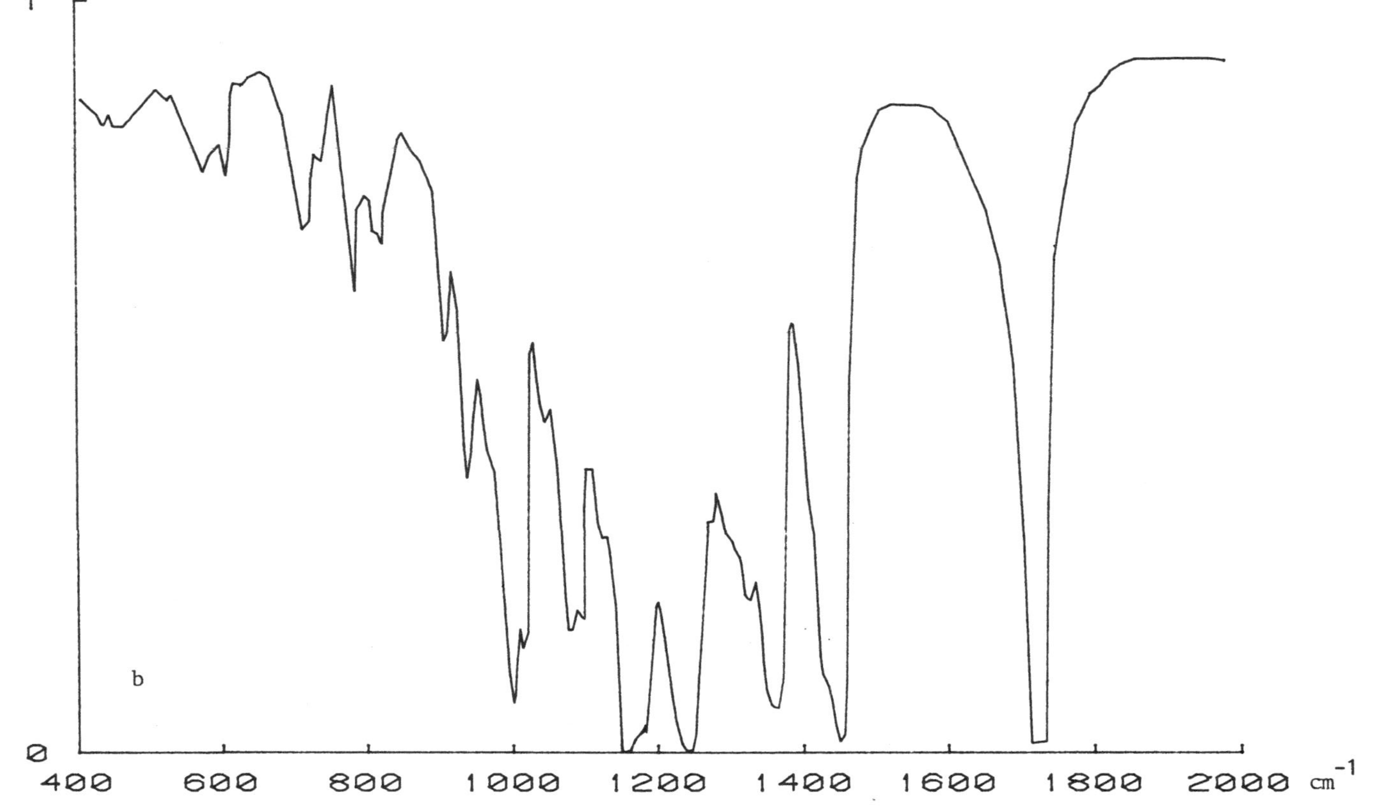

1
0
b

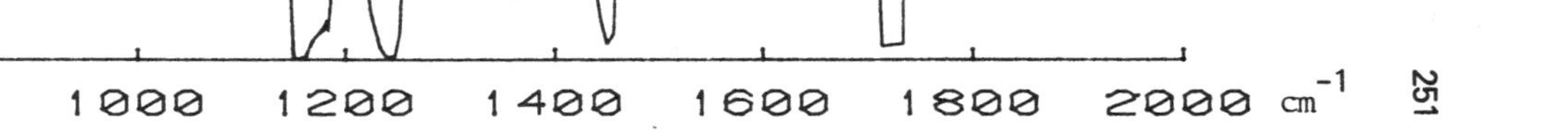

400
600
800
1000
1200
1400
1600
1800
2000
cm^{-1}

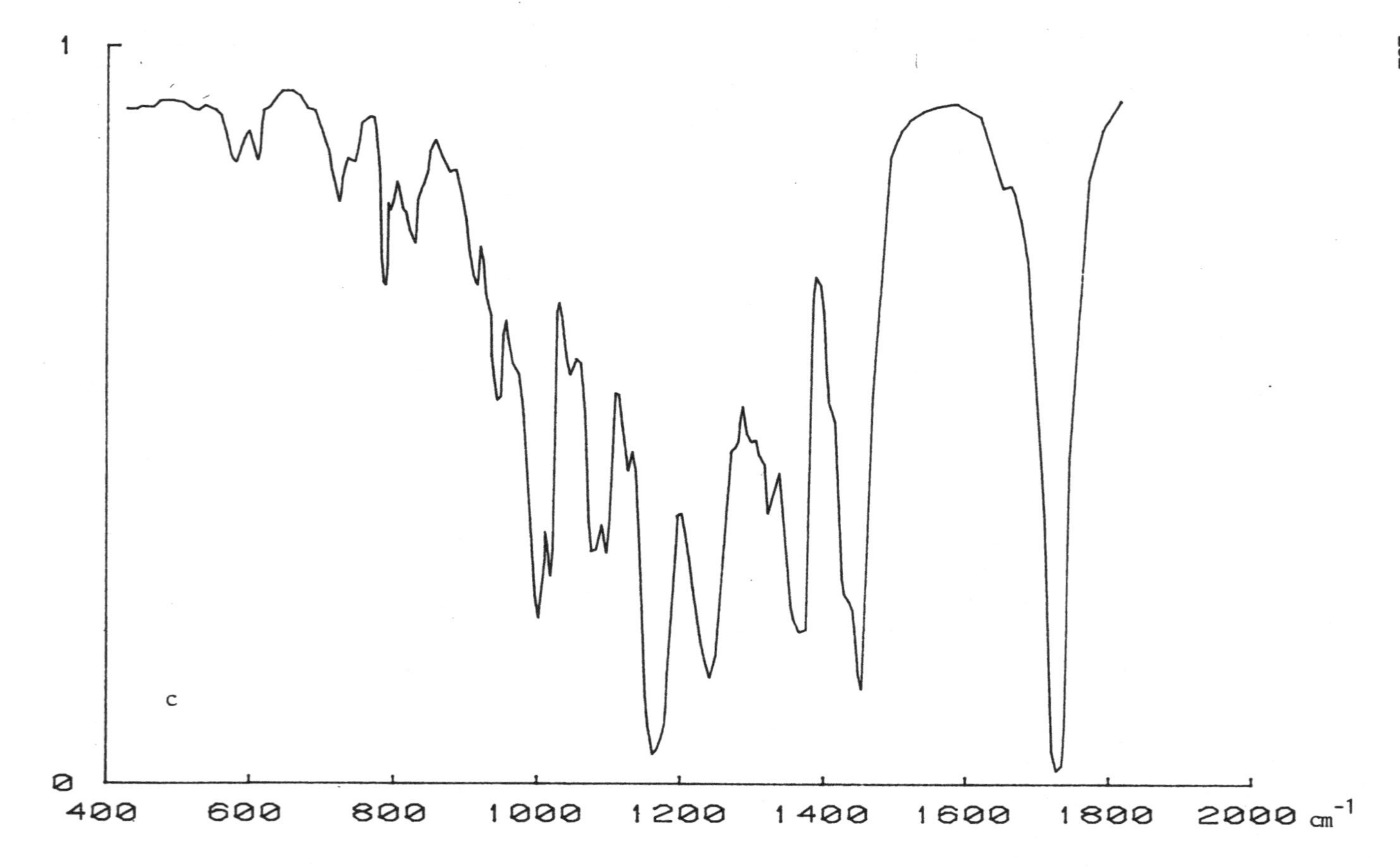

1
0
c
400
600
800
1000
1200
1400
1600
1800
2000 cm-1

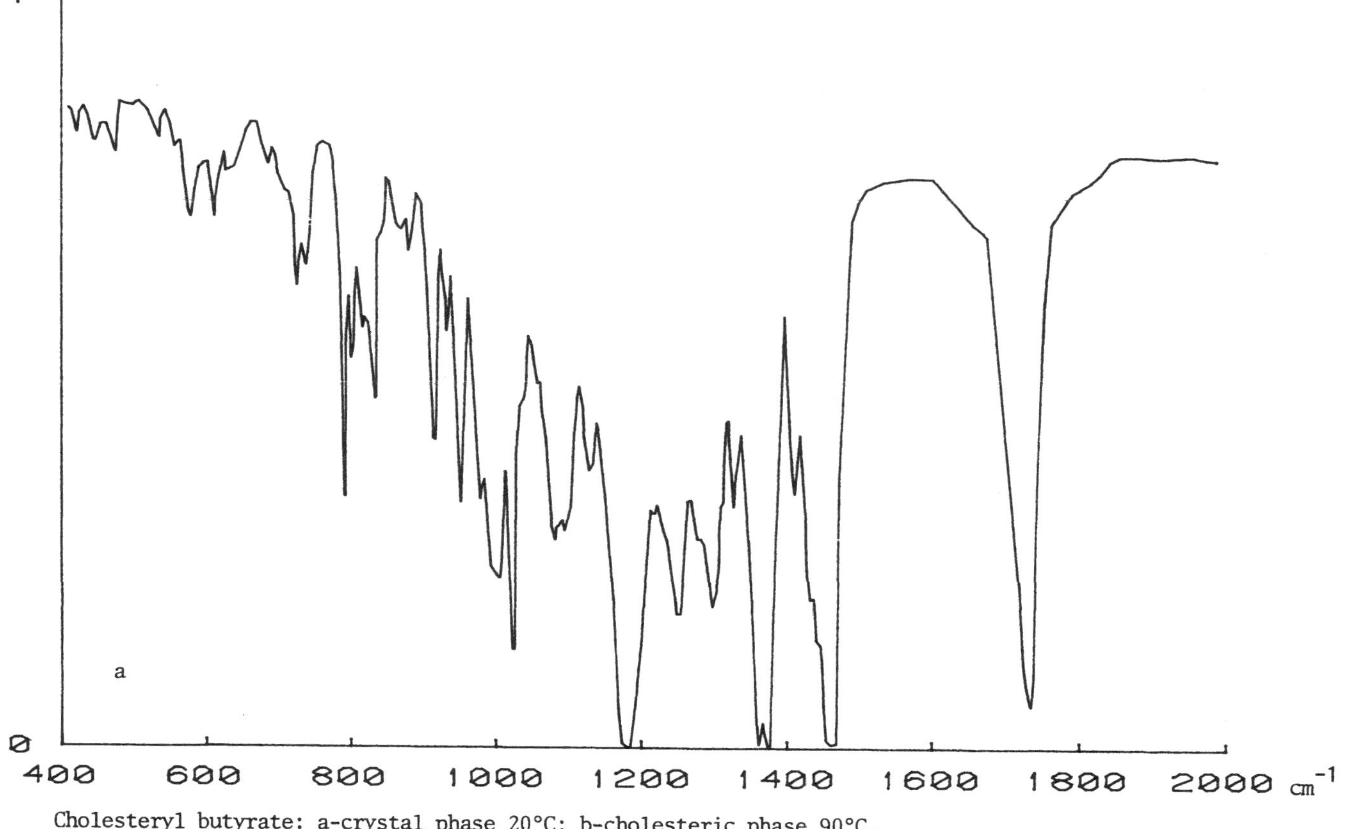

Cholesteryl butyrate: a-crystal phase 20°C; b-cholesteric phase 90°C.

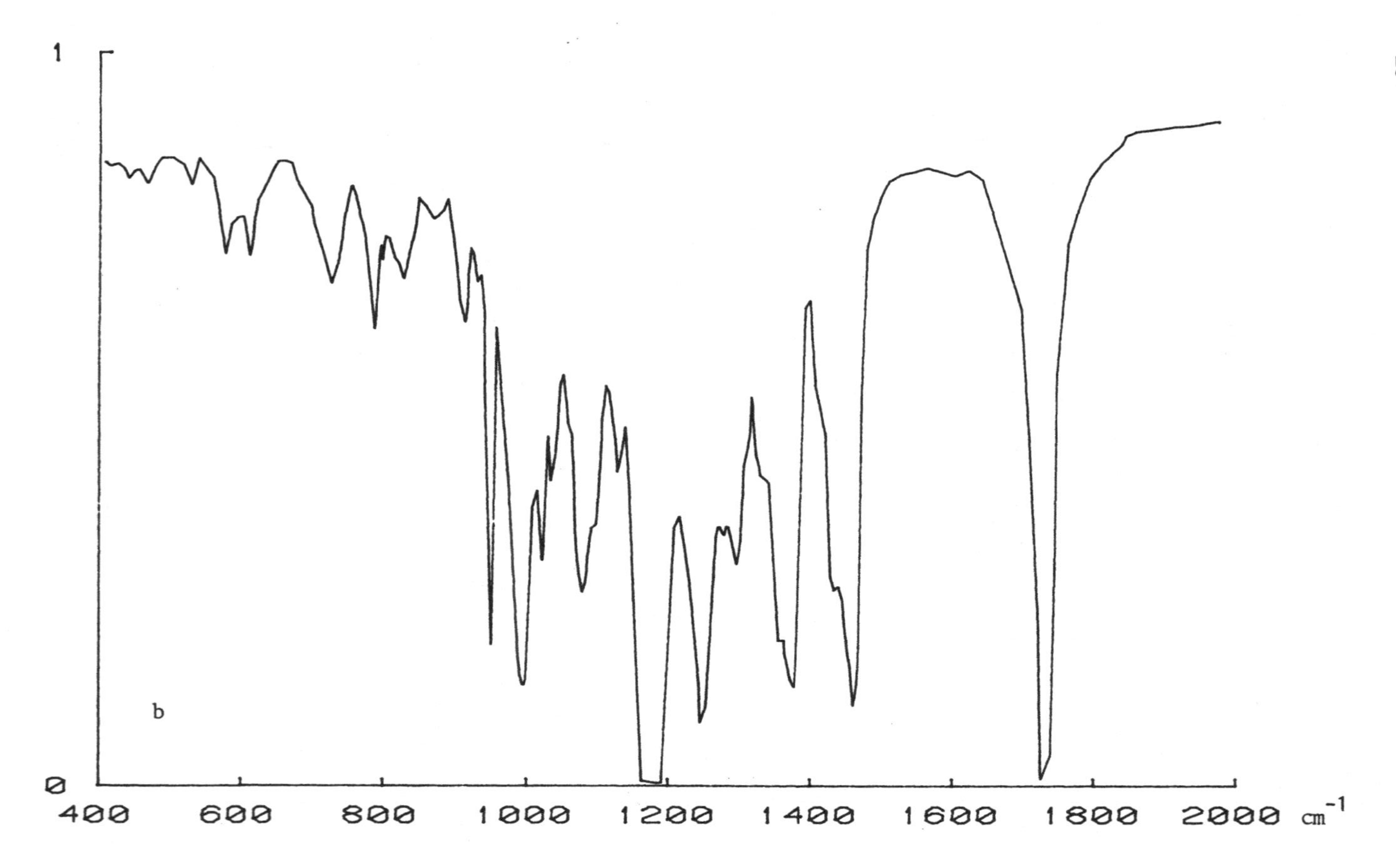
1
0
400
600
800
1000
1200
1400
1600
1800
2000
cm^{-1}
b

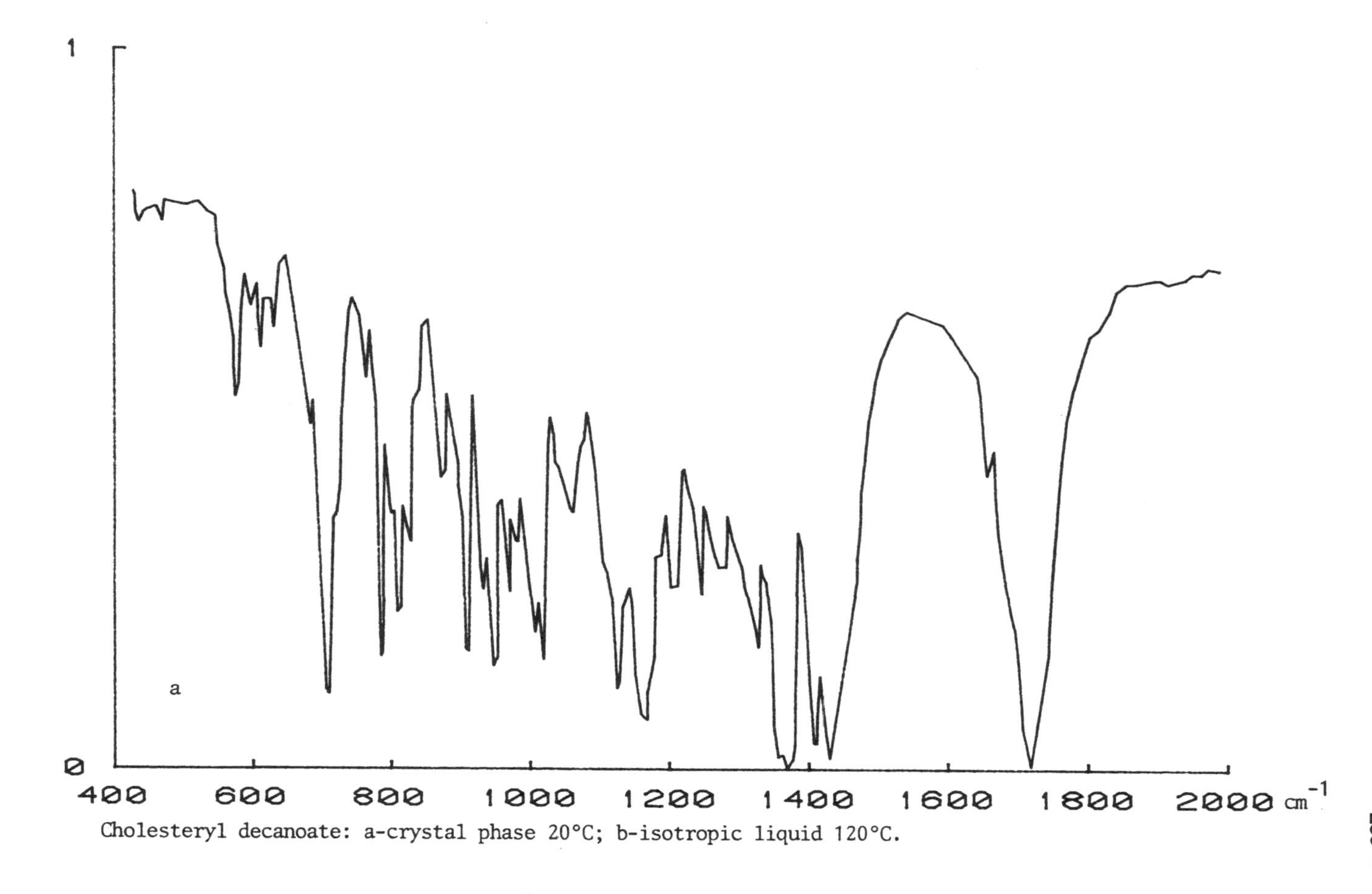

Cholesteryl decanoate: a-crystal phase 20°C; b-isotropic liquid 120°C.

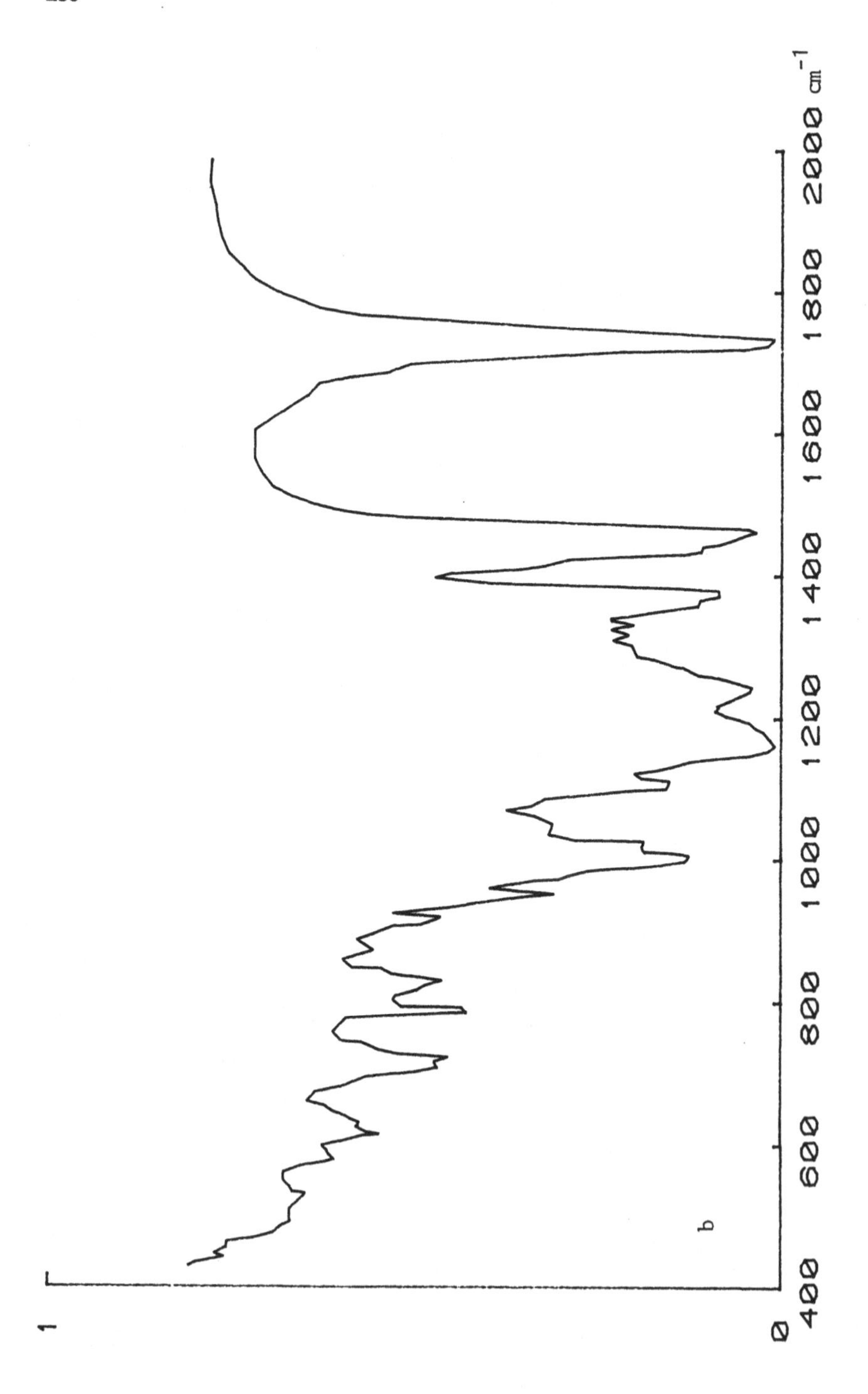
1
0
b
400
600
800
1000
1200
1400
1600
1800
2000 cm-1

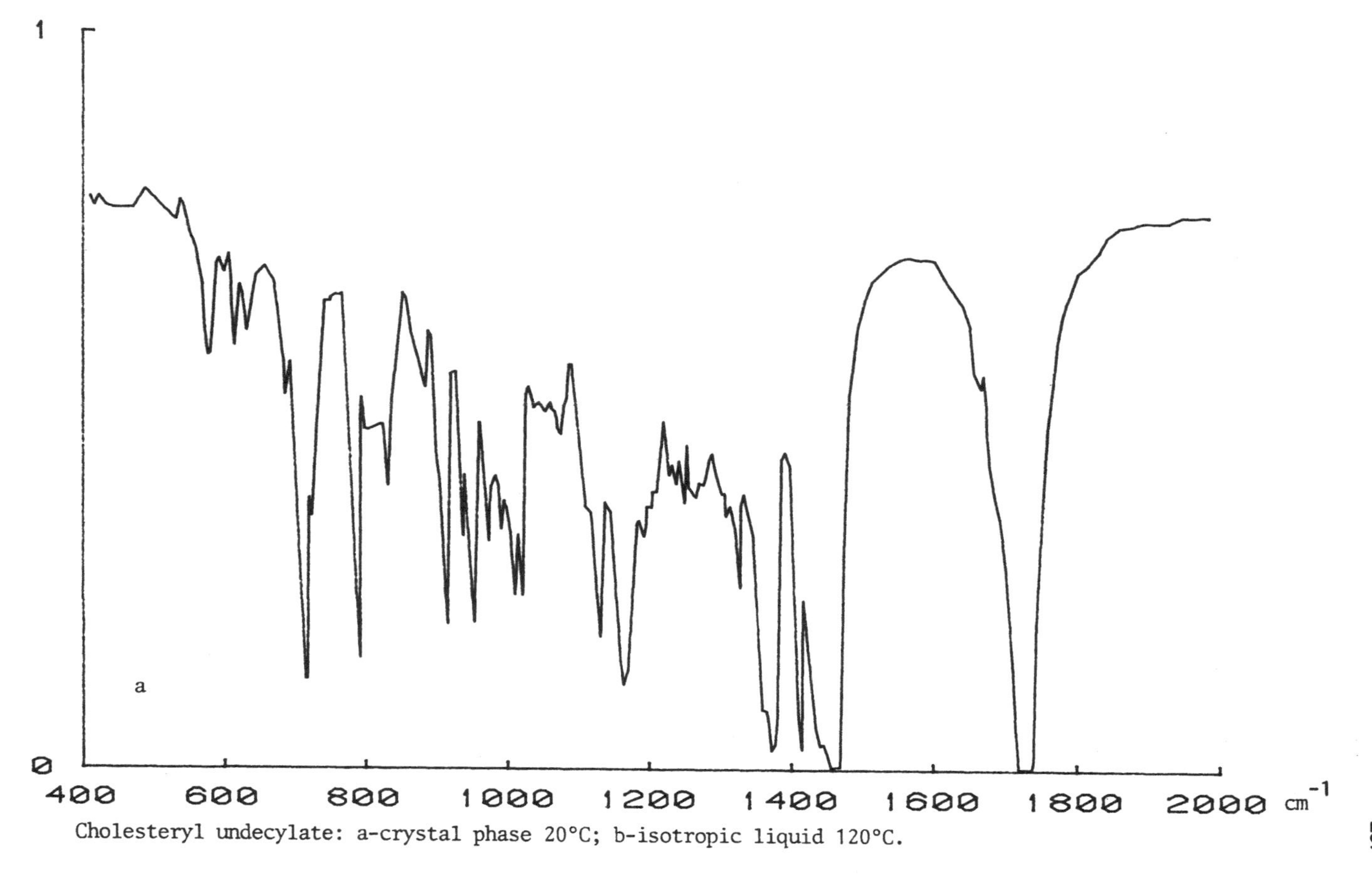

Cholesteryl undecylate: a-crystal phase 20°C; b-isotropic liquid 120°C.

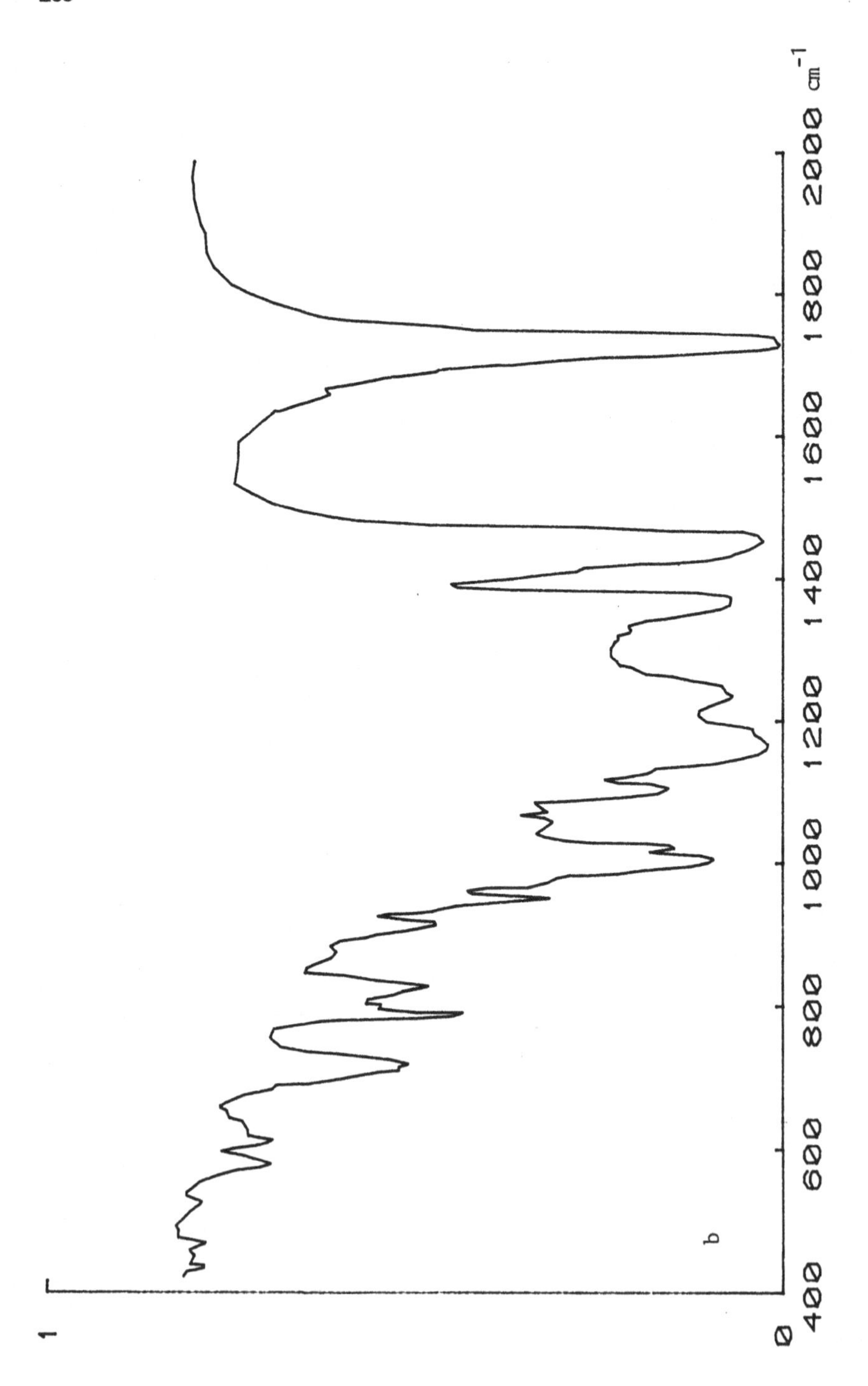

1
0
400
600
800
1000
1200
1400
1600
1800
2000
cm⁻¹
b

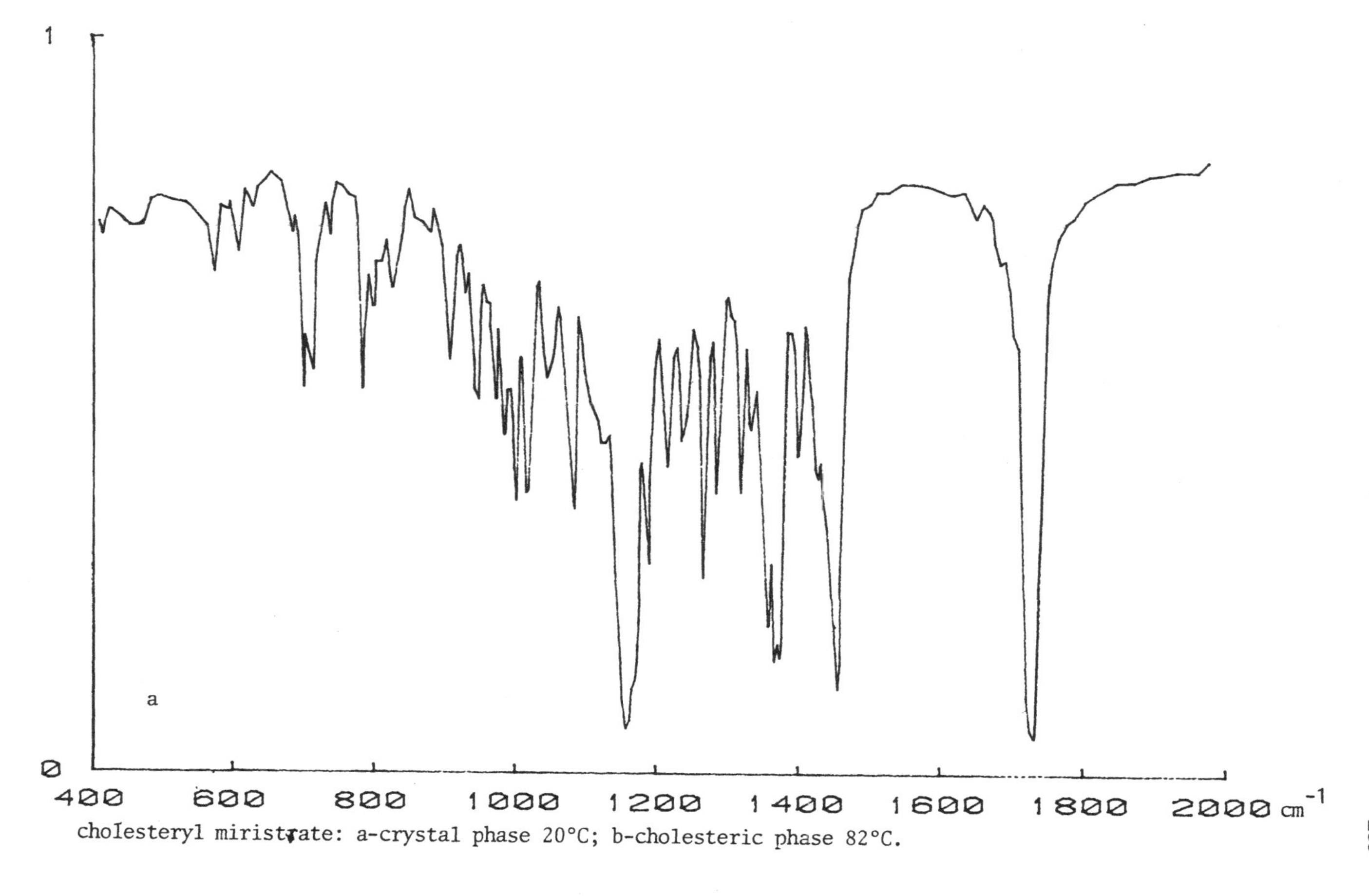

cholesteryl miristrate: a-crystal phase 20°C; b-cholesteric phase 82°C.

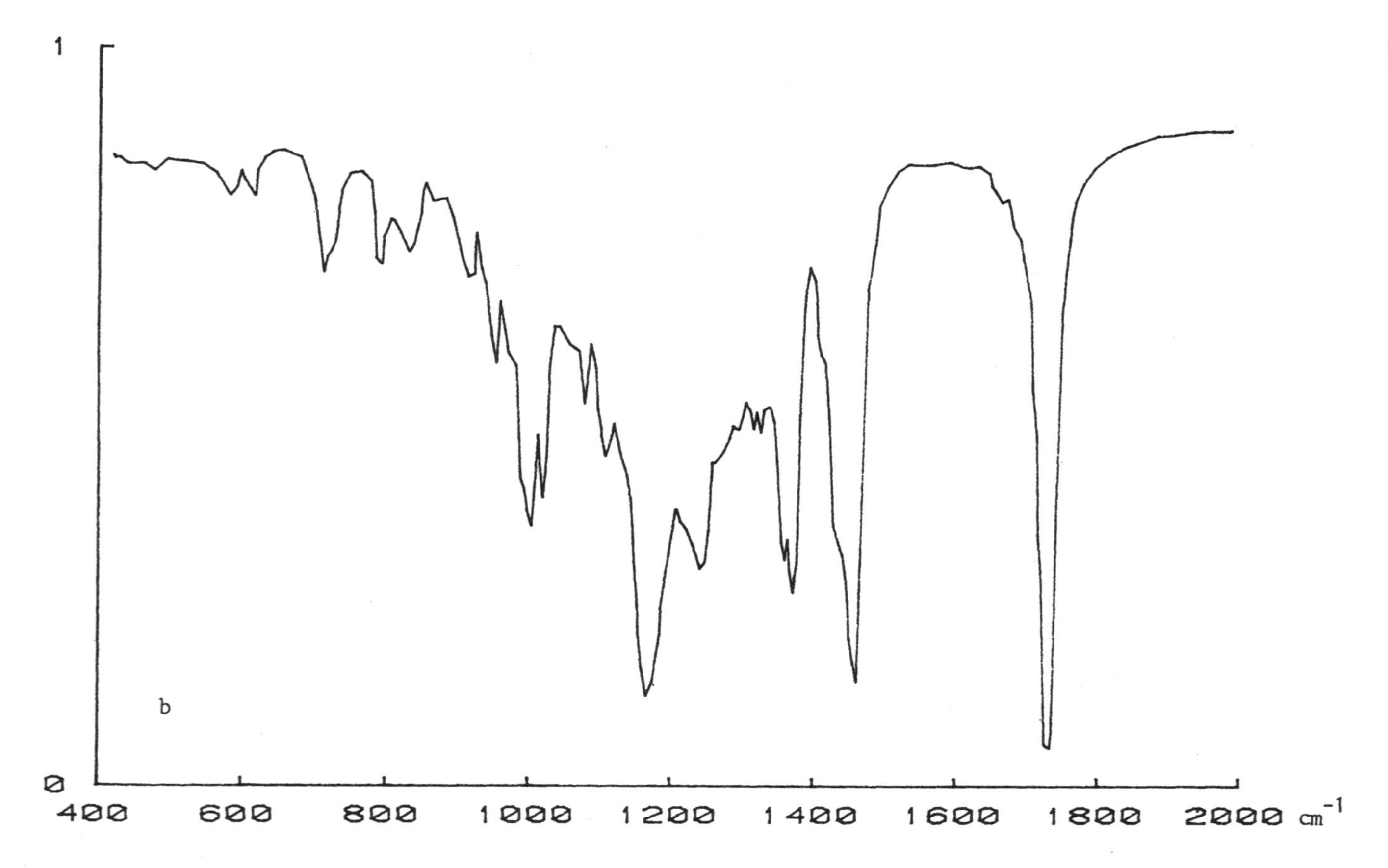
1
0
b
400
600
800
1000
1200
1400
1600
1800
2000 cm^{-1}

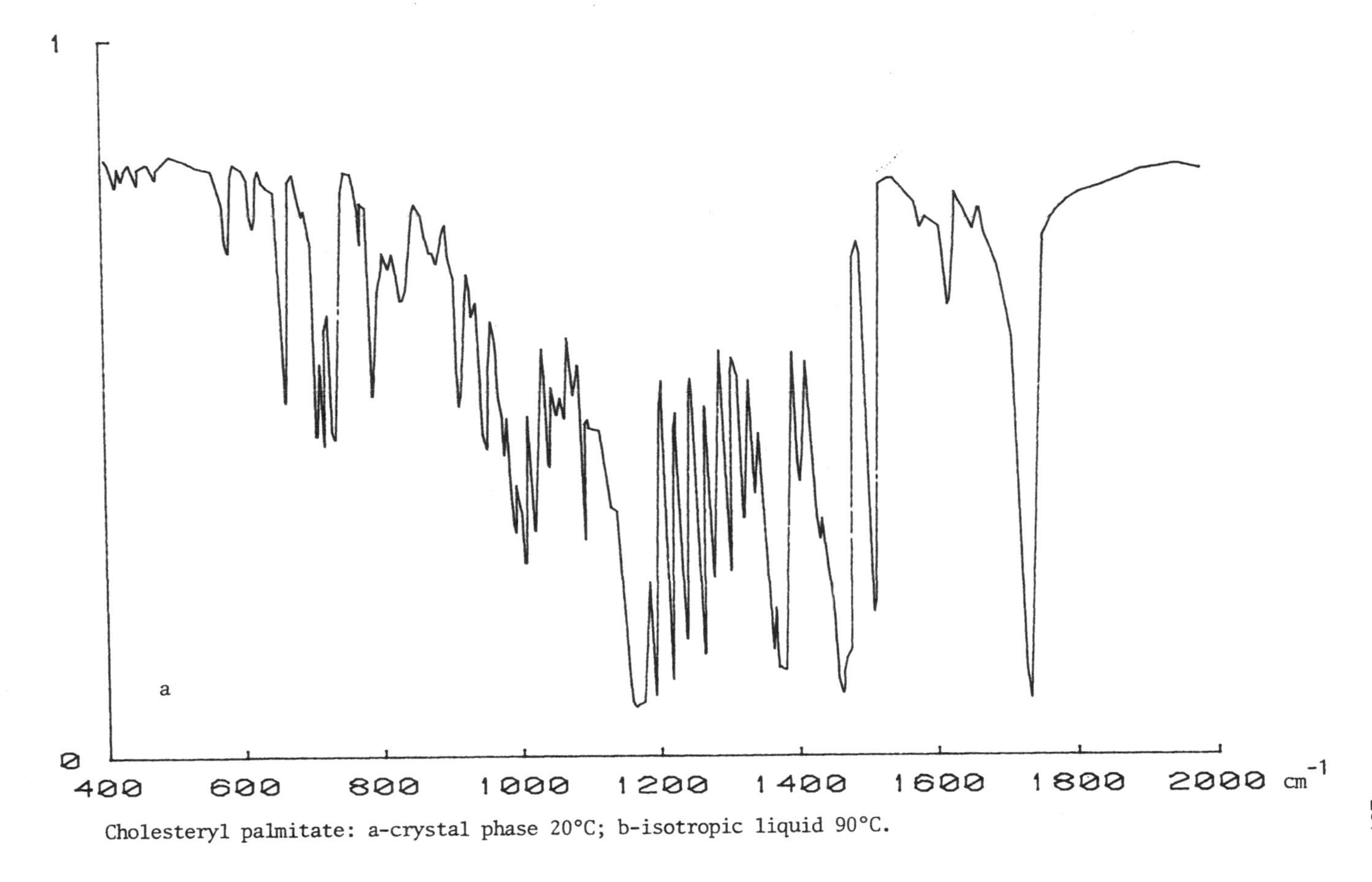

Cholesteryl palmitate: a-crystal phase 20°C; b-isotropic liquid 90°C.

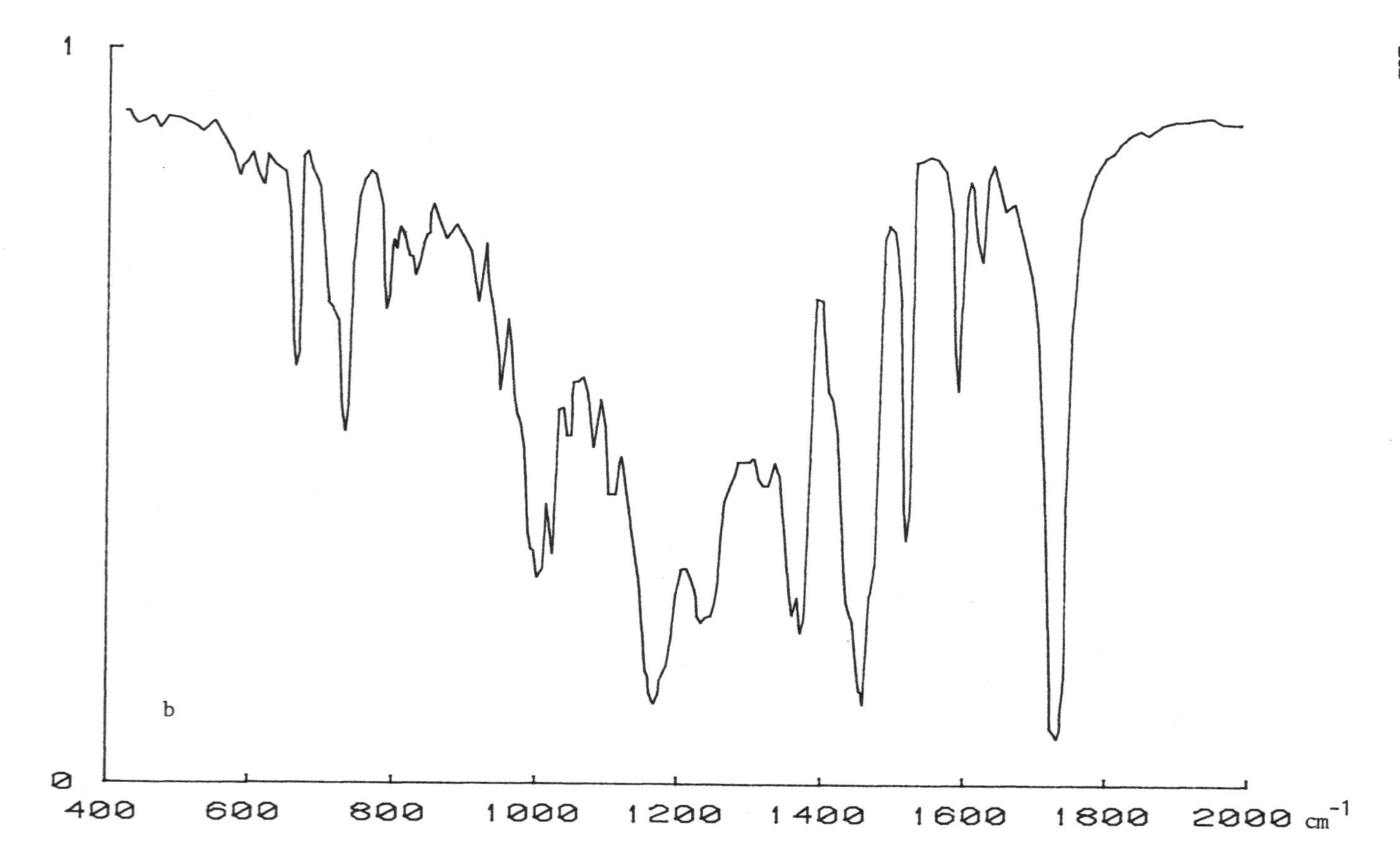

1
0
b
400
600
800
1000
1200
1400
1600
1800
2000 cm-1

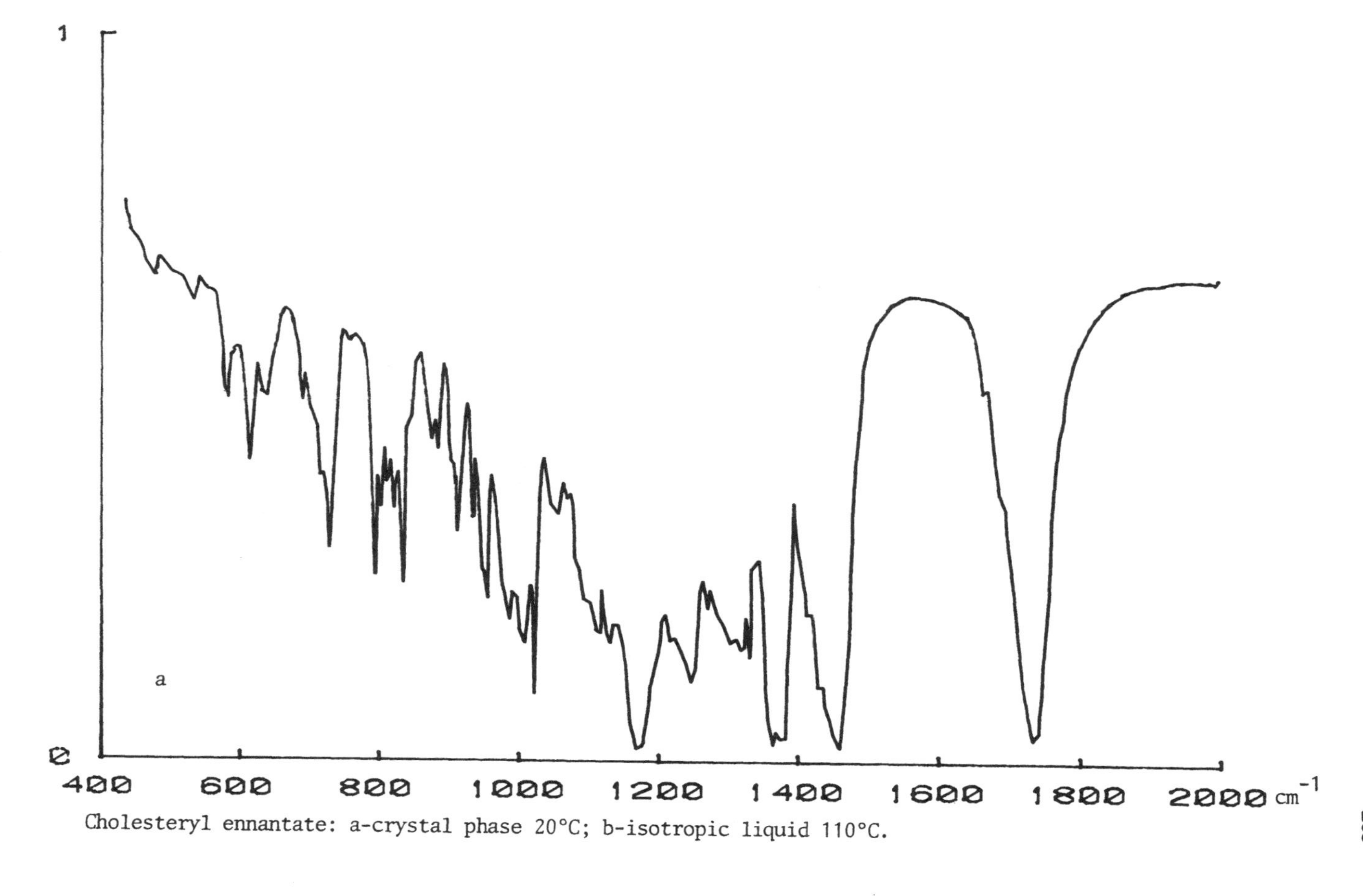

Cholesteryl ennantate: a-crystal phase 20°C; b-isotropic liquid 110°C.

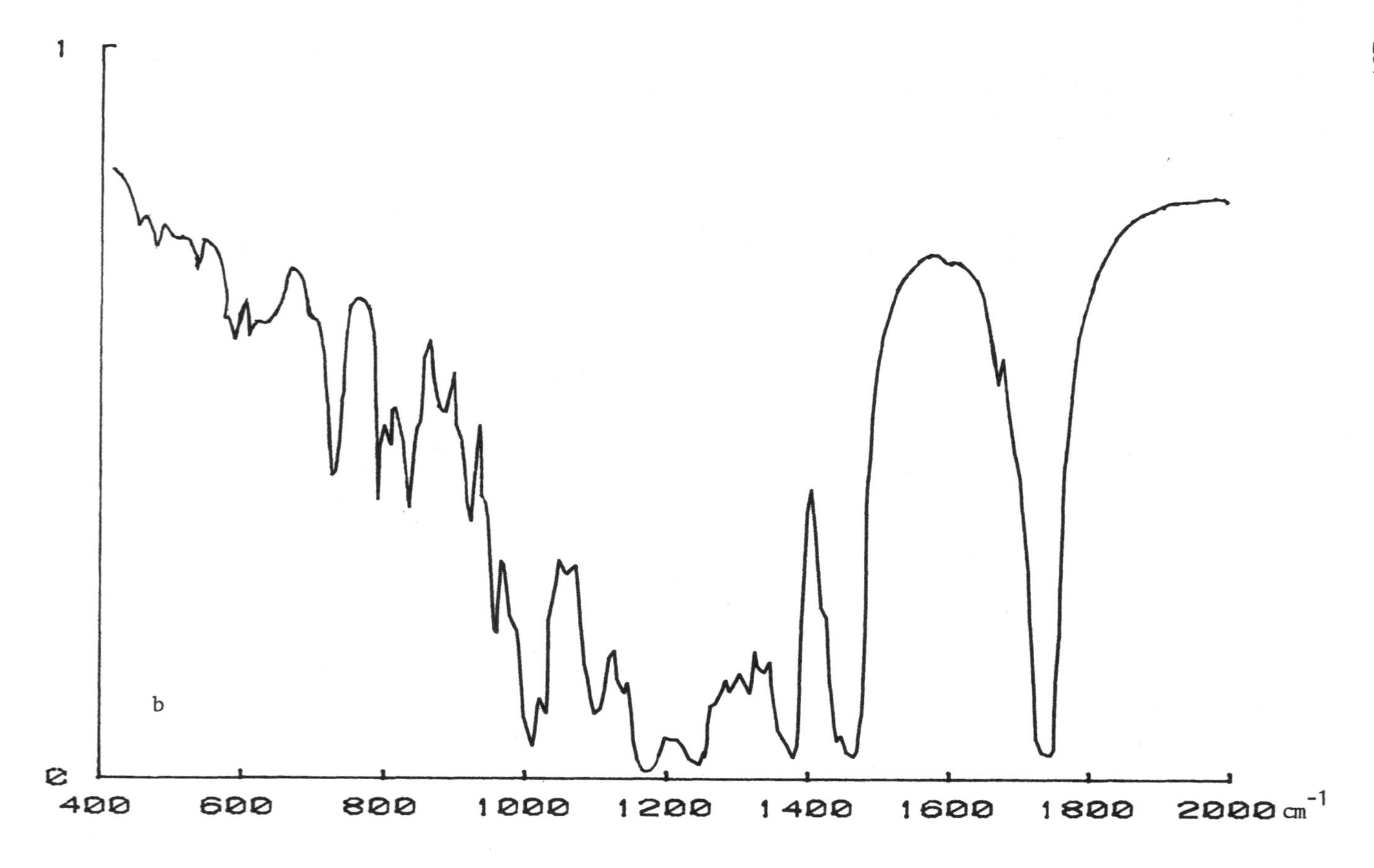

1
0
b
400
600
800
1000
1200
1400
1600
1800
2000 cm-1

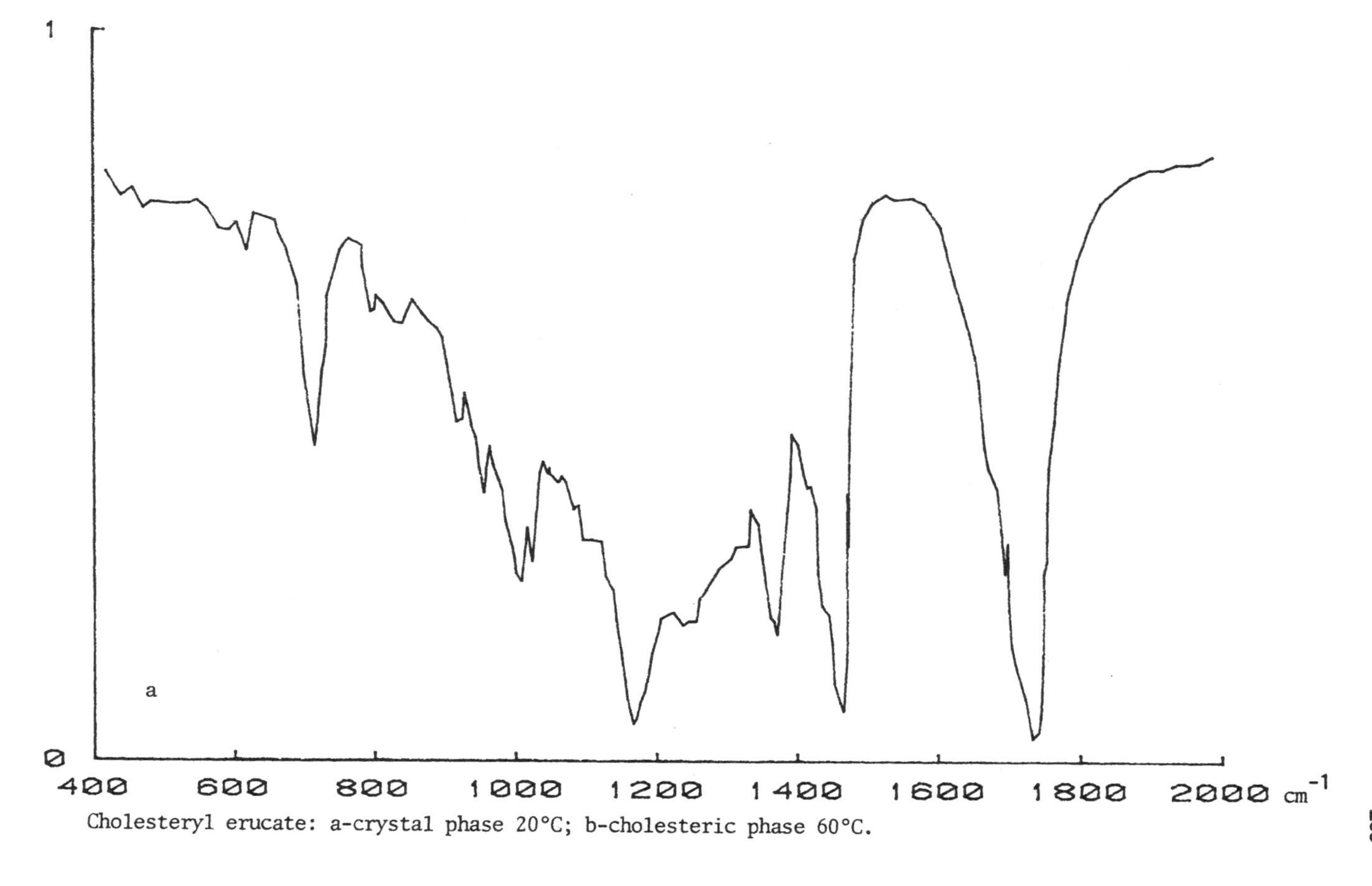

Cholesteryl erucate: a-crystal phase 20°C; b-cholesteric phase 60°C.

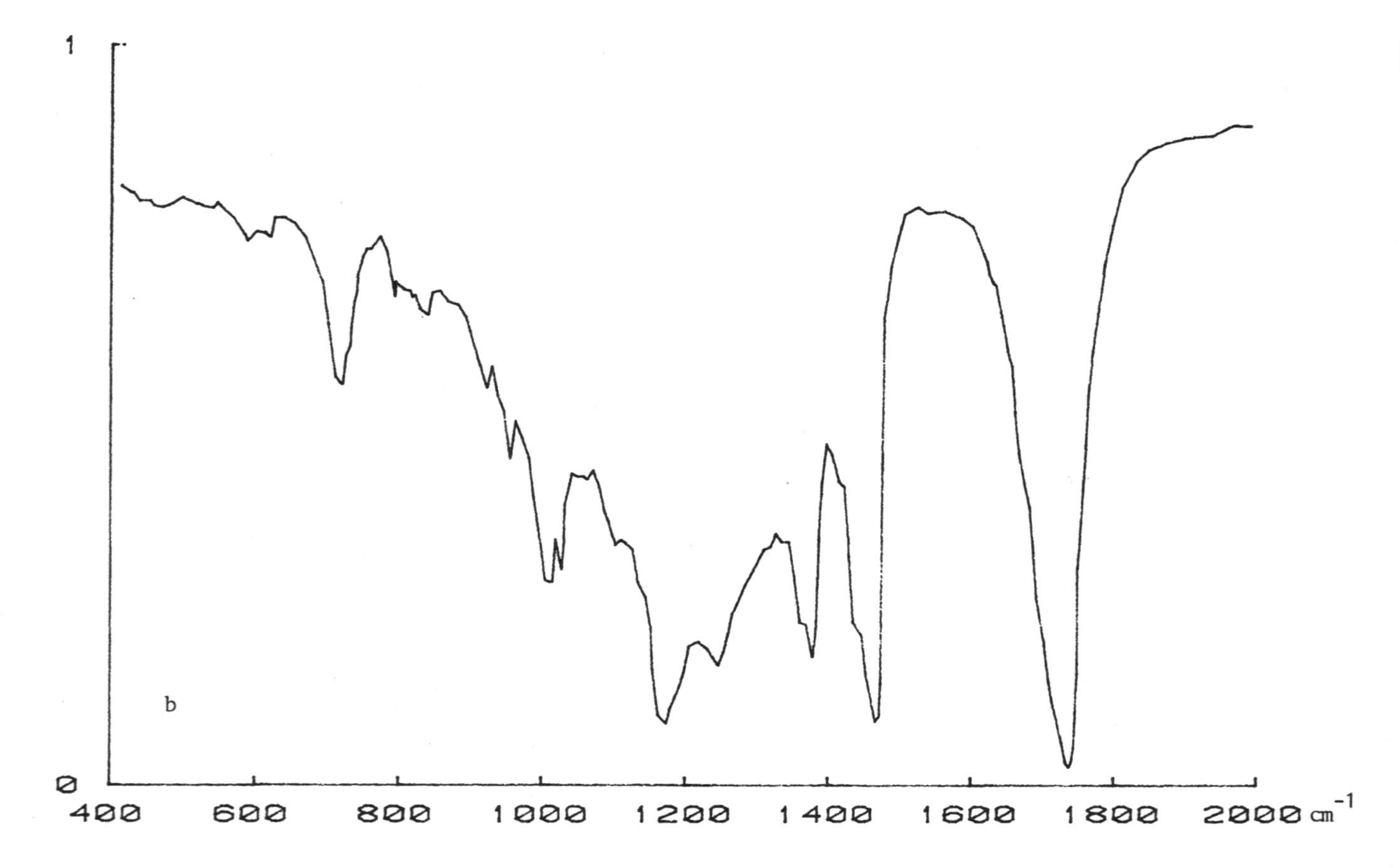

1
0
b
400
600
800
1000
1200
1400
1600
1800
2000 cm-1

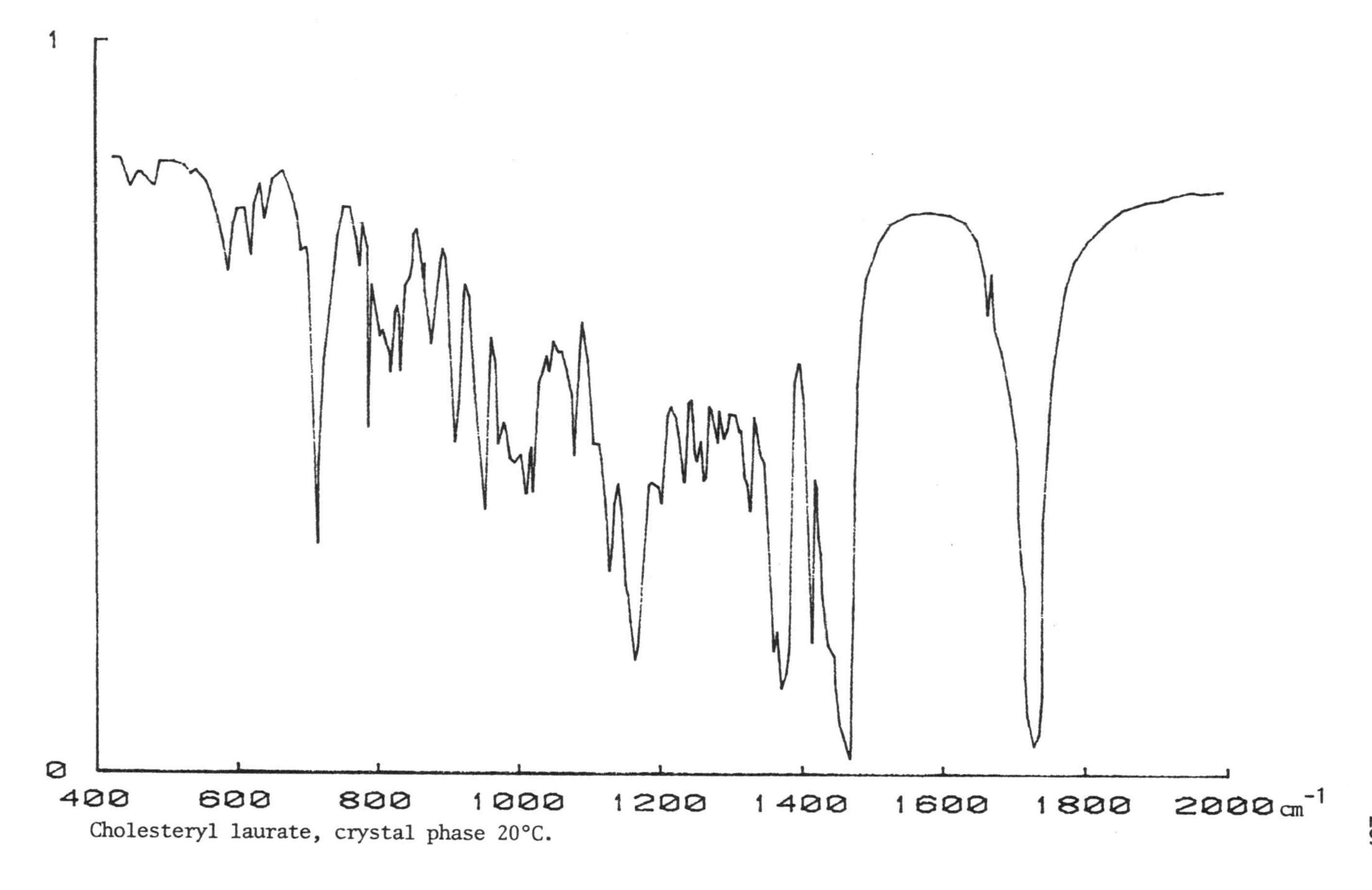

Cholesteryl laurate, crystal phase 20°C.

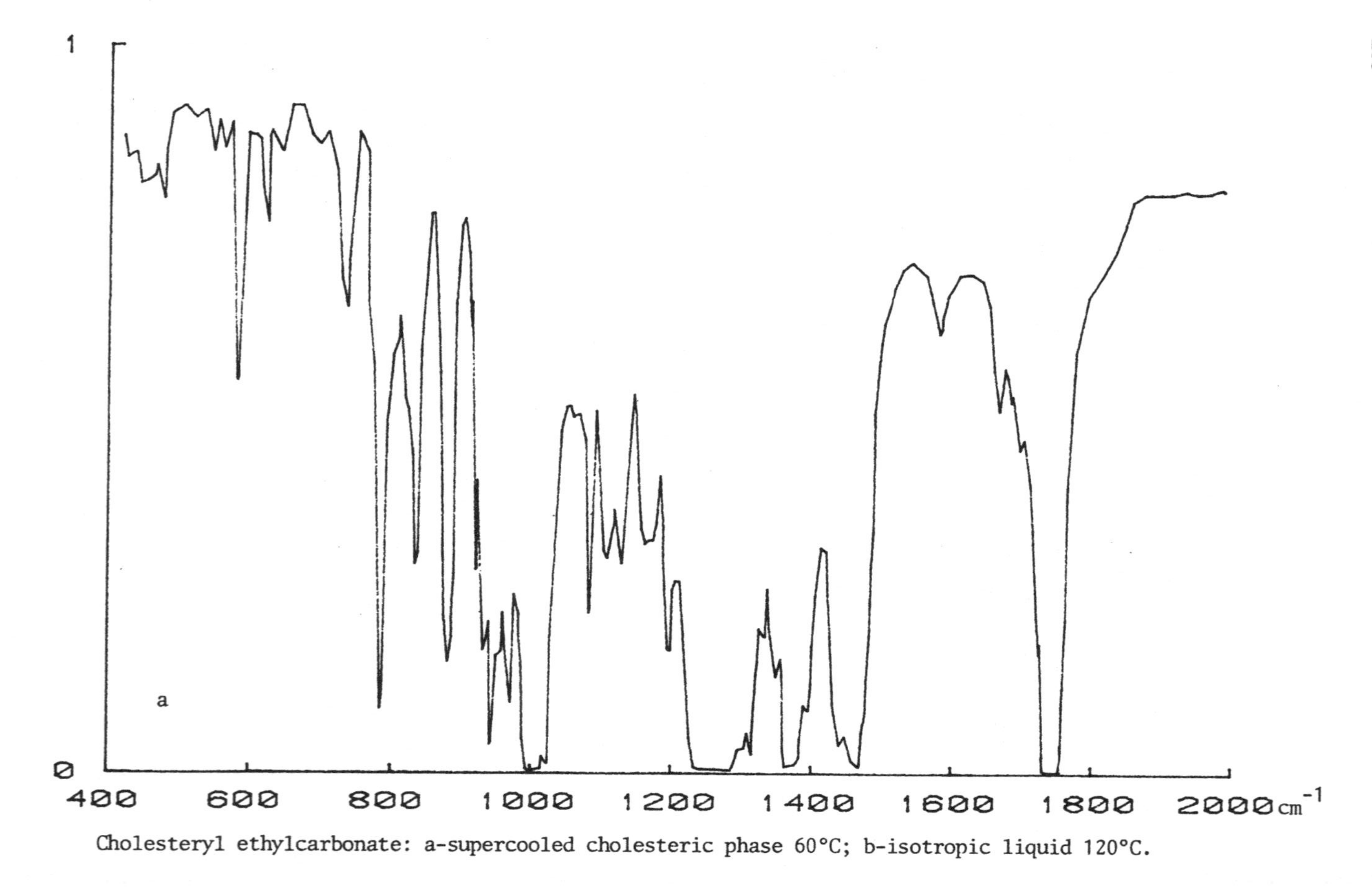

Cholesteryl ethylcarbonate: a-supercooled cholesteric phase 60°C; b-isotropic liquid 120°C.

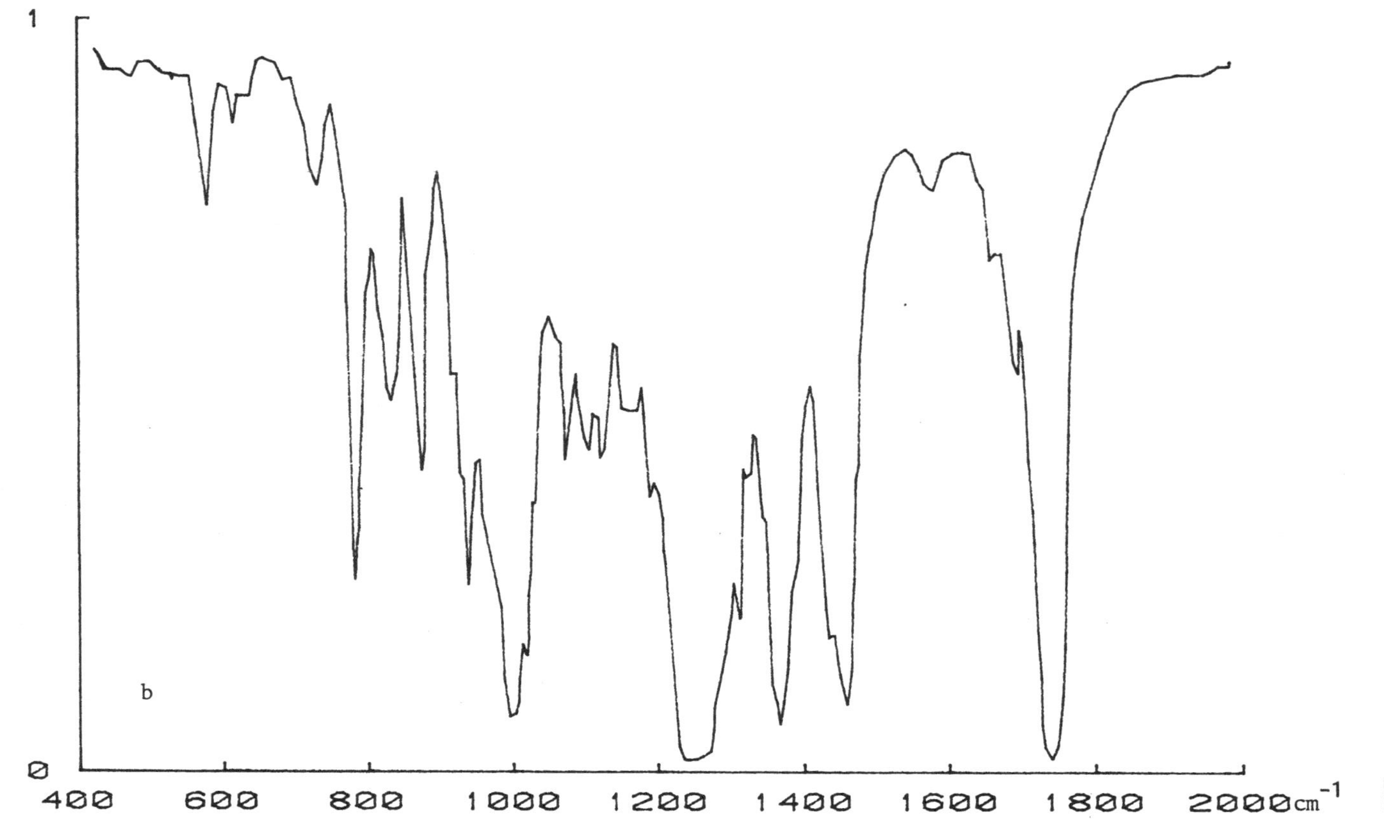

1
0
b
400
600
800
1000
1200
1400
1600
1800
2000cm-1

RAMAN SPECTRA

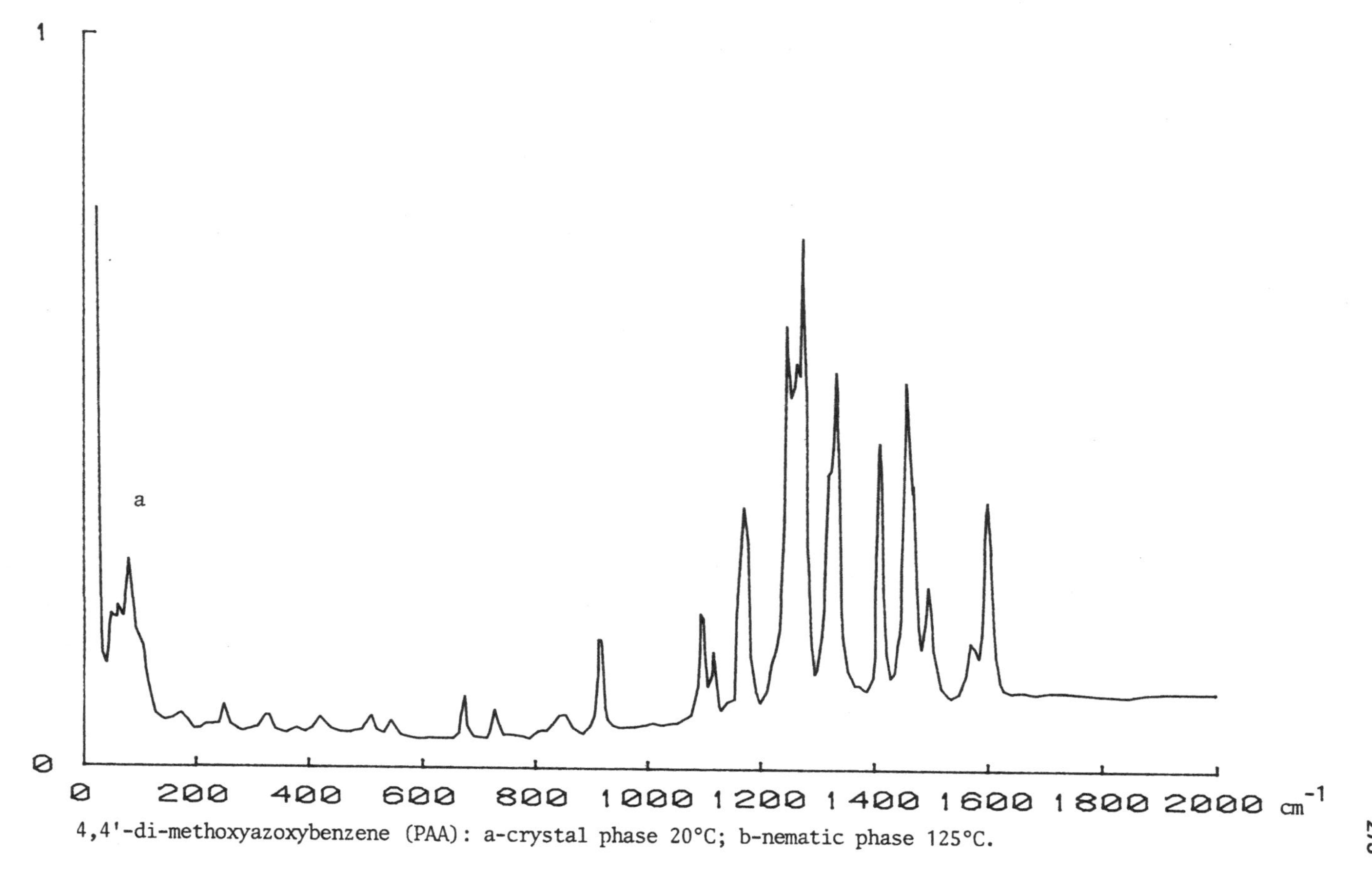

4,4'-di-methoxyazoxybenzene (PAA): a-crystal phase 20°C; b-nematic phase 125°C.

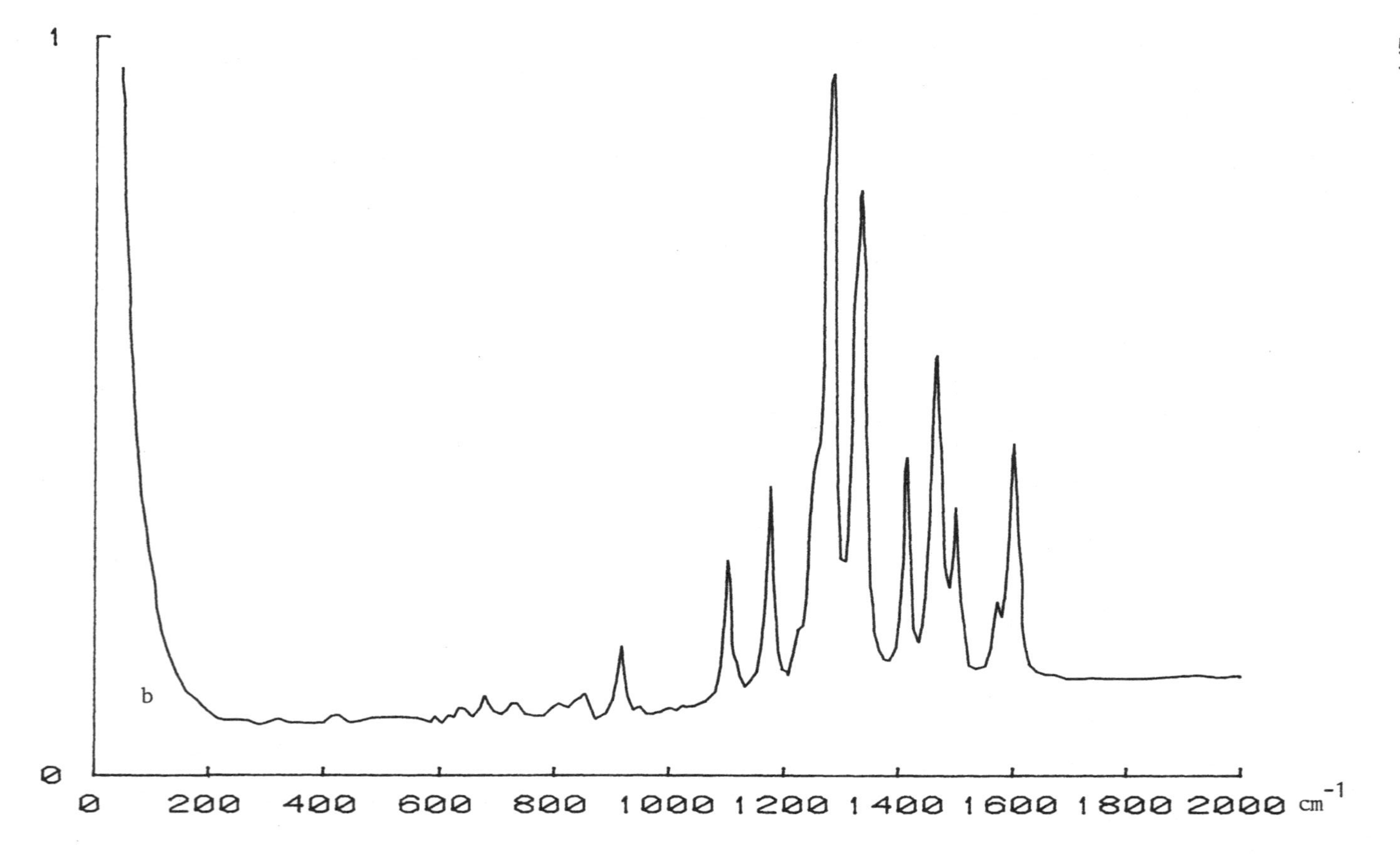

1
0
b
0
200
400
600
800
1000
1200
1400
1600
1800
2000 cm-1

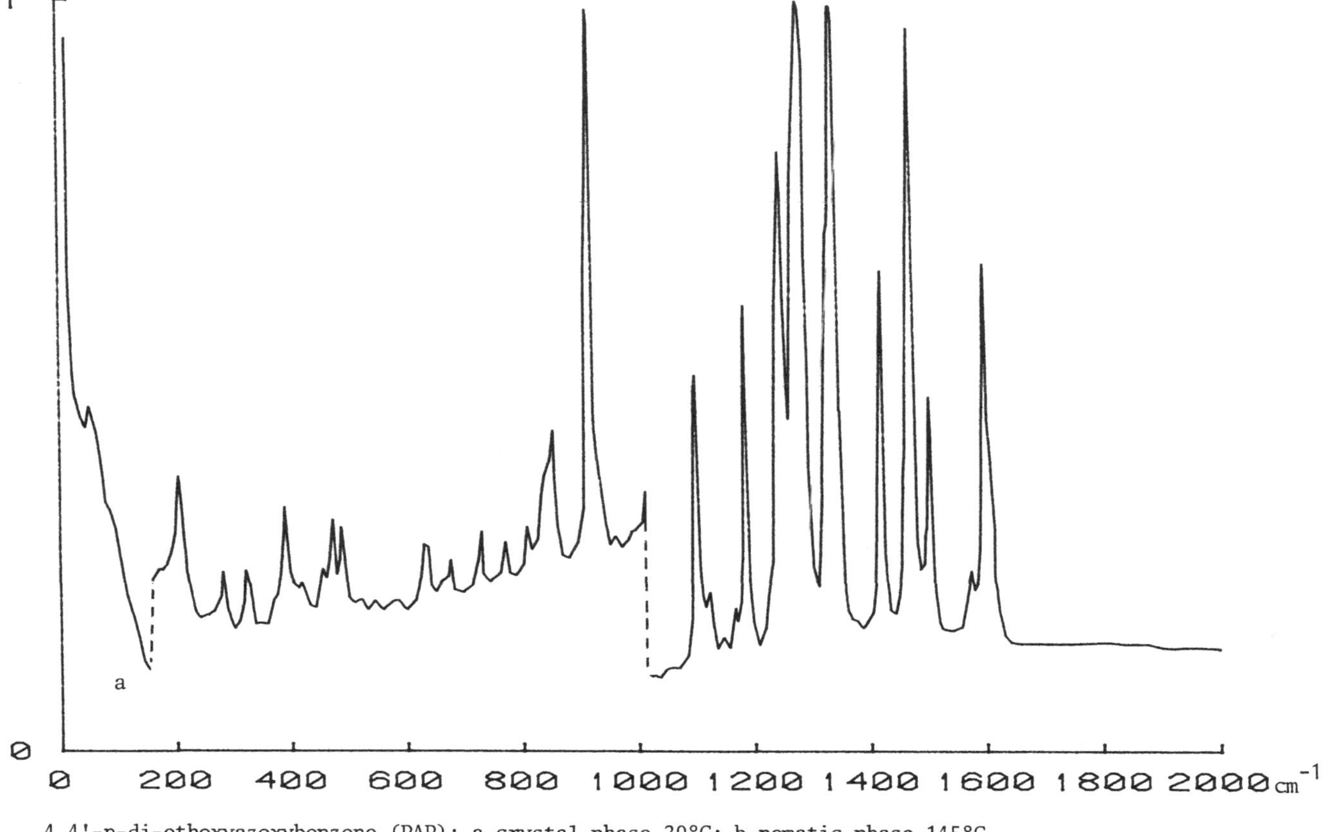

4,4'-n-di-ethoxyazoxybenzene (PAP): a-crystal phase 20°C; b-nematic phase 145°C.

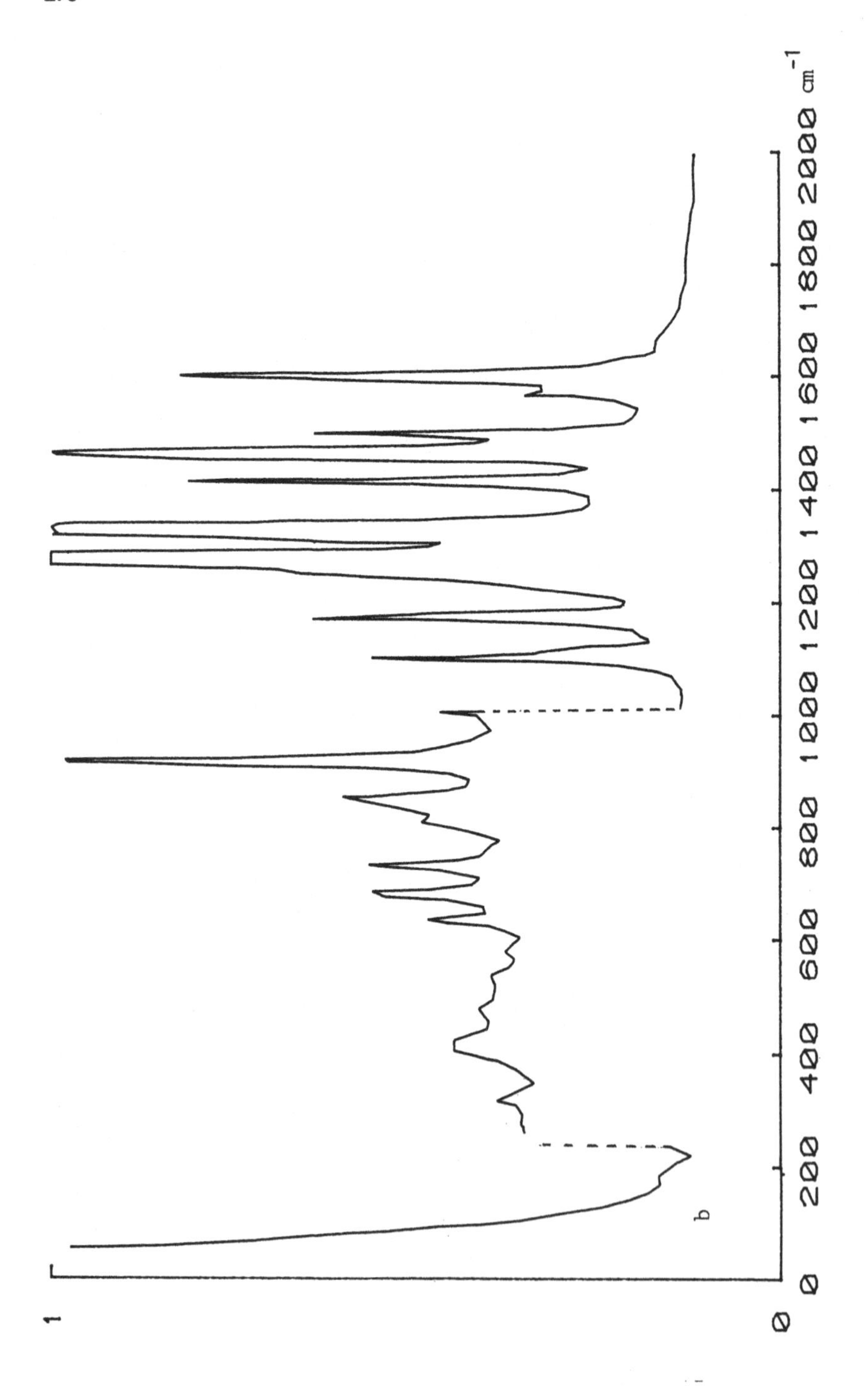

1
0
b
0 200 400 600 800 1000 1200 1400 1600 1800 2000 cm⁻¹

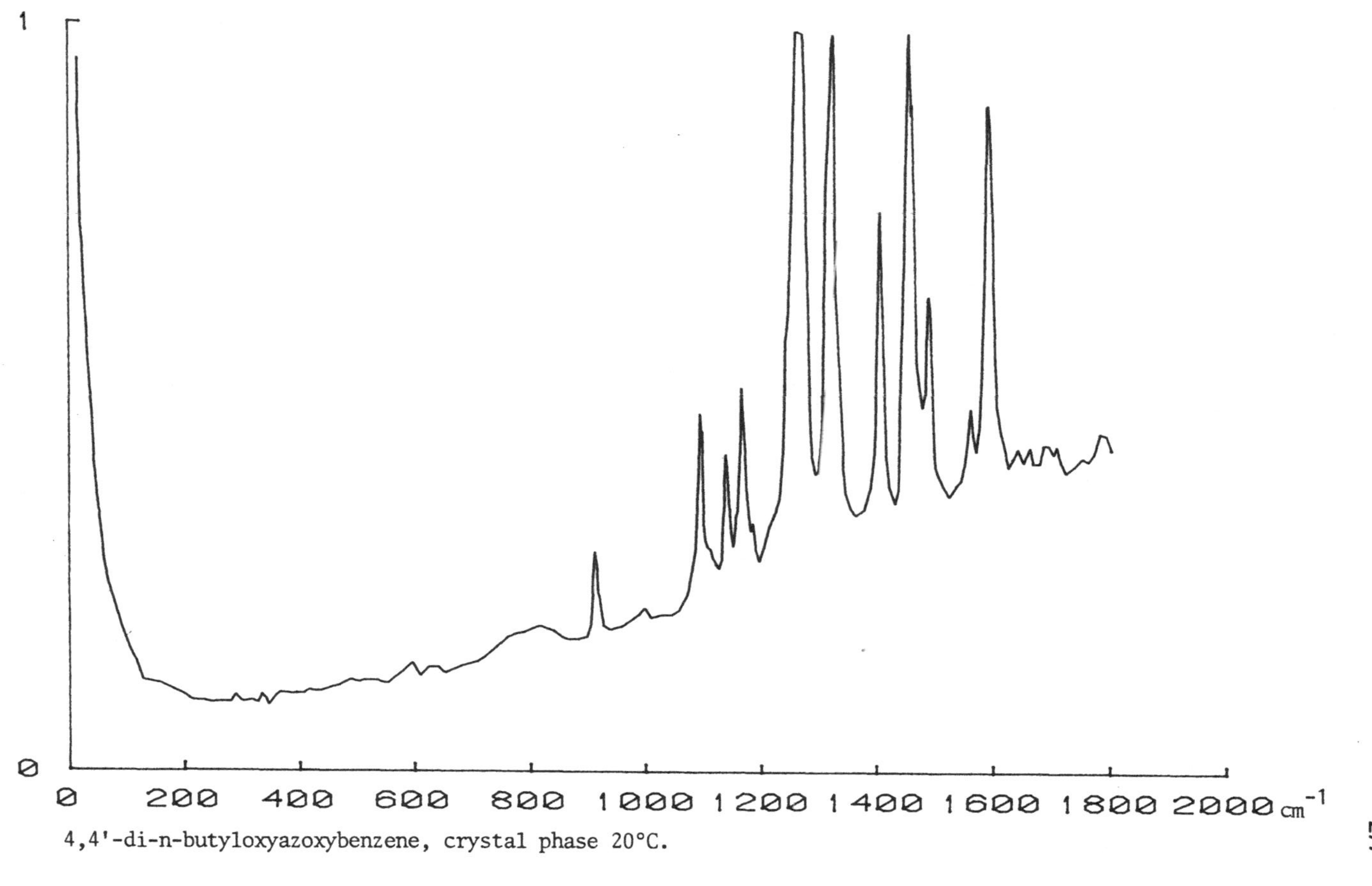

4,4'-di-n-butyloxyazoxybenzene, crystal phase 20°C.

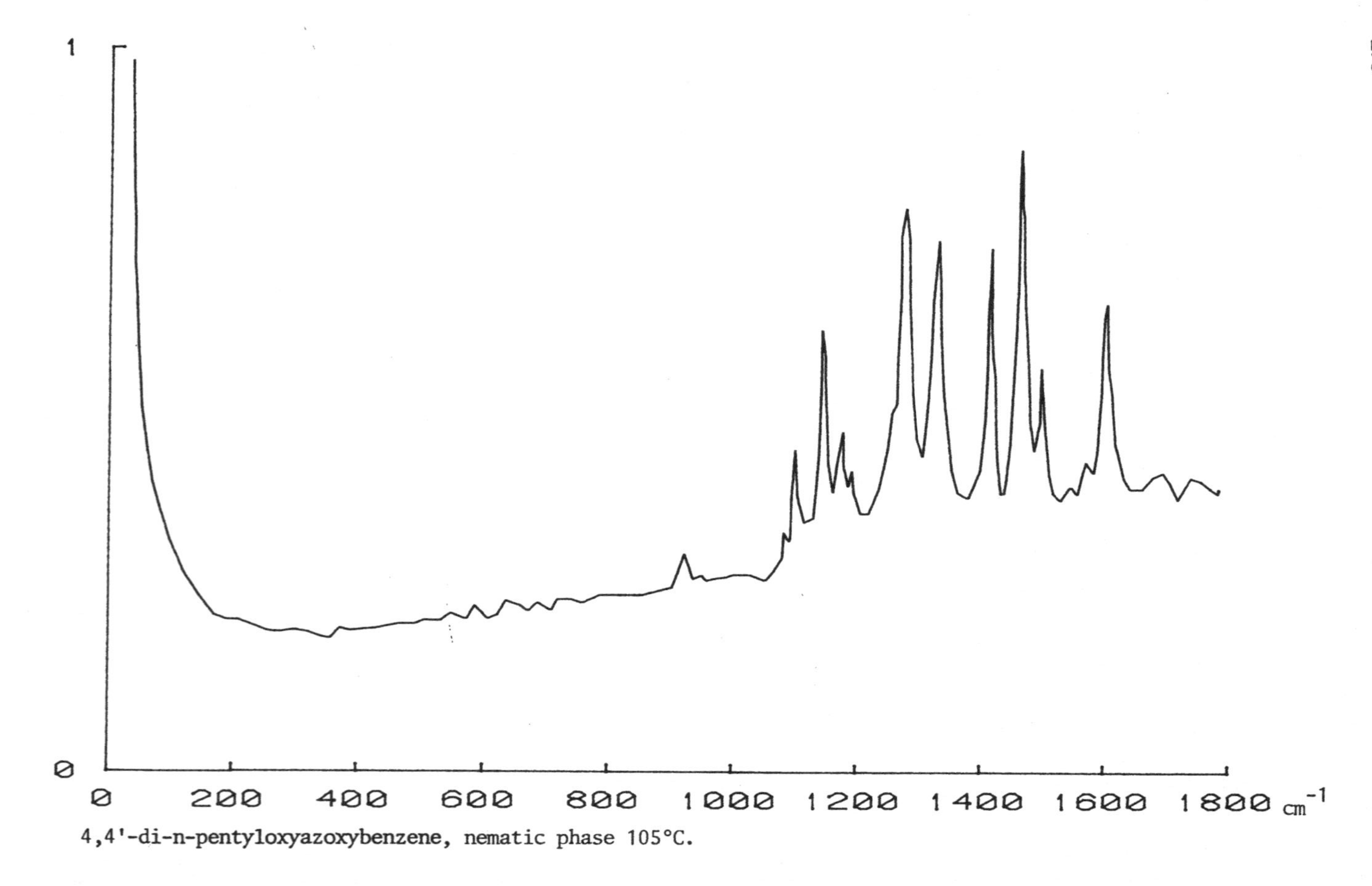

4,4'-di-n-pentyloxyazoxybenzene, nematic phase 105°C.

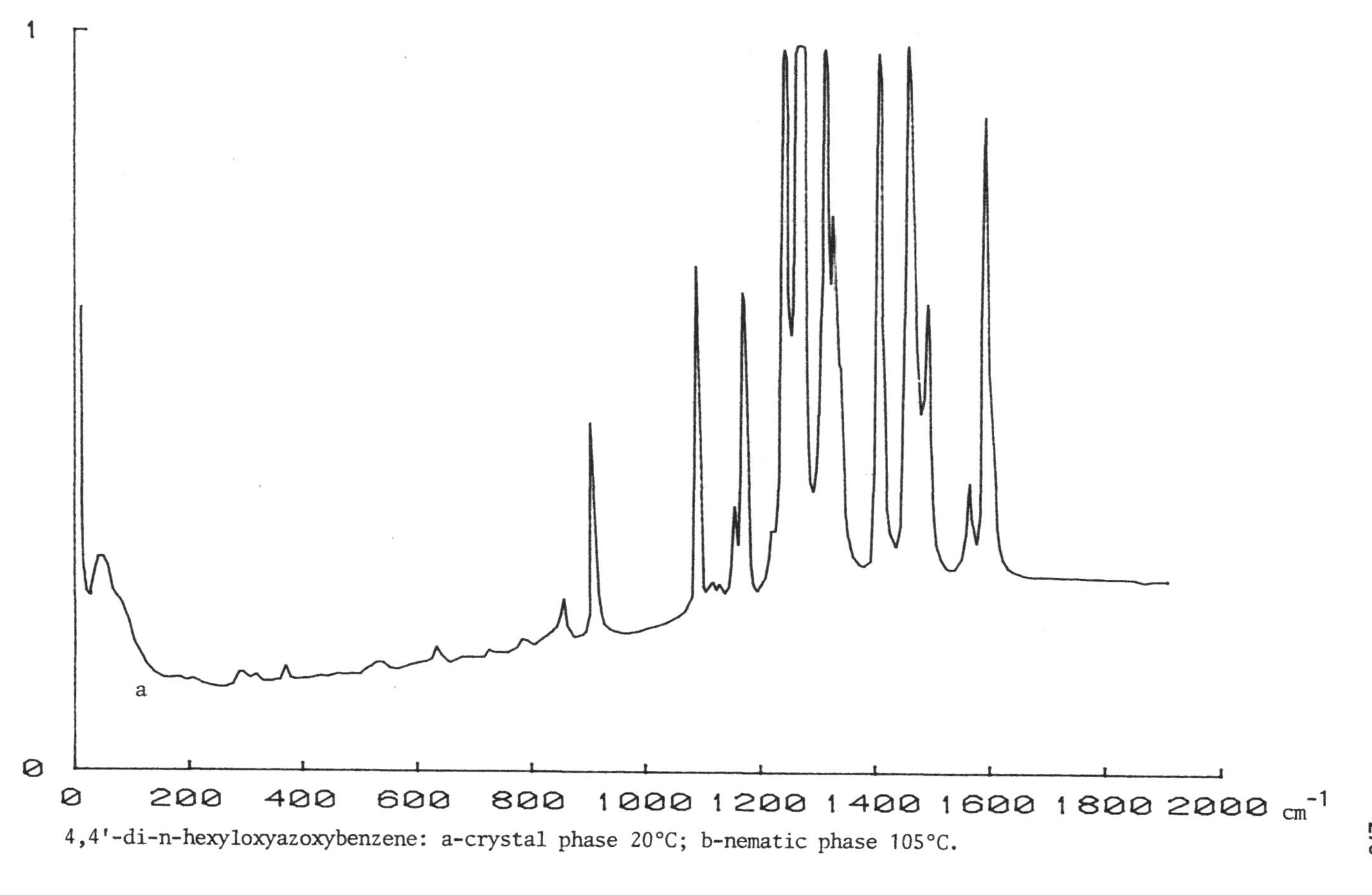

4,4'-di-n-hexyloxyazoxybenzene: a-crystal phase 20°C; b-nematic phase 105°C.

1
0
b
0 200 400 600 800 1000 1200 1400 1600 1800 2000 cm-1

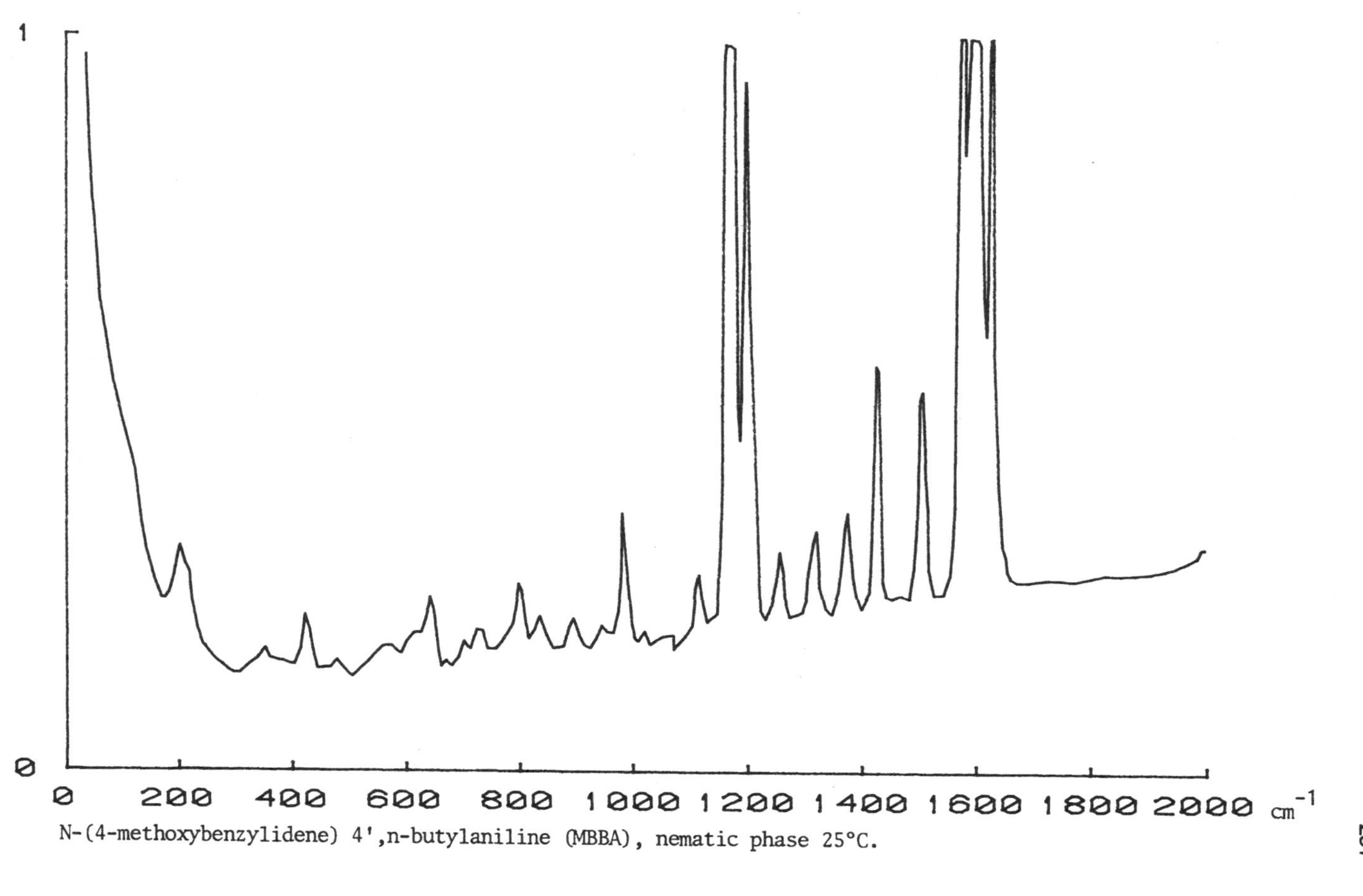

N-(4-methoxybenzylidene) 4',n-butylaniline (MBBA), nematic phase 25°C.

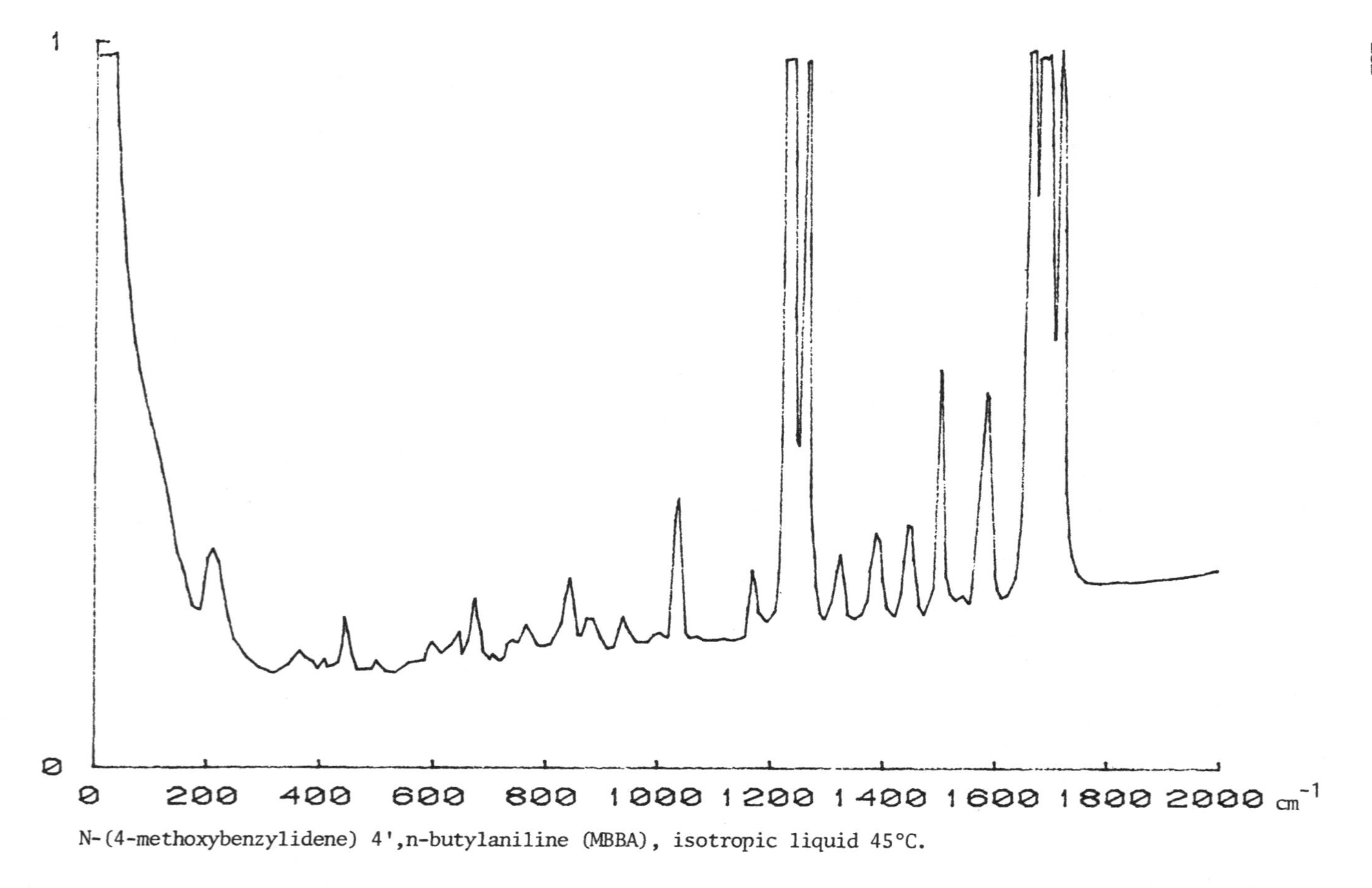

N-(4-methoxybenzylidene) 4',n-butylaniline (MBBA), isotropic liquid 45°C.

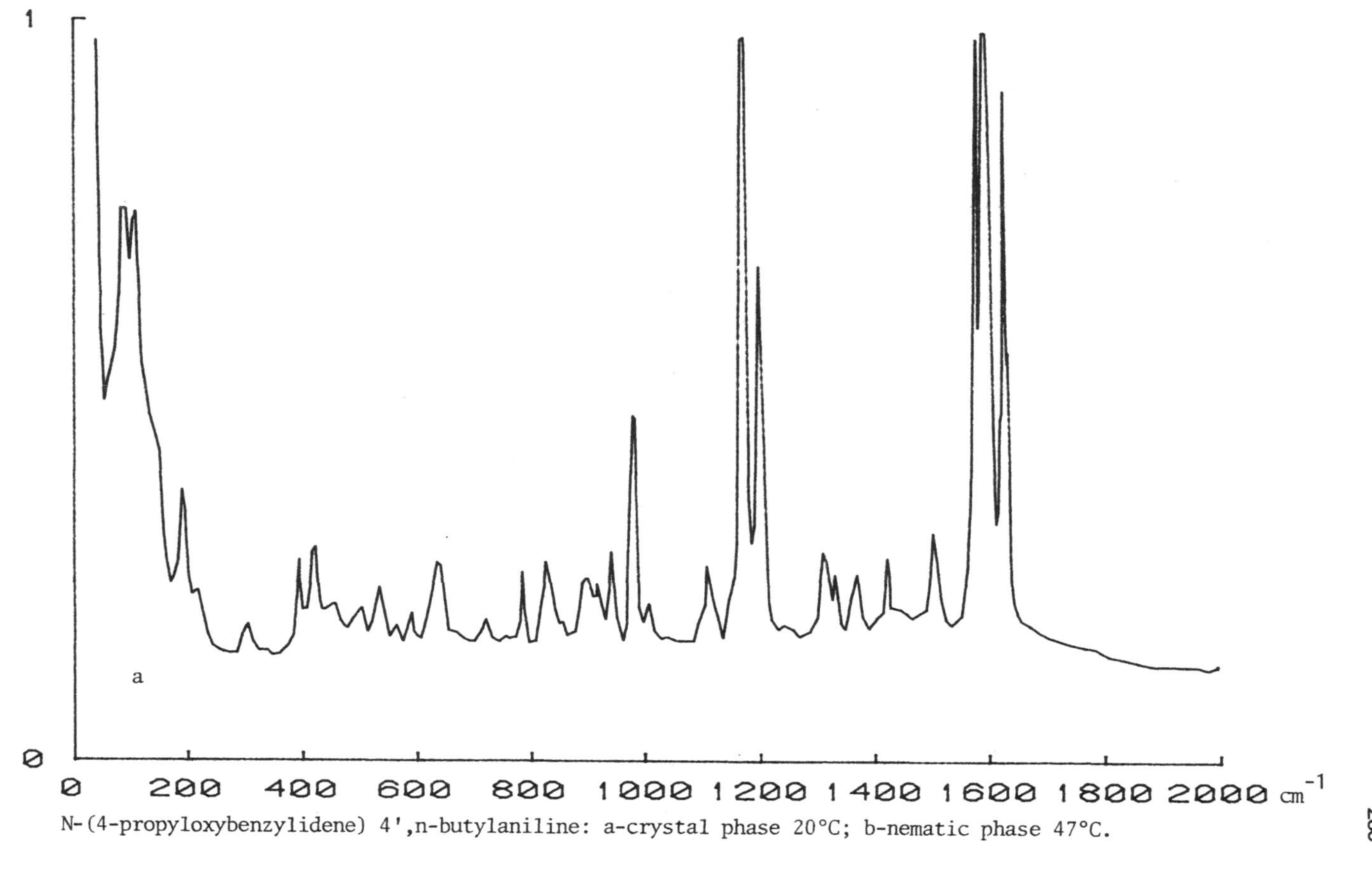

N-(4-propyloxybenzylidene) 4',n-butylaniline: a-crystal phase 20°C; b-nematic phase 47°C.

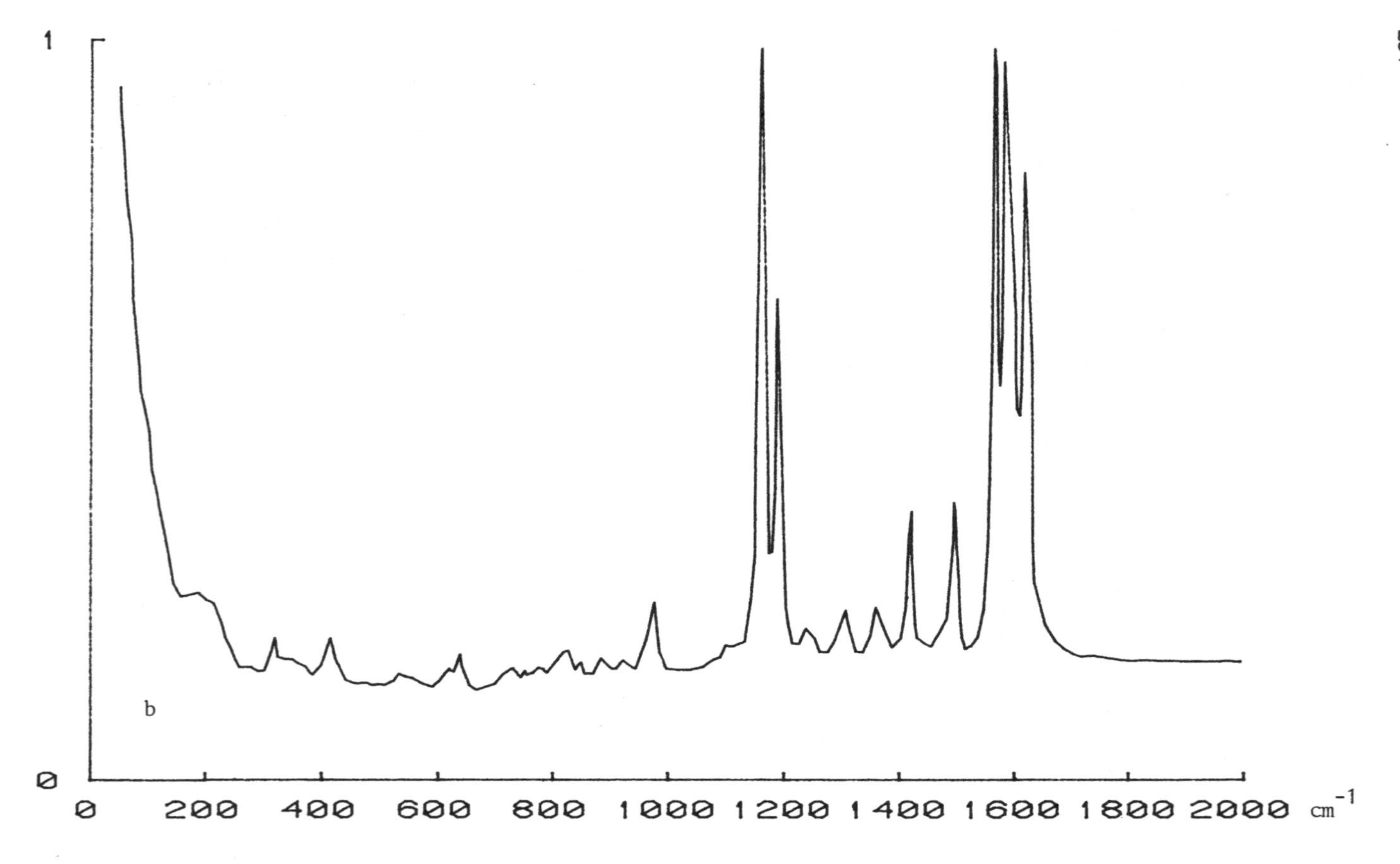
1
0
b
0
200
400
600
800
1000
1200
1400
1600
1800
2000
cm⁻¹

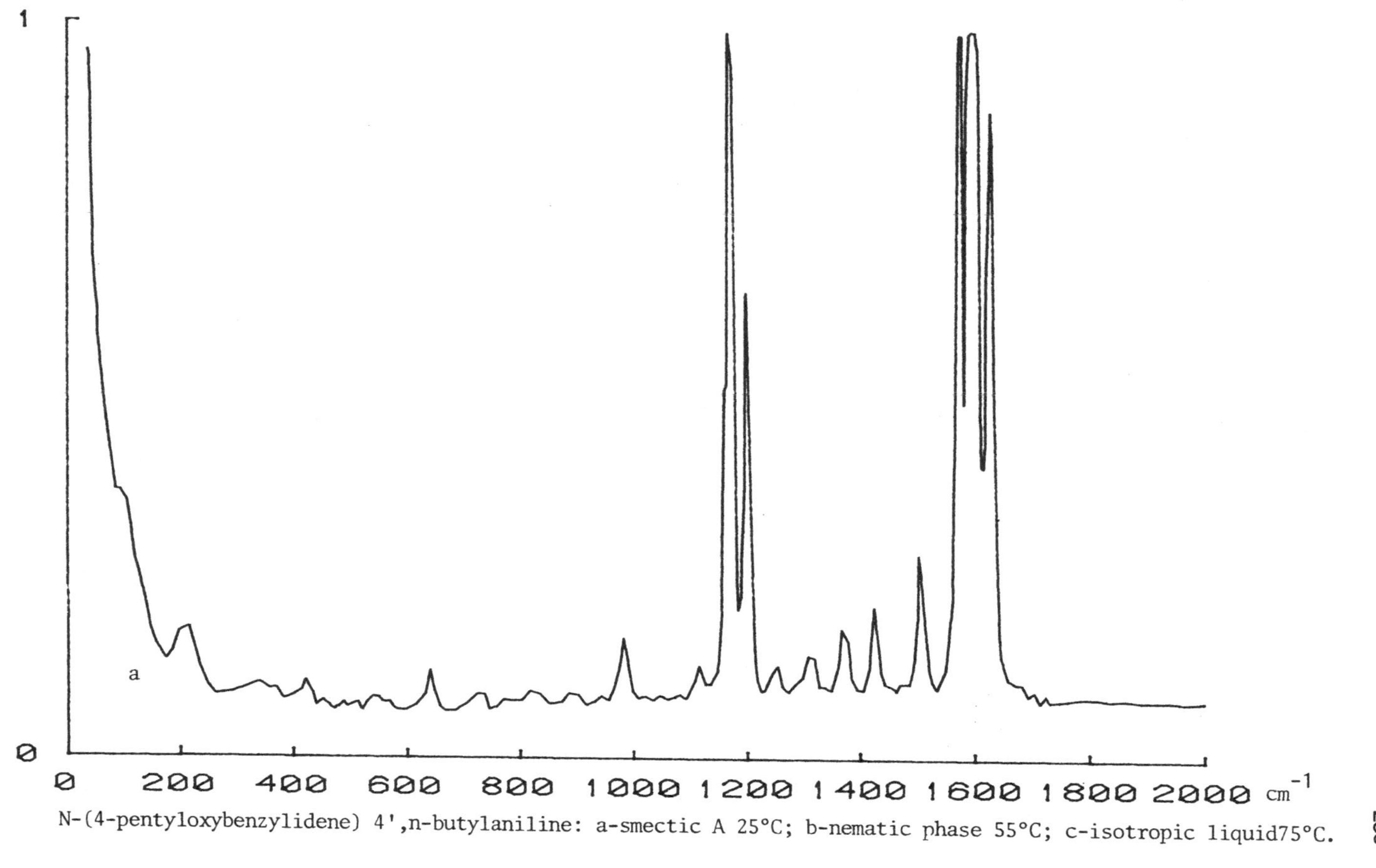

N-(4-pentyloxybenzylidene) 4',n-butylaniline: a-smectic A 25°C; b-nematic phase 55°C; c-isotropic liquid75°C.

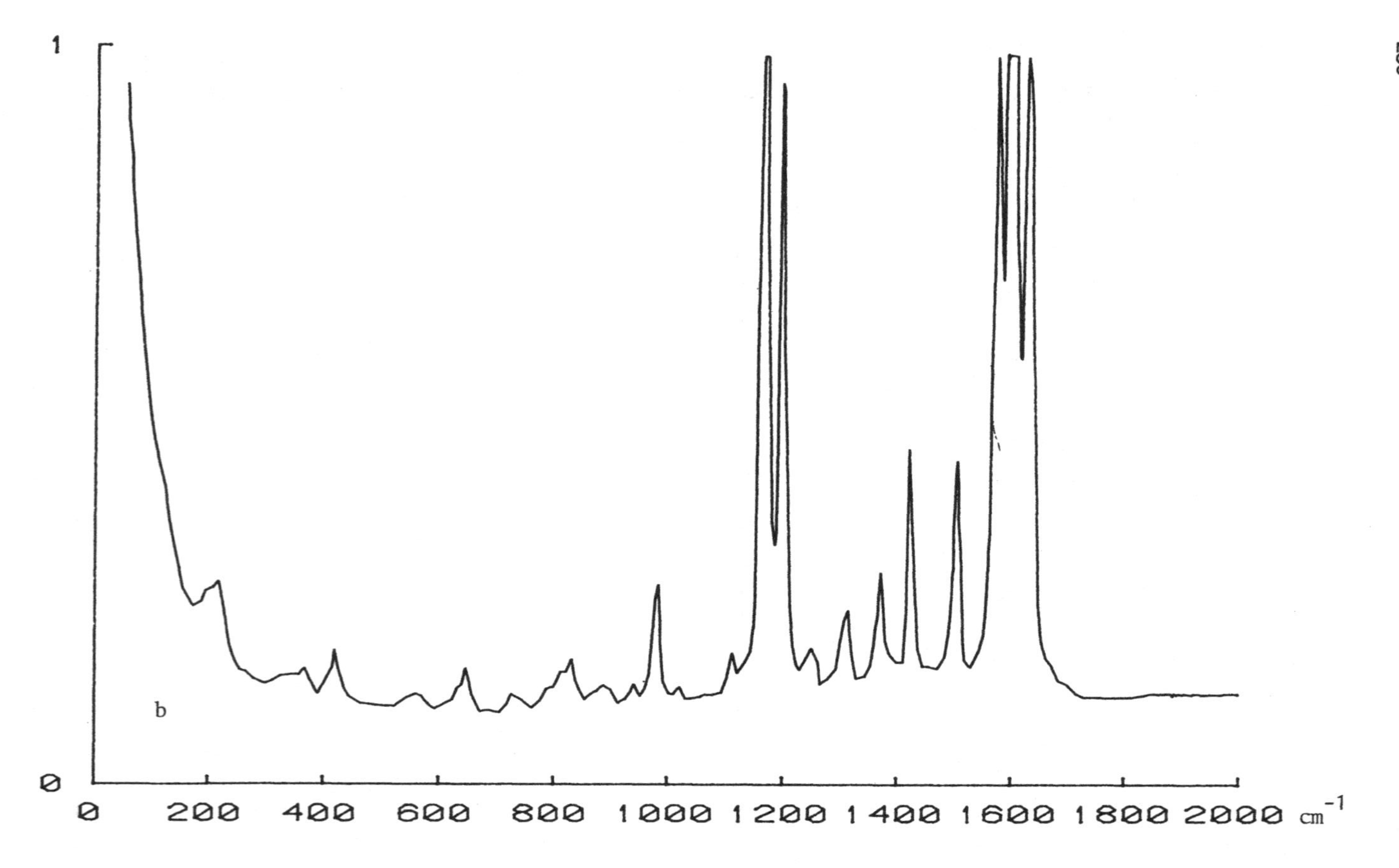

1
0
0
200
400
600
800
1000
1200
1400
1600
1800
2000
cm^{-1}
b

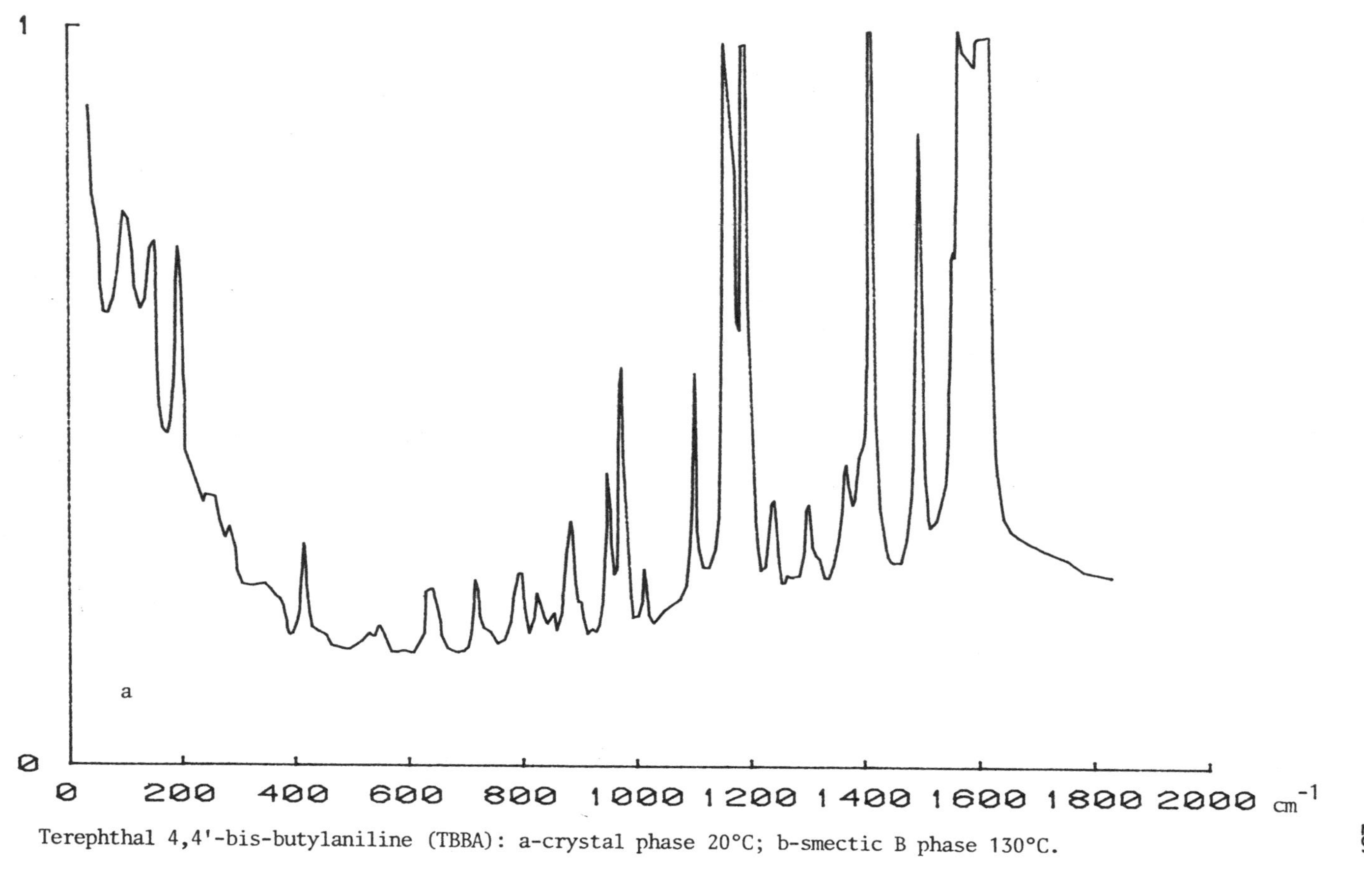

Terephthal 4,4'-bis-butylaniline (TBBA): a-crystal phase 20°C; b-smectic B phase 130°C.

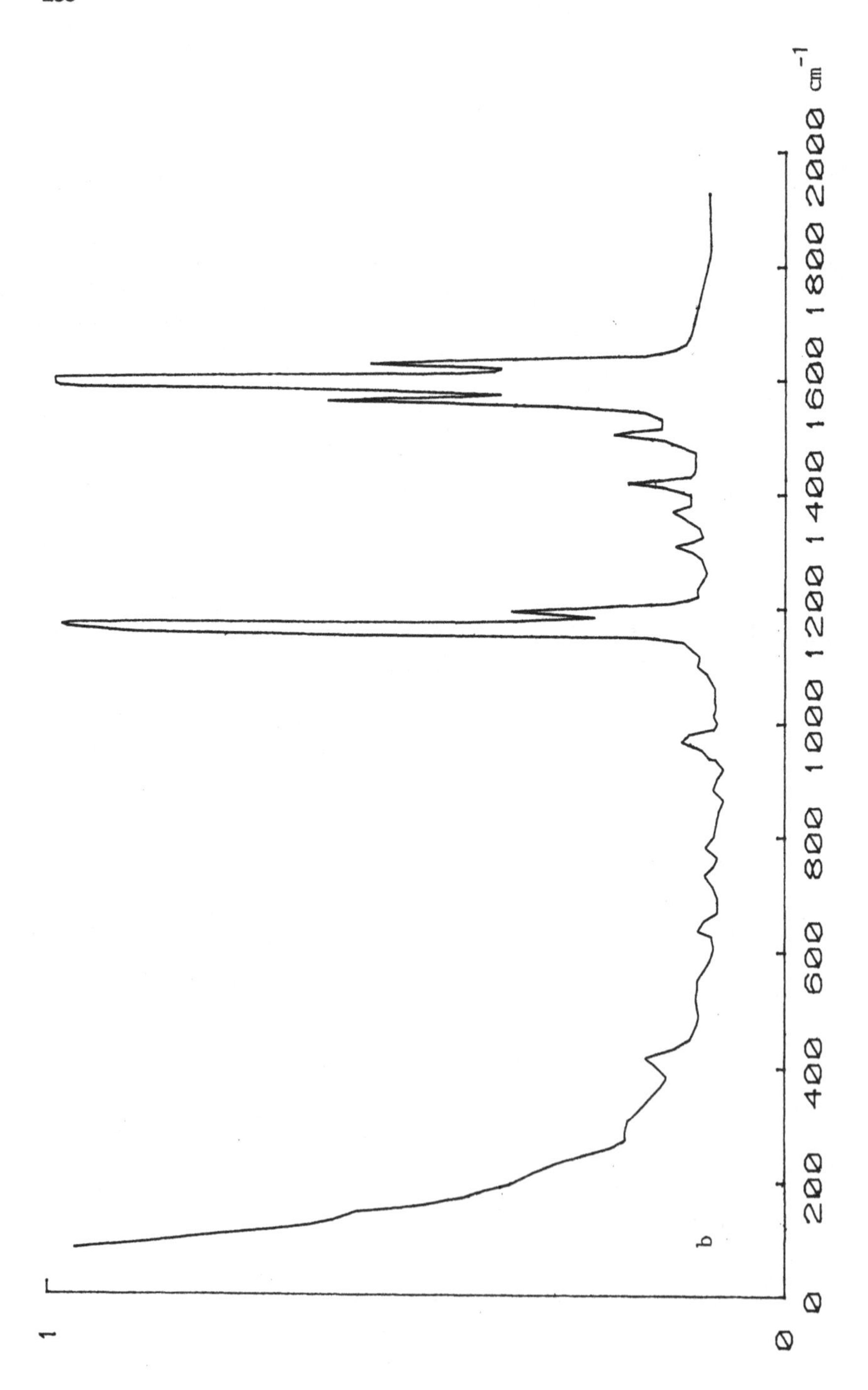

1
0
b
0
200
400
600
800
1000
1200
1400
1600
1800
2000 cm-1

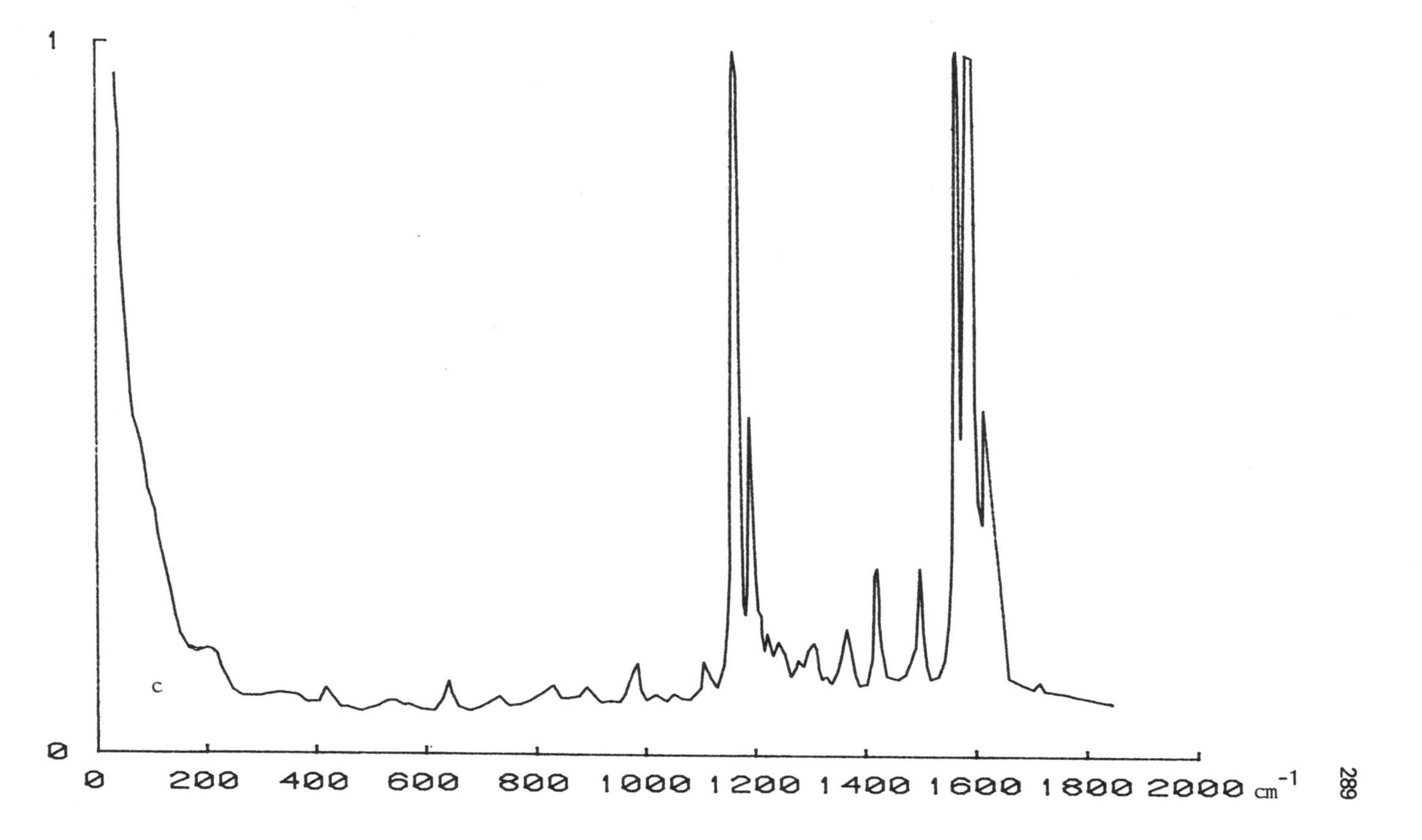

1
0
c
0
200
400
600
800
1000
1200
1400
1600
1800
2000 cm⁻¹

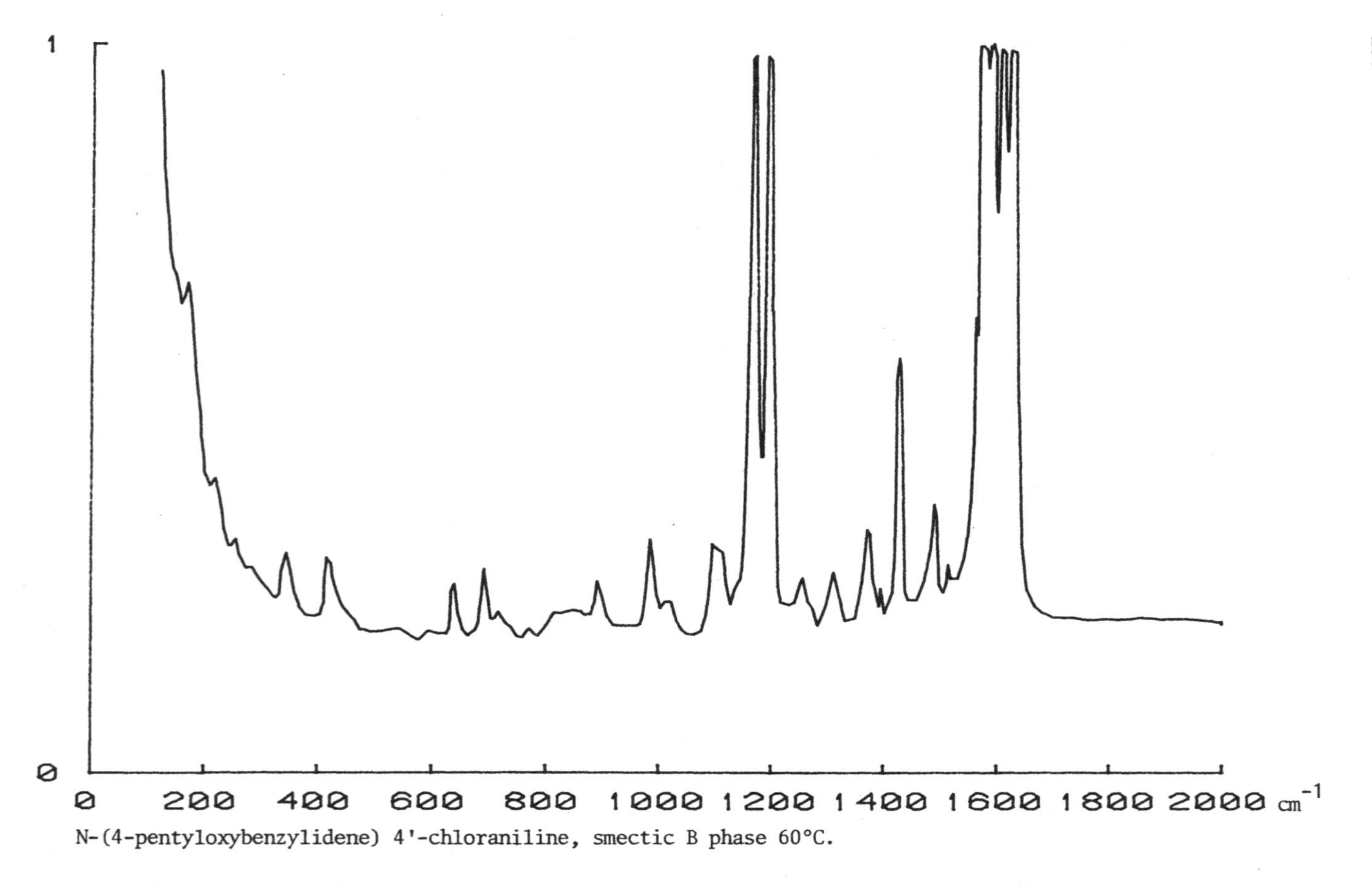

N-(4-pentyloxybenzylidene) 4'-chloraniline, smectic B phase 60°C.

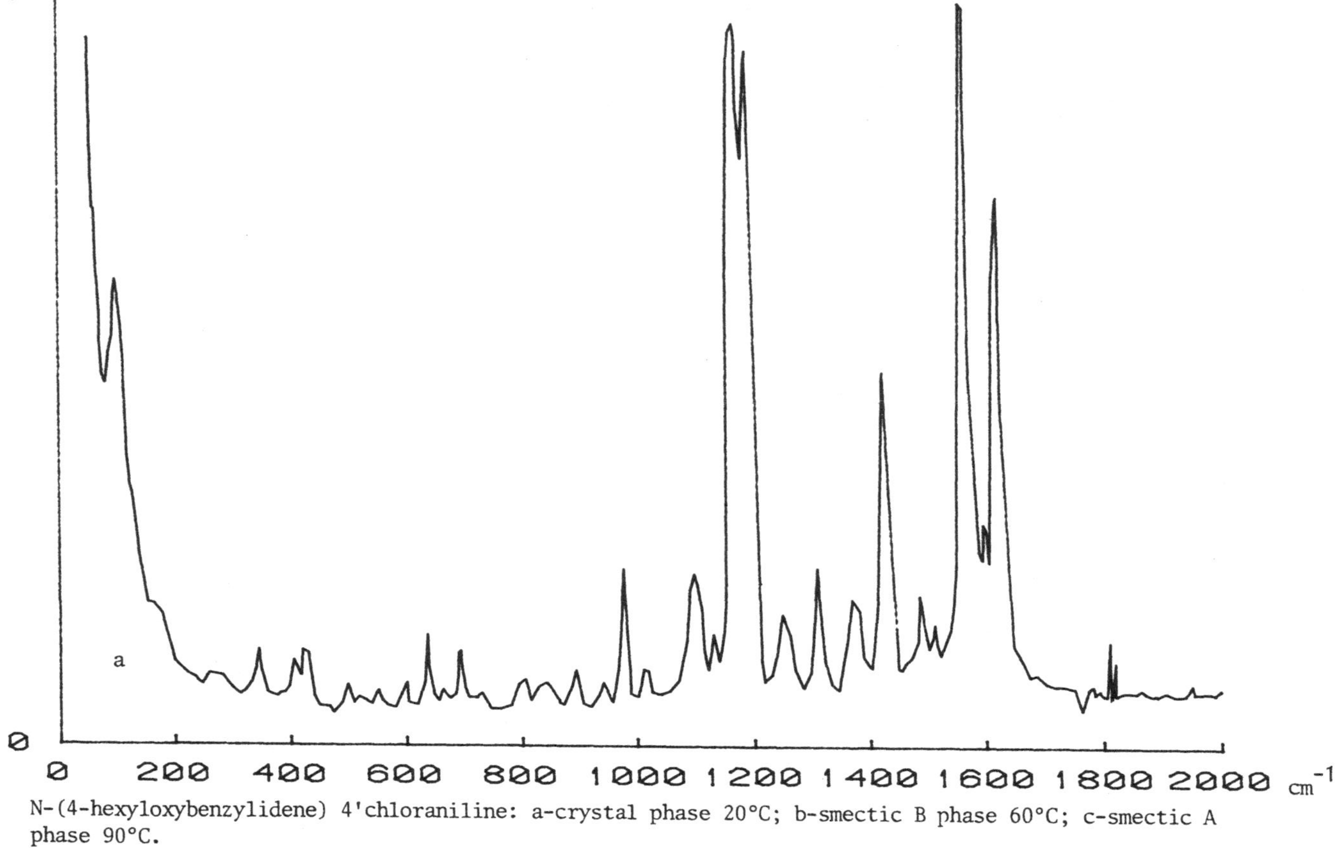

N-(4-hexyloxybenzylidene) 4'chloraniline: a-crystal phase 20°C; b-smectic B phase 60°C; c-smectic A phase 90°C.

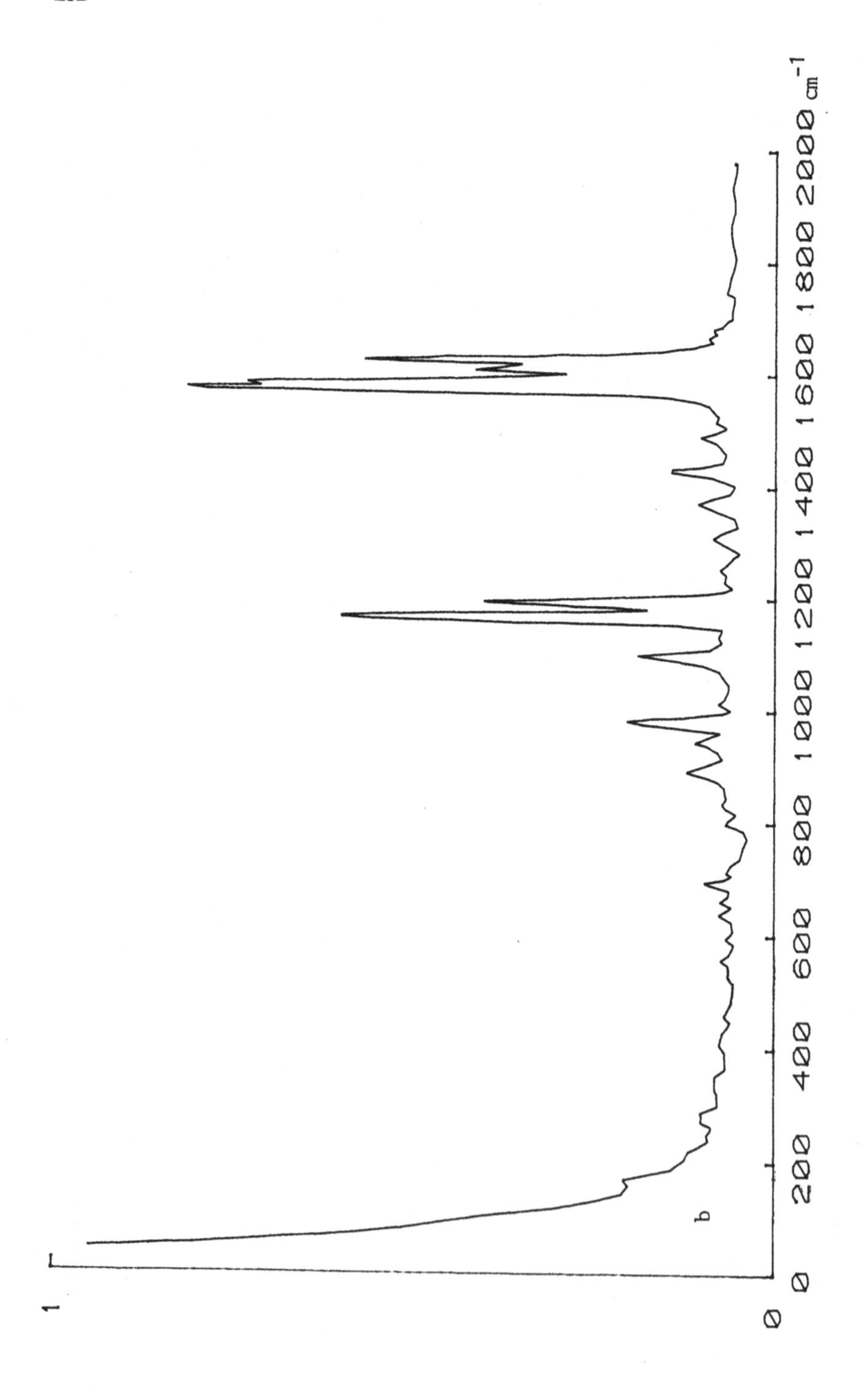
1
0
b
0
200
400
600
800
1000
1200
1400
1600
1800
2000
cm-1

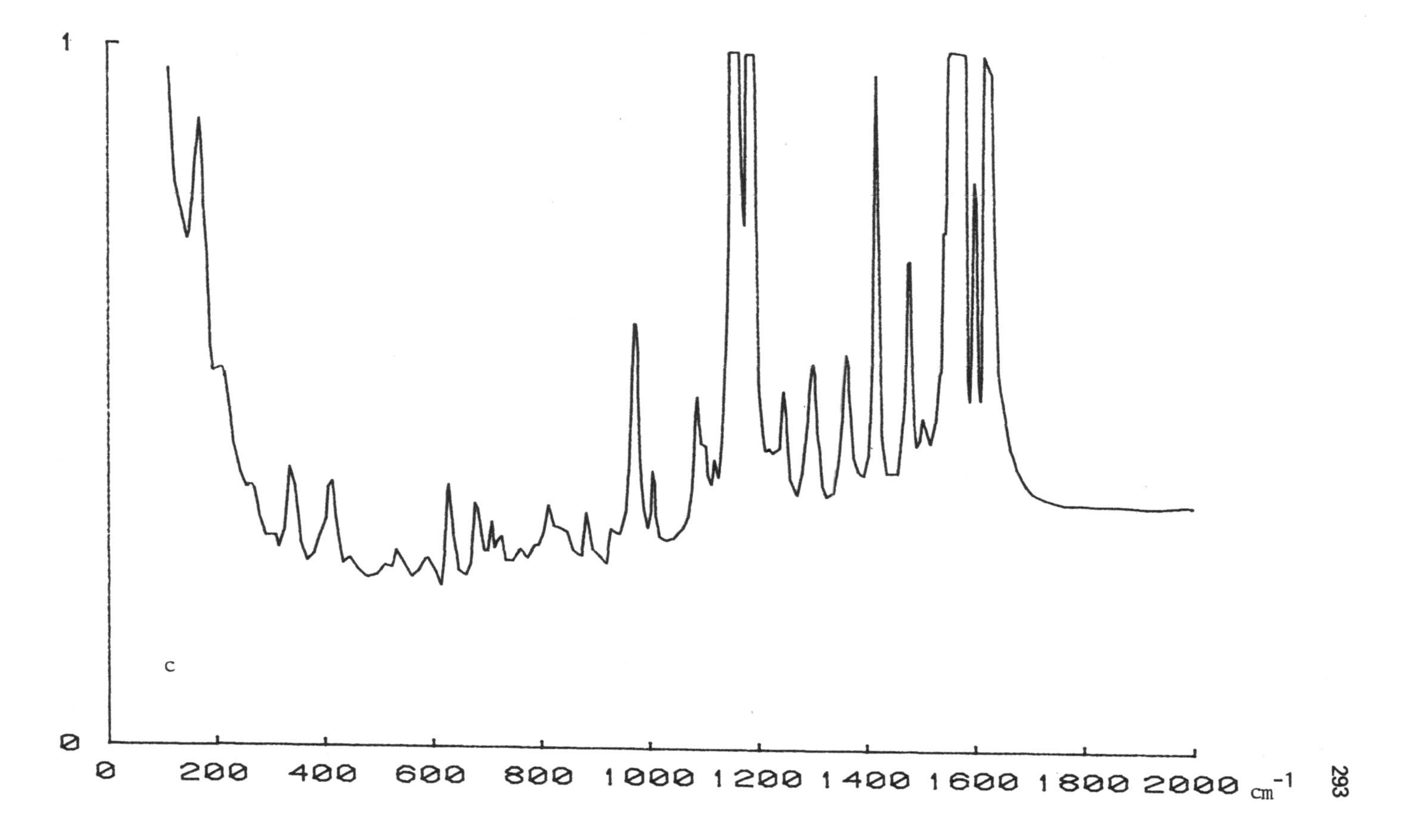

1
0
c
0
200
400
600
800
1000
1200
1400
1600
1800
2000
cm-1

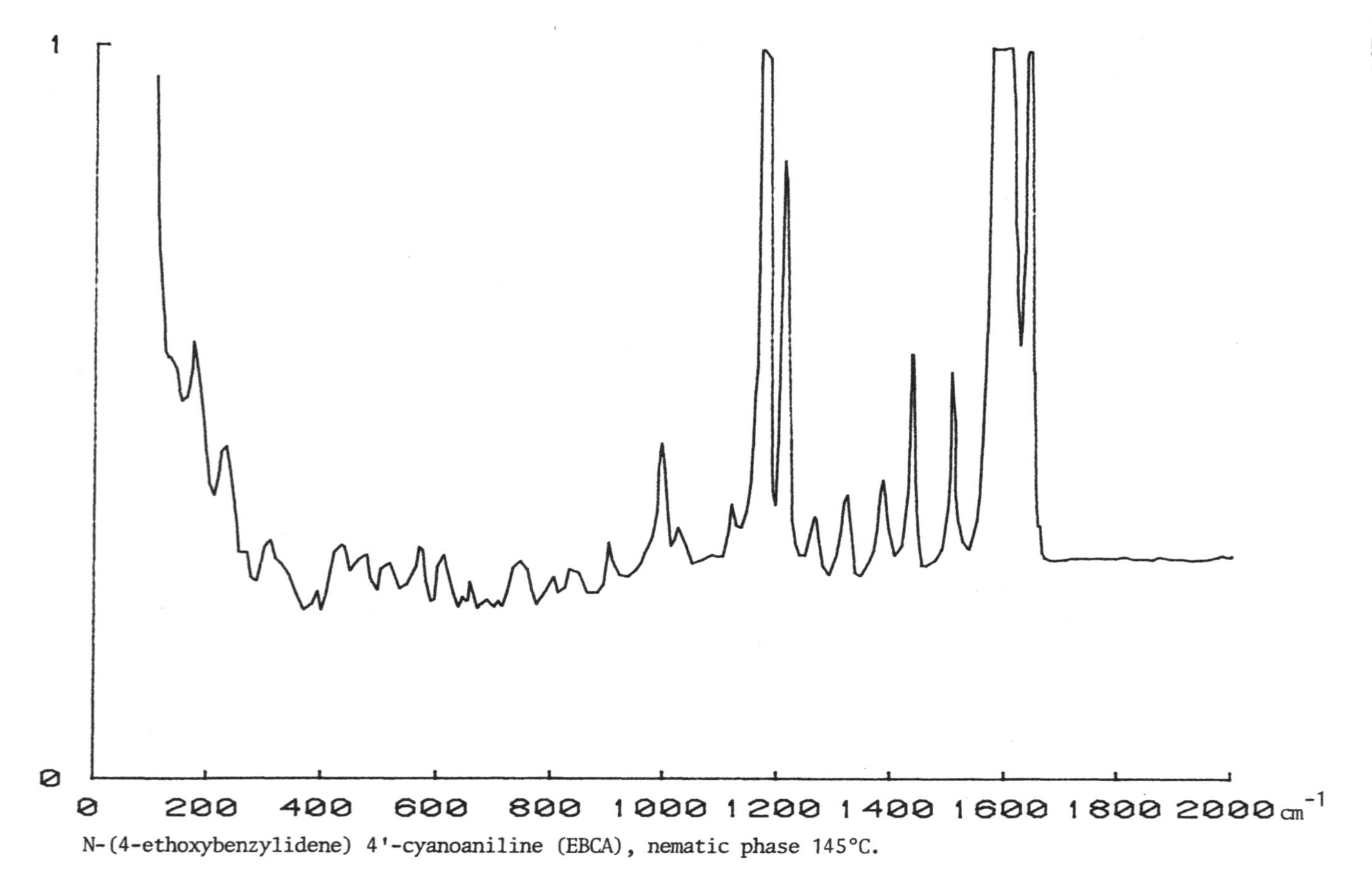

N-(4-ethoxybenzylidene) 4'-cyanoaniline (EBCA), nematic phase 145°C.

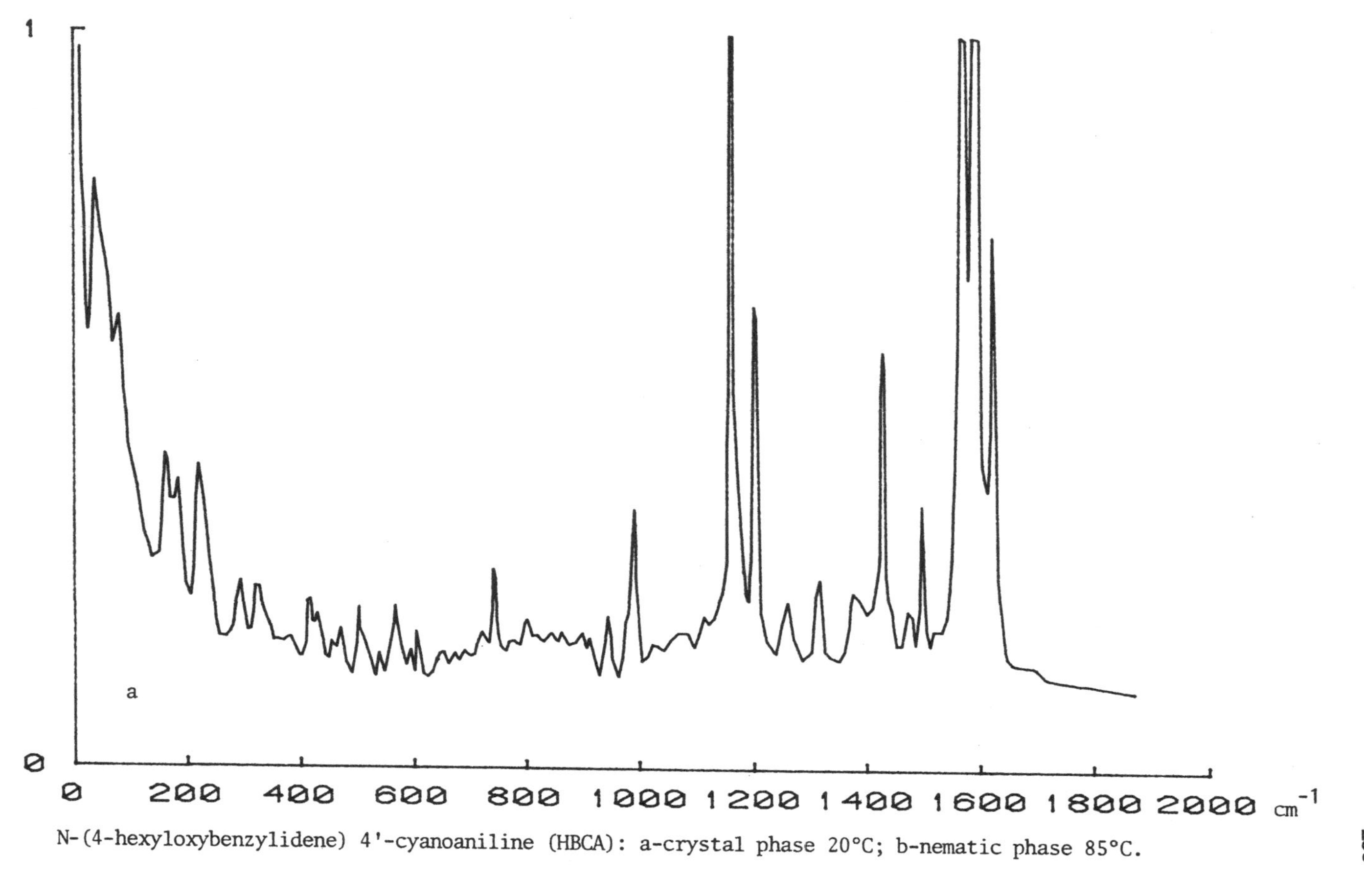

N-(4-hexyloxybenzylidene) 4'-cyanoaniline (HBCA): a-crystal phase 20°C; b-nematic phase 85°C.

1
0
b
0
200
400
600
800
1000
1200
1400
1600
1800
2000
cm-1

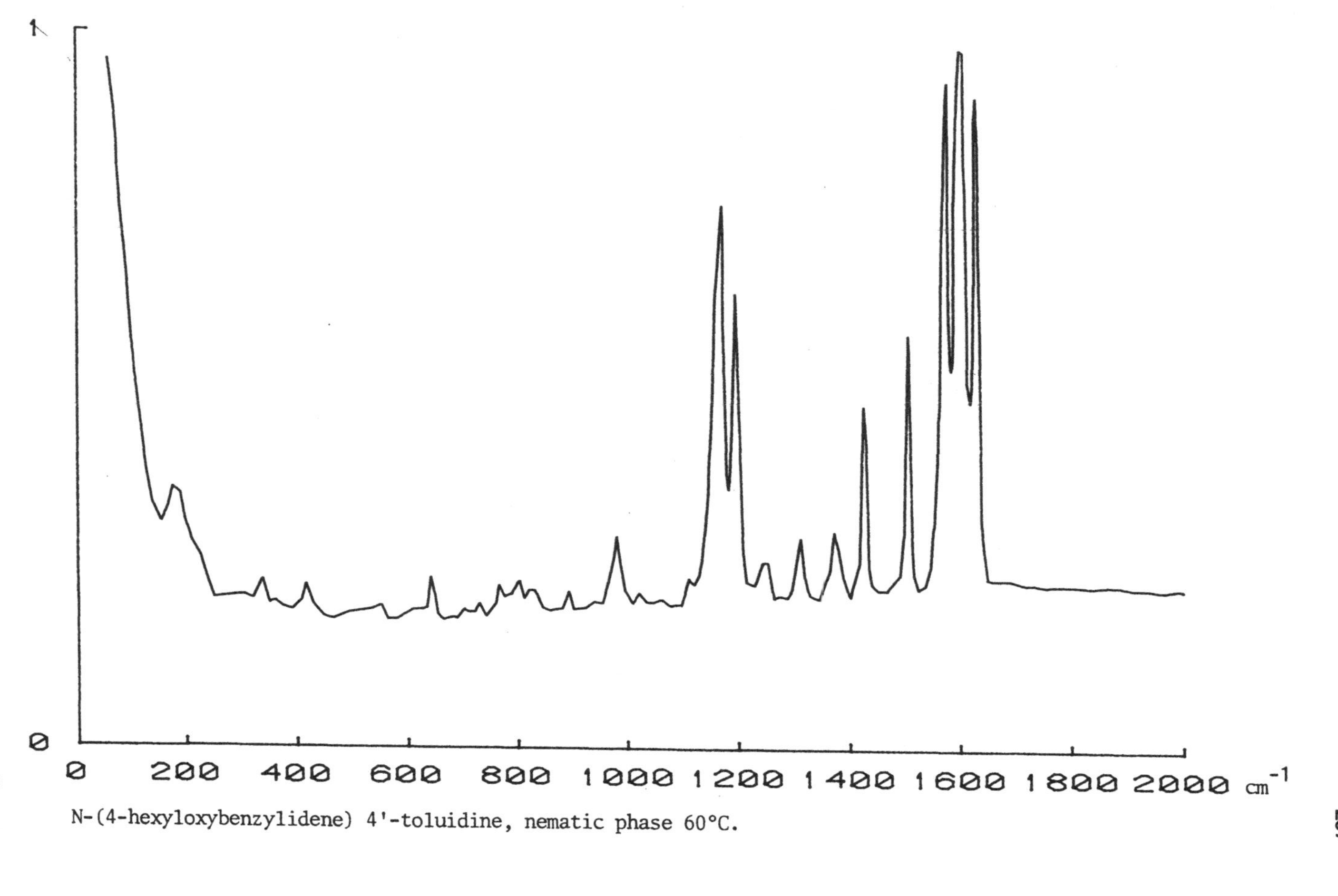

N-(4-hexyloxybenzylidene) 4'-toluidine, nematic phase 60°C.

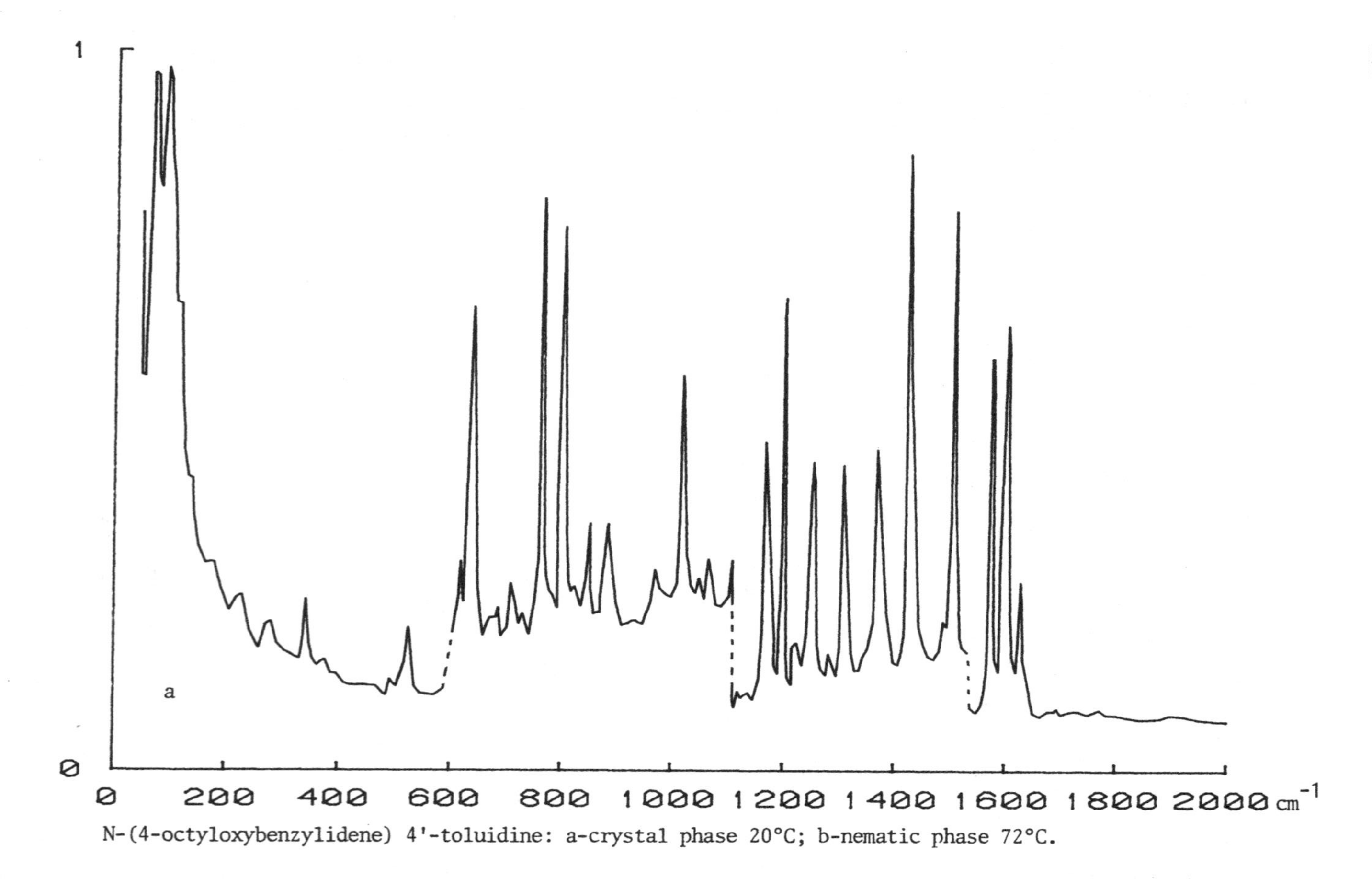

N-(4-octyloxybenzylidene) 4'-toluidine: a-crystal phase 20°C; b-nematic phase 72°C.

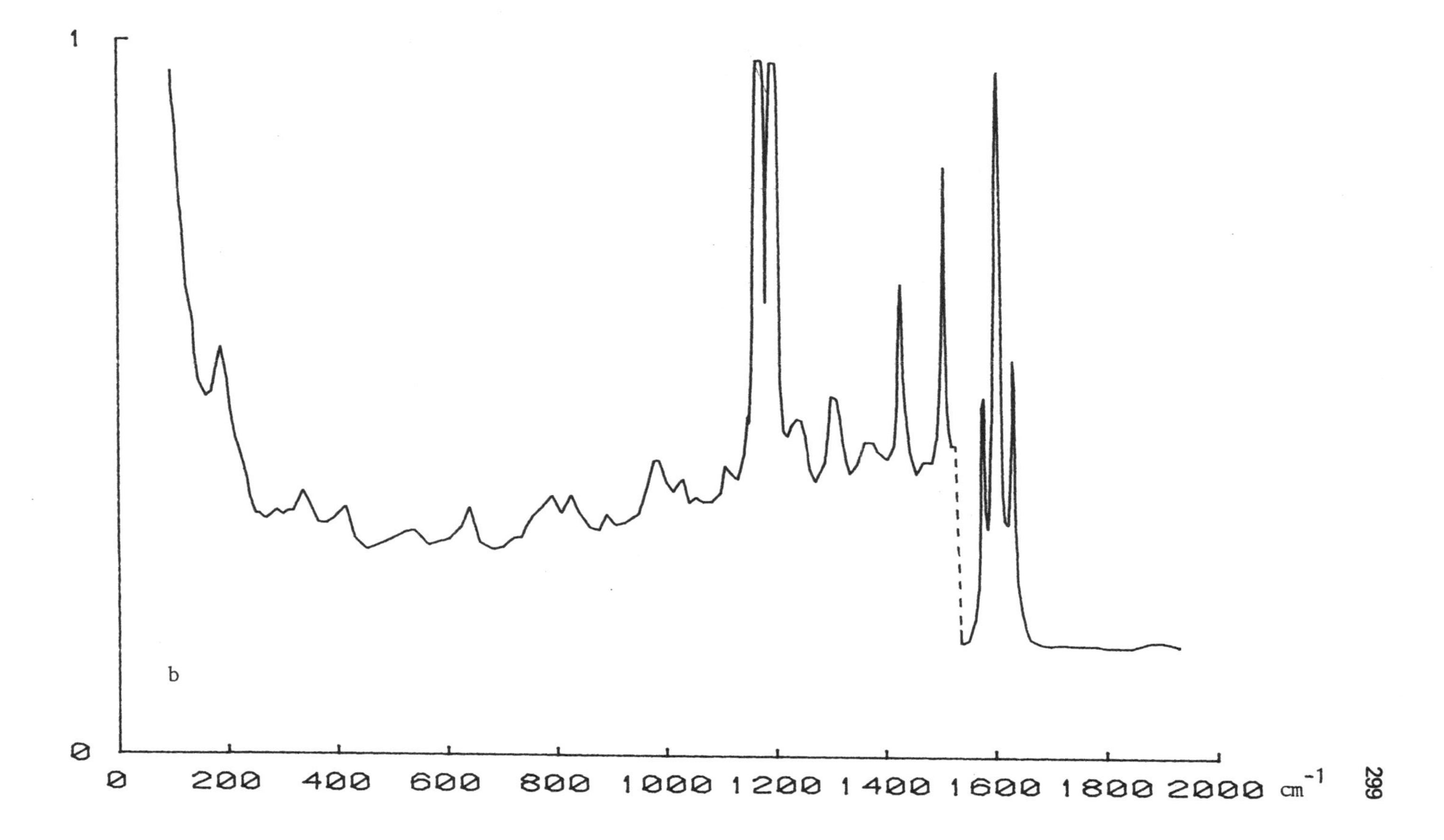

1
0
b
0
200
400
600
800
1000
1200
1400
1600
1800
2000 cm-1

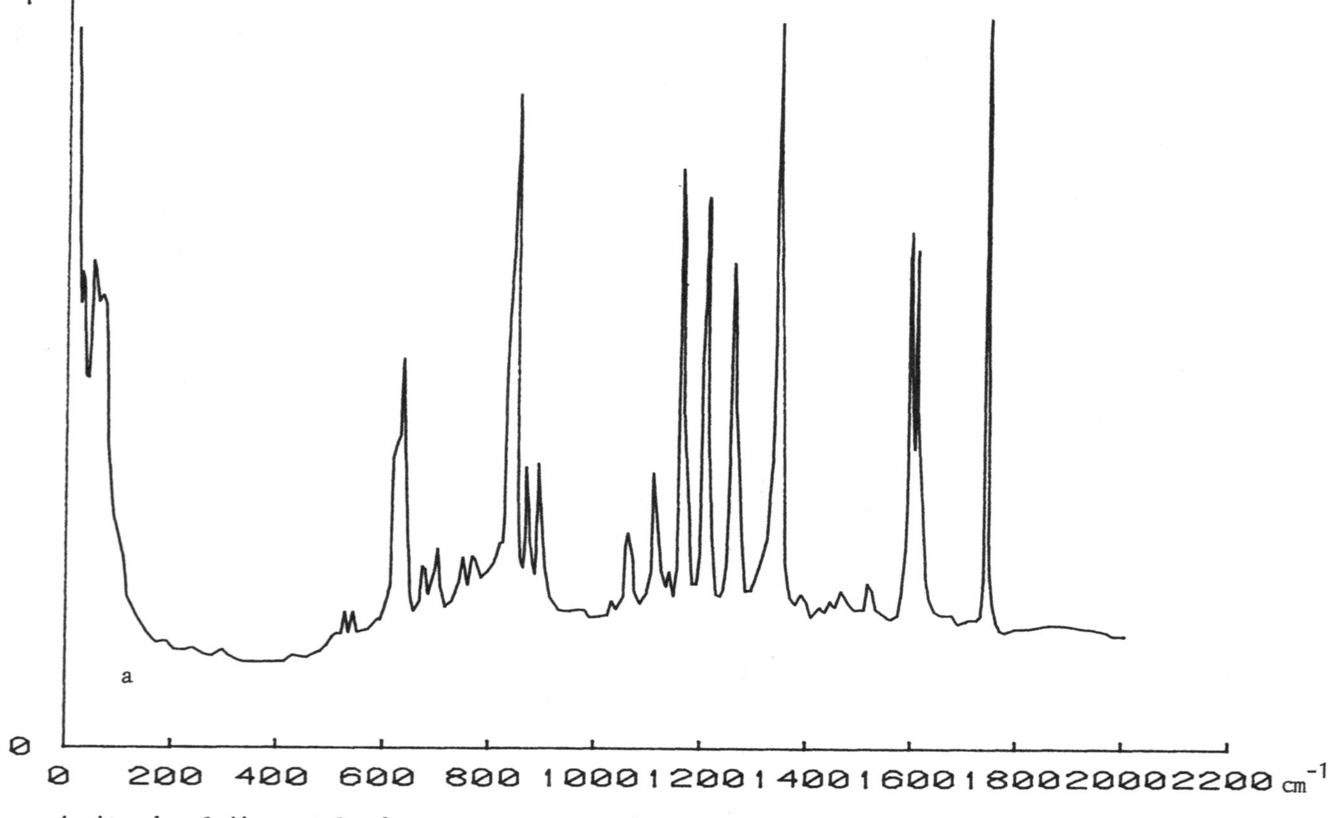

4-nitrophenol 4',n-octyloxybenzoate: a-crystal phase 20°C; b-nematic phase 63°C.

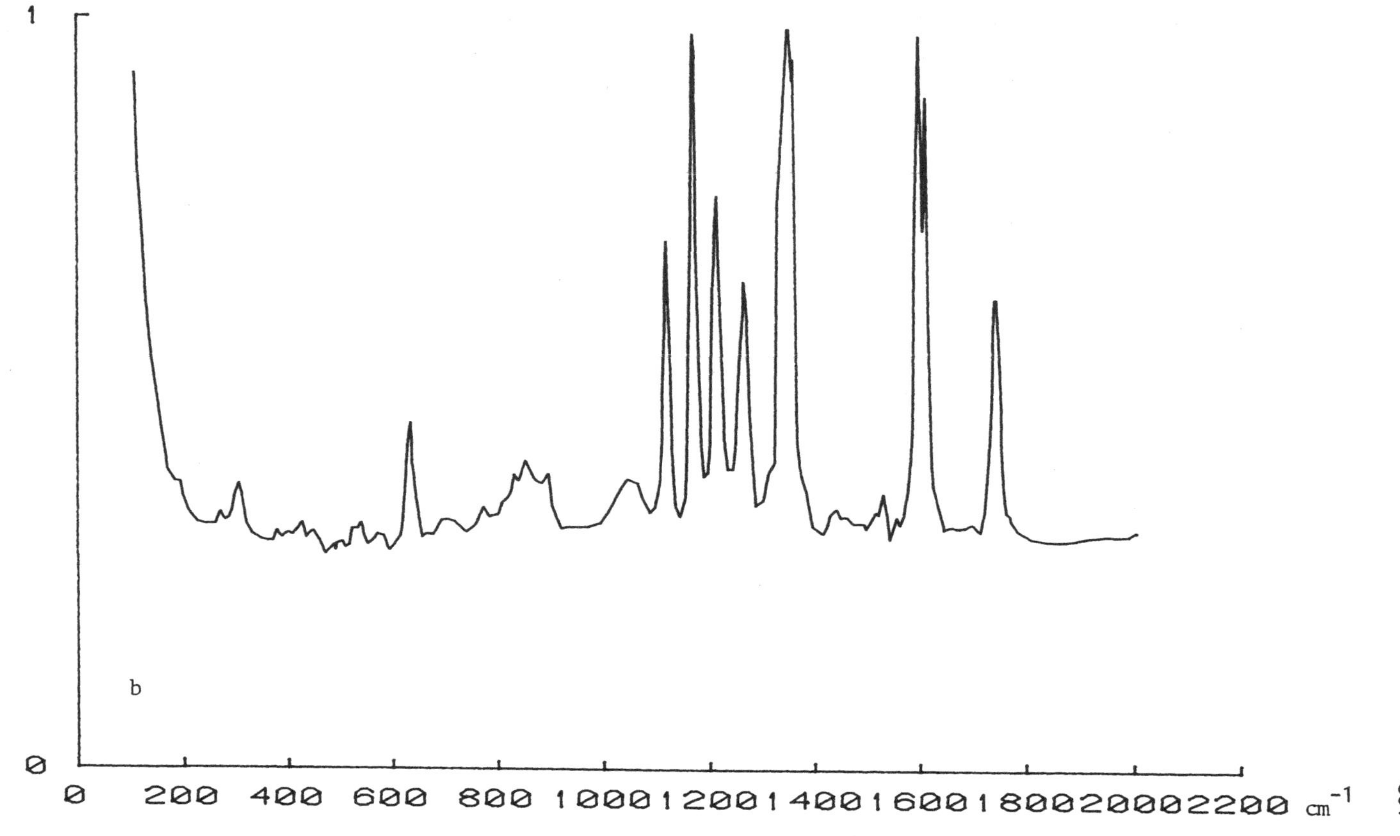
1
b
0
0
200
400
600
800
1000
1200
1400
1600
1800
2000
2200
cm^{-1}

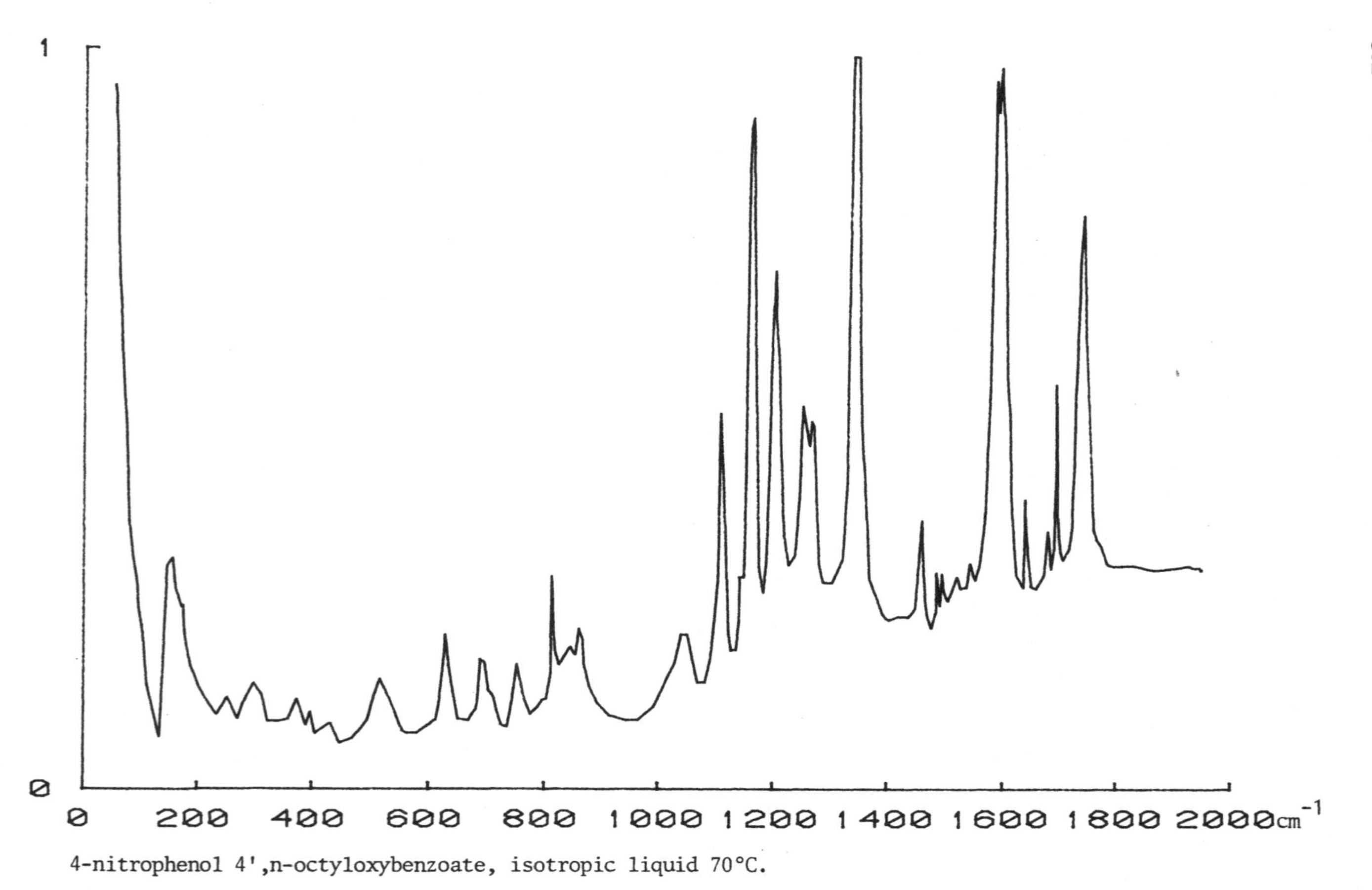

4-nitrophenol 4',n-octyloxybenzoate, isotropic liquid 70°C.

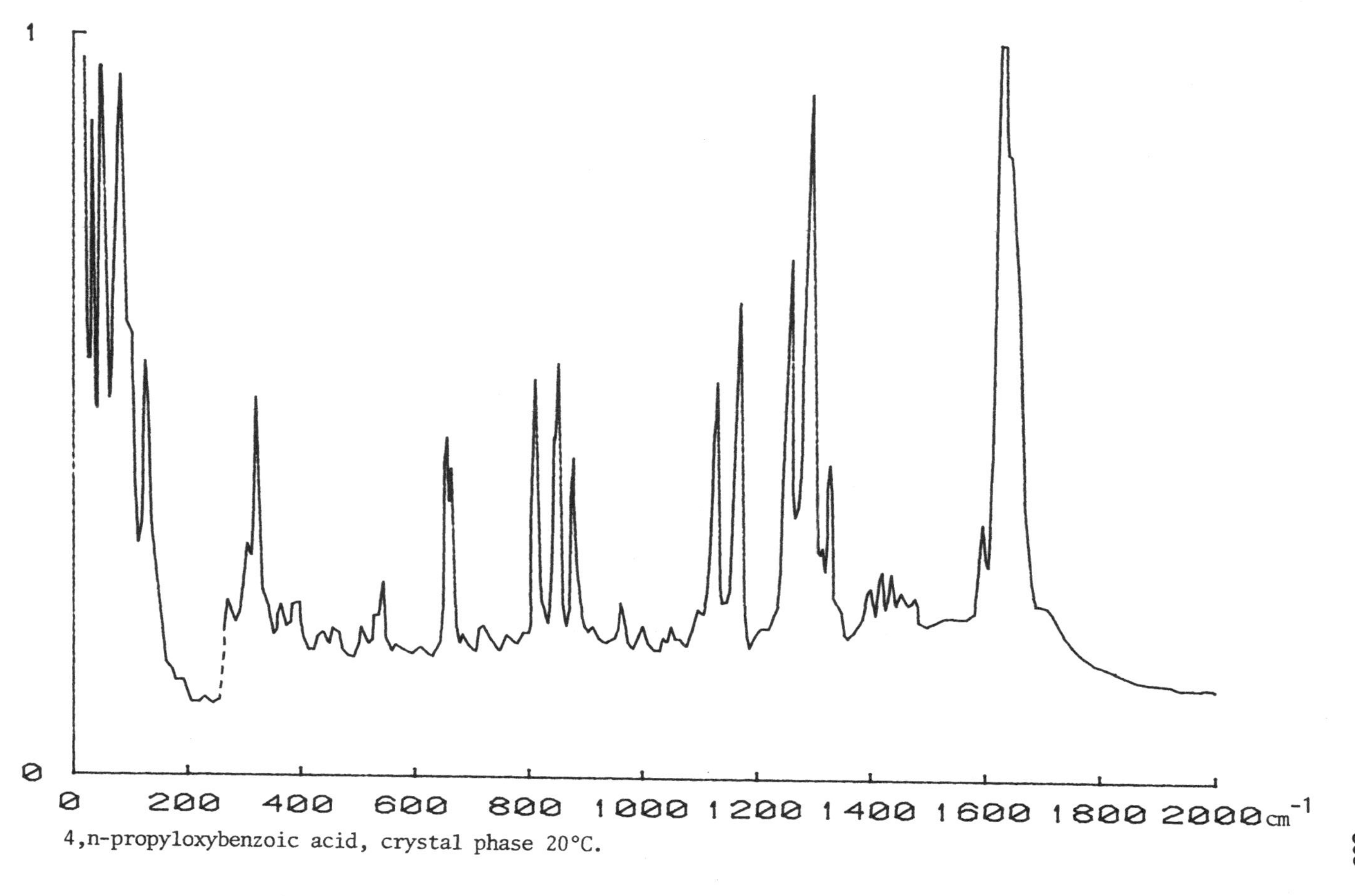

4,n-propyloxybenzoic acid, crystal phase 20°C.

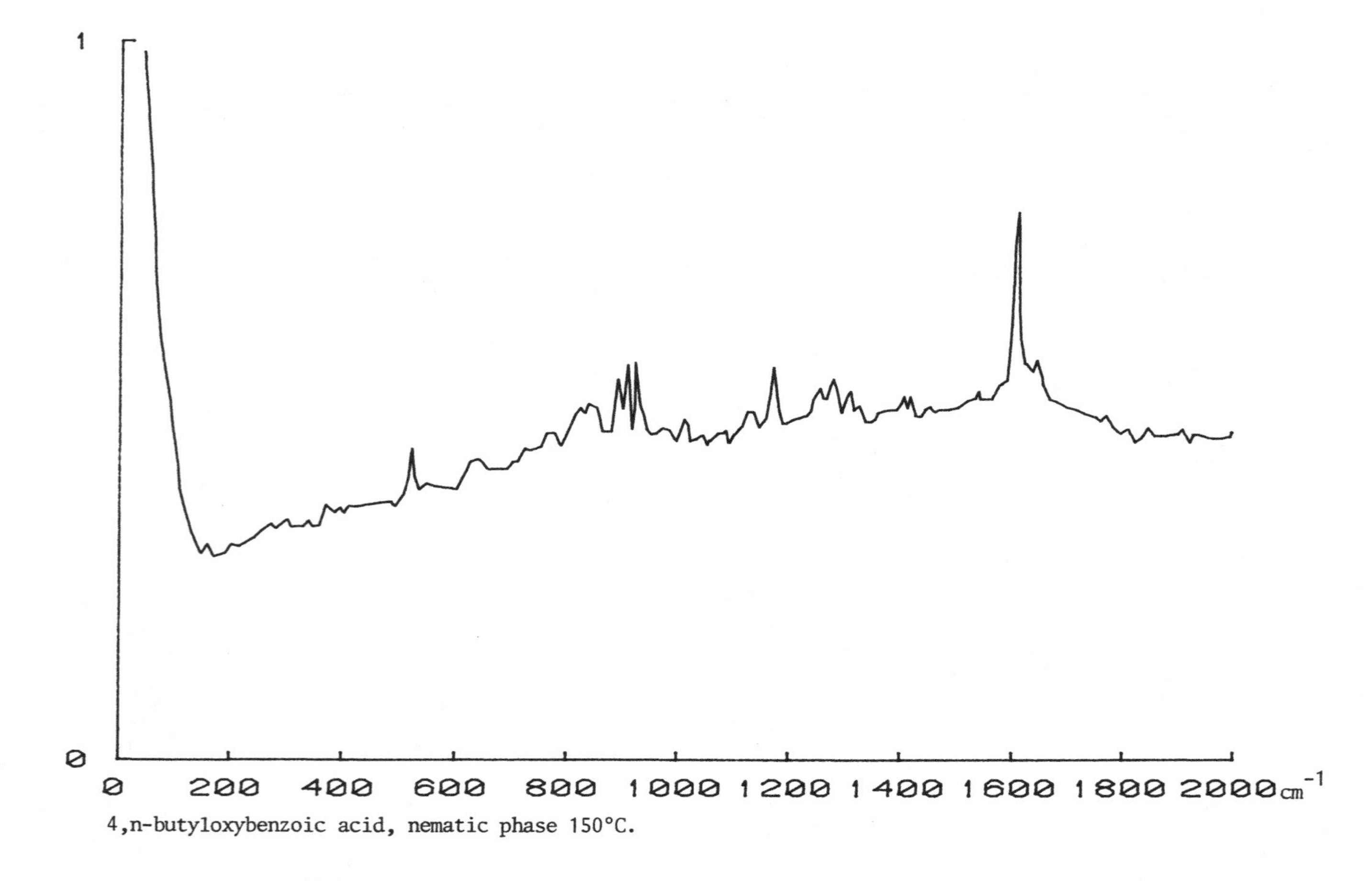

4,n-butyloxybenzoic acid, nematic phase 150°C.

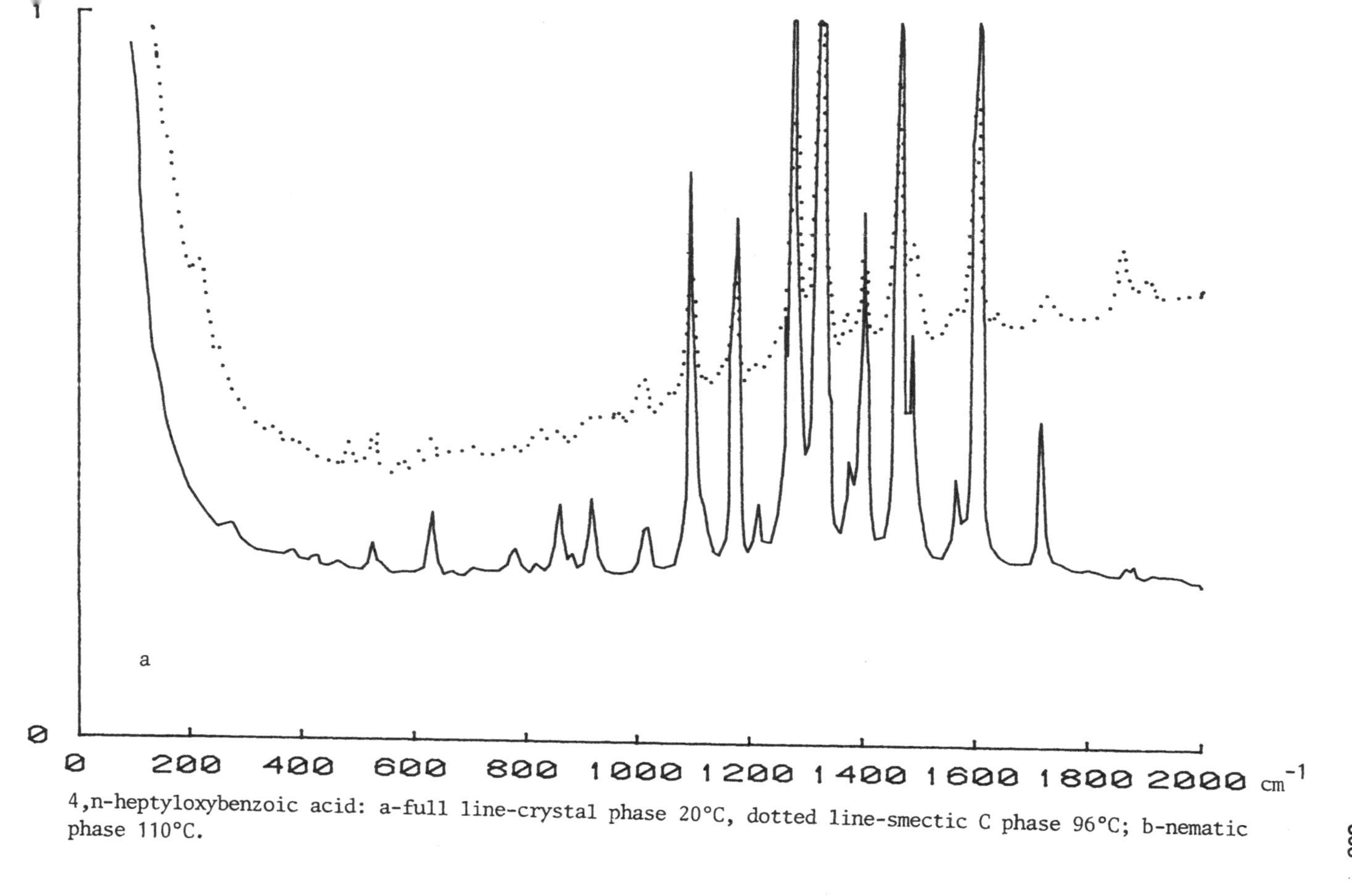

4,n-heptyloxybenzoic acid: a-full line-crystal phase 20°C, dotted line-smectic C phase 96°C; b-nematic phase 110°C.

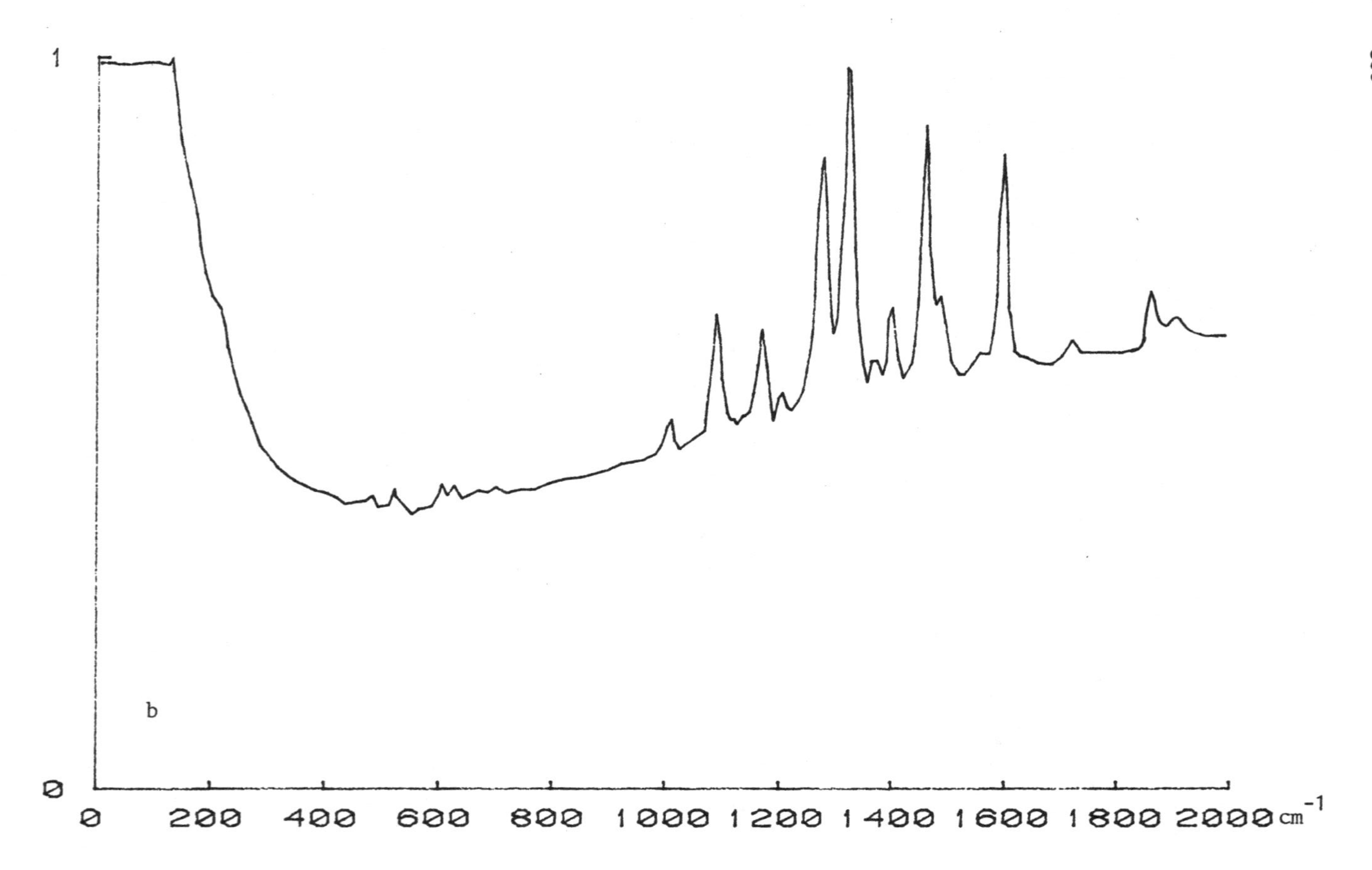
1
0
b
0 200 400 600 800 1000 1200 1400 1600 1800 2000 cm-1

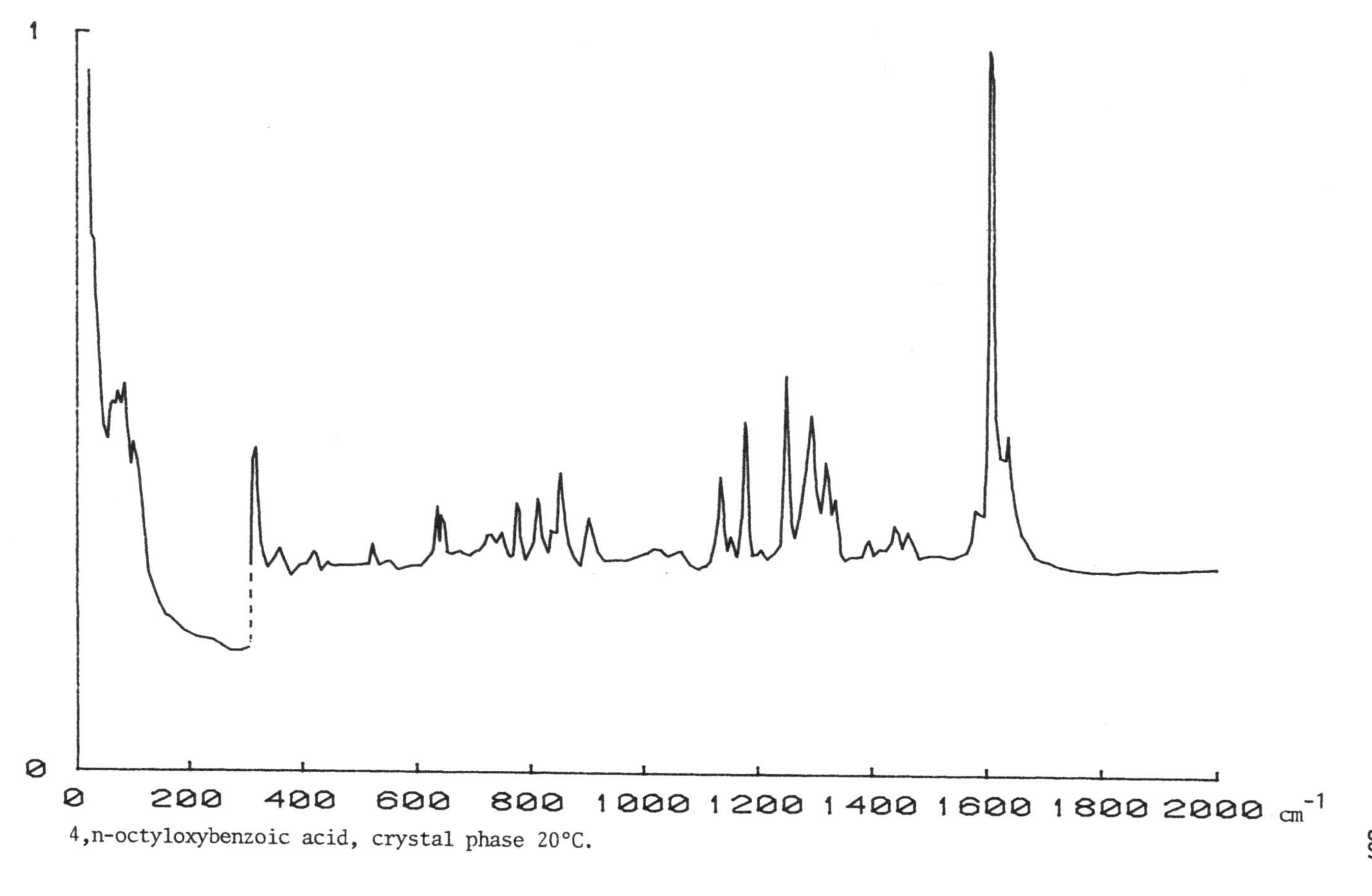

4,n-octyloxybenzoic acid, crystal phase 20°C.

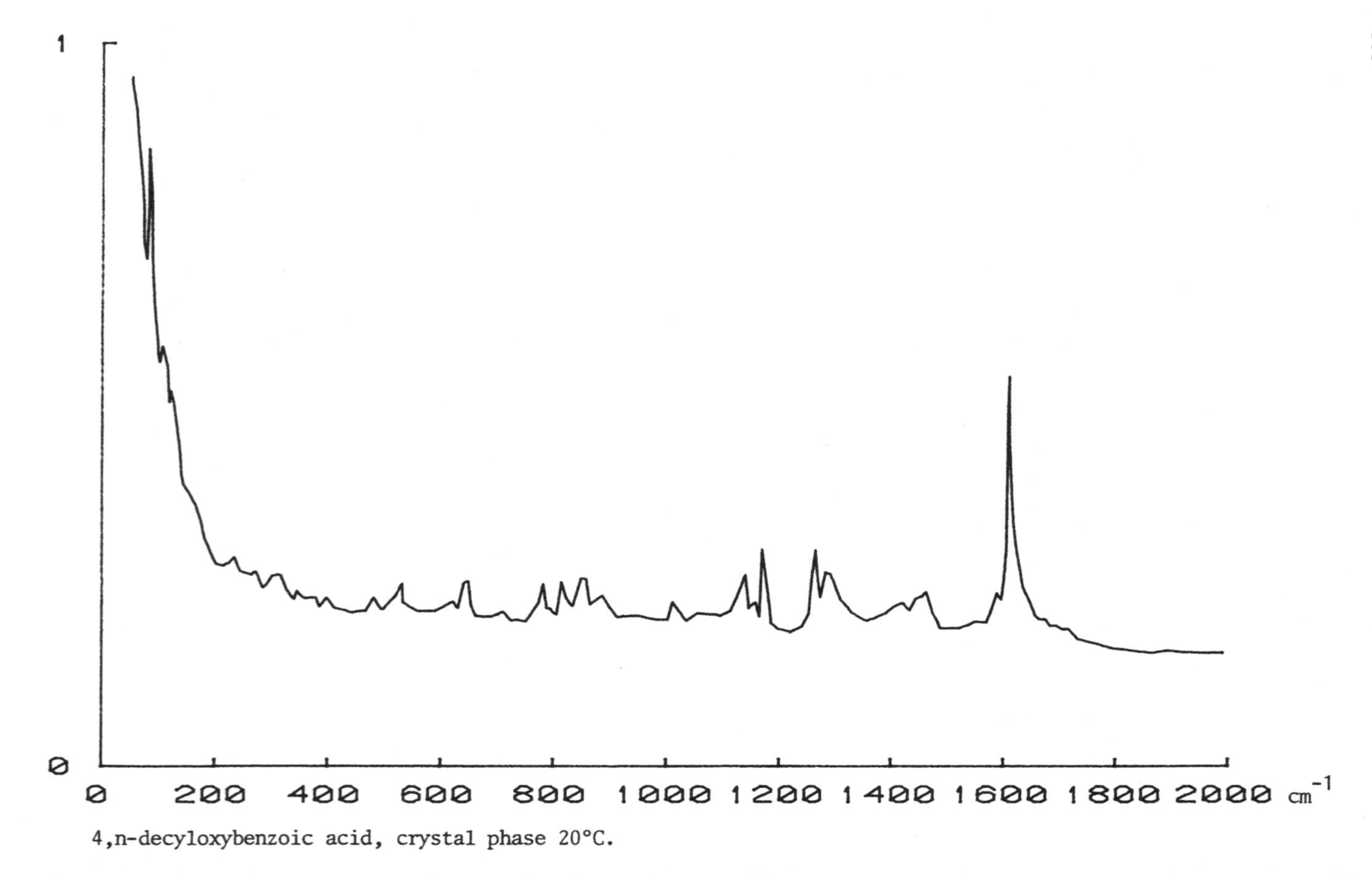

4,n-decyloxybenzoic acid, crystal phase 20°C.

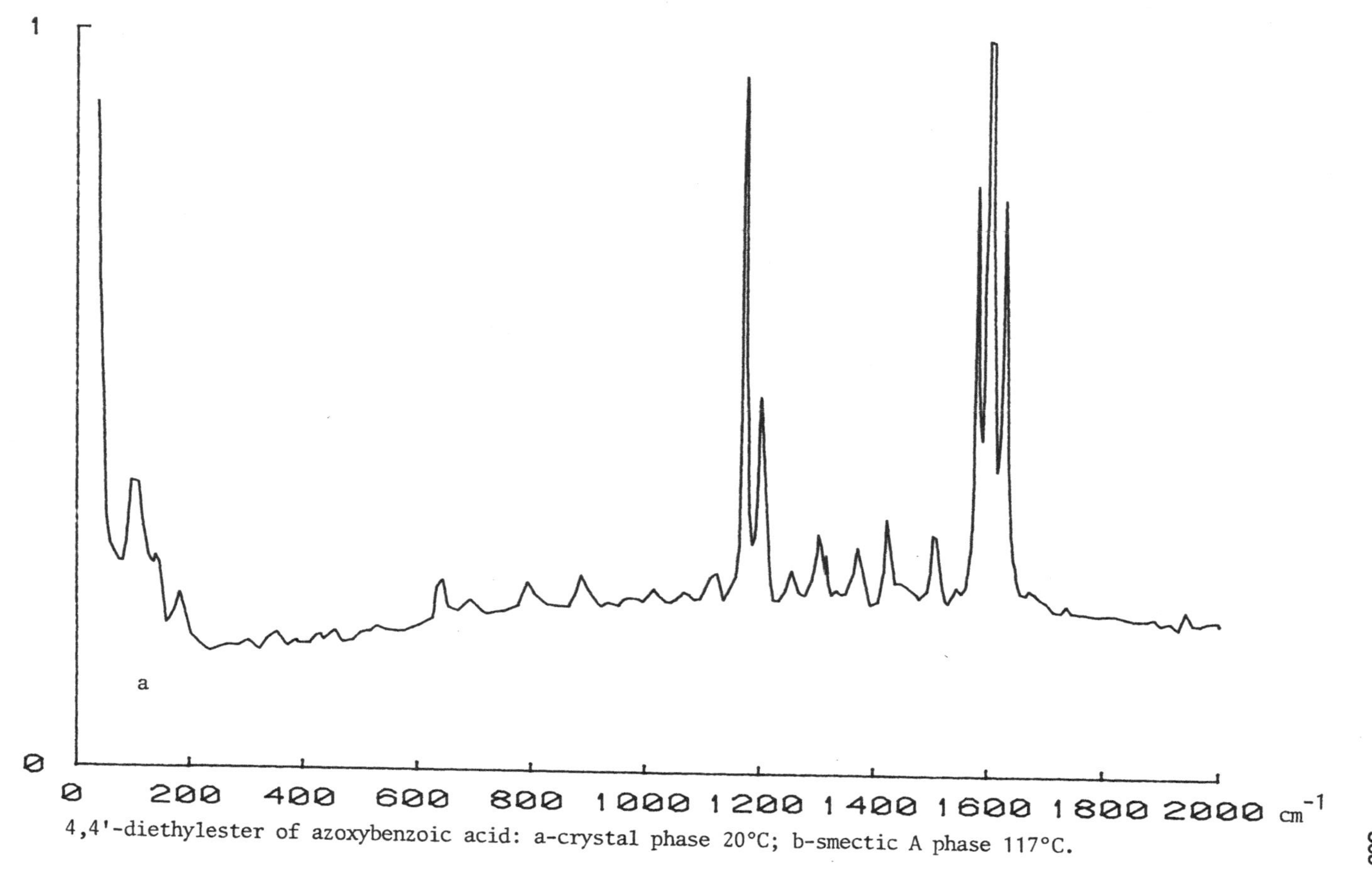

4,4'-diethylester of azoxybenzoic acid: a-crystal phase 20°C; b-smectic A phase 117°C.

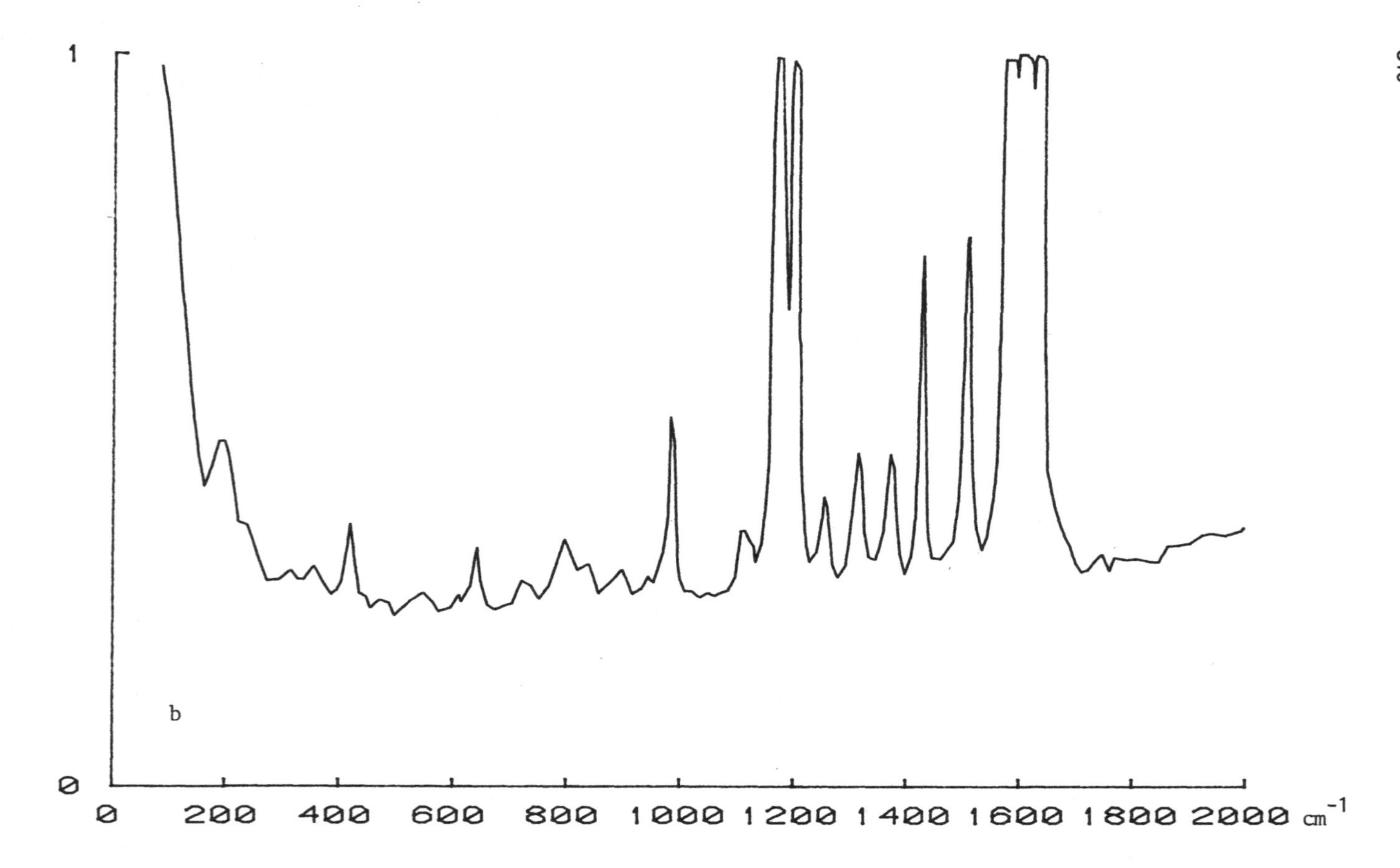
1
0
b
0
200
400
600
800
1000
1200
1400
1600
1800
2000
cm^{-1}

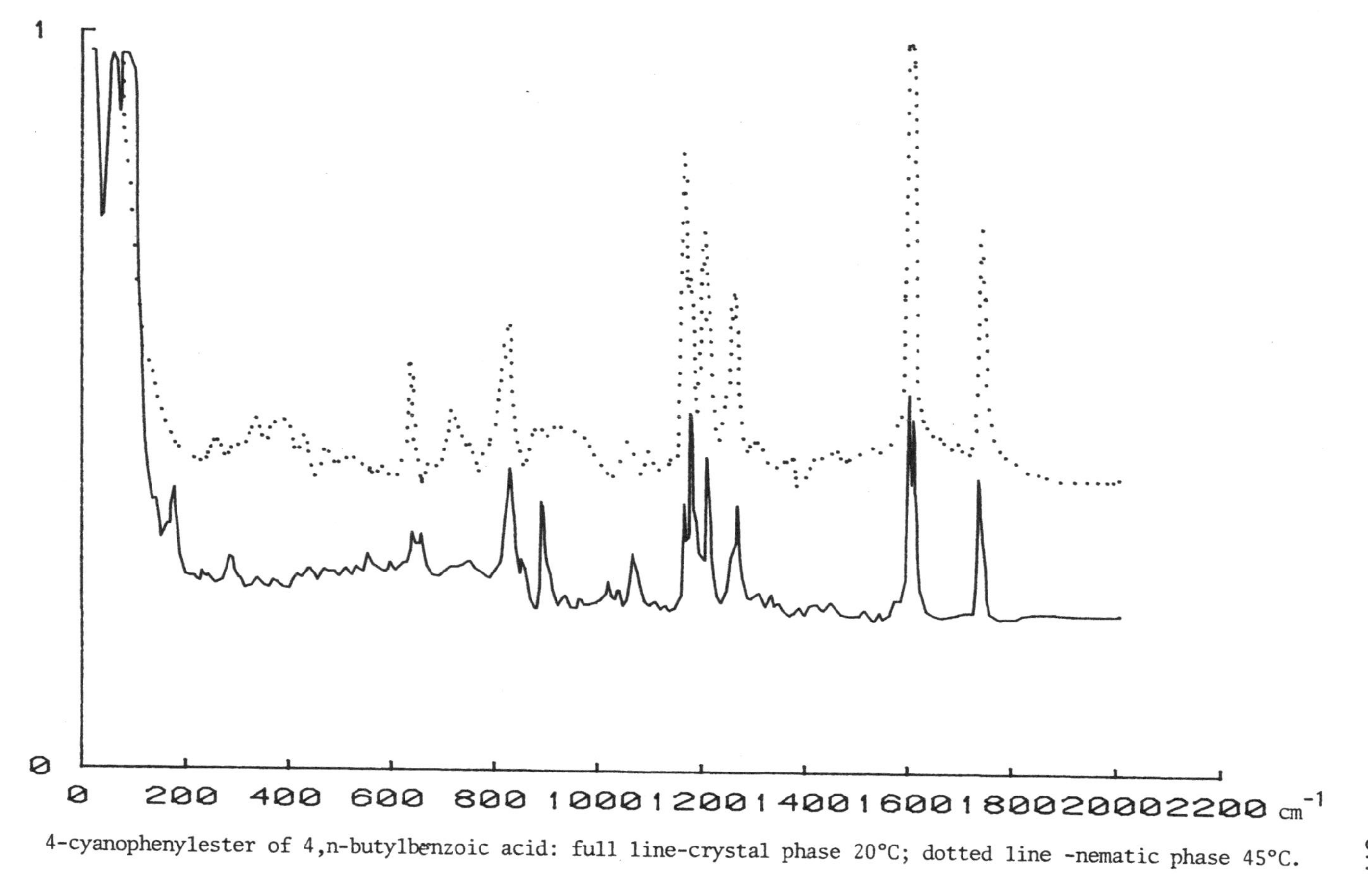

4-cyanophenylester of 4,n-butylbenzoic acid: full line-crystal phase 20°C; dotted line -nematic phase 45°C.

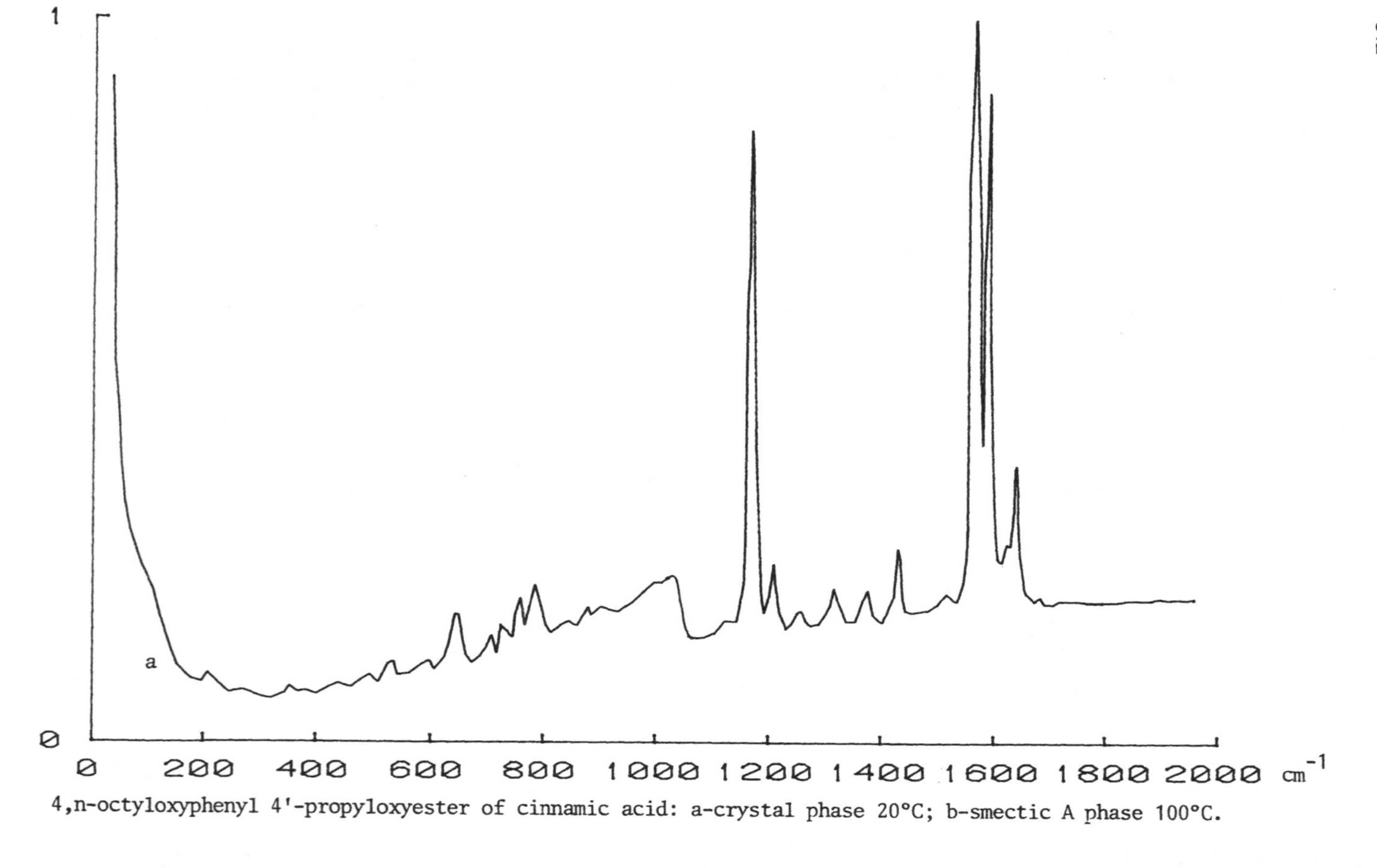

4,n-octyloxyphenyl 4'-propyloxyester of cinnamic acid: a-crystal phase 20°C; b-smectic A phase 100°C.

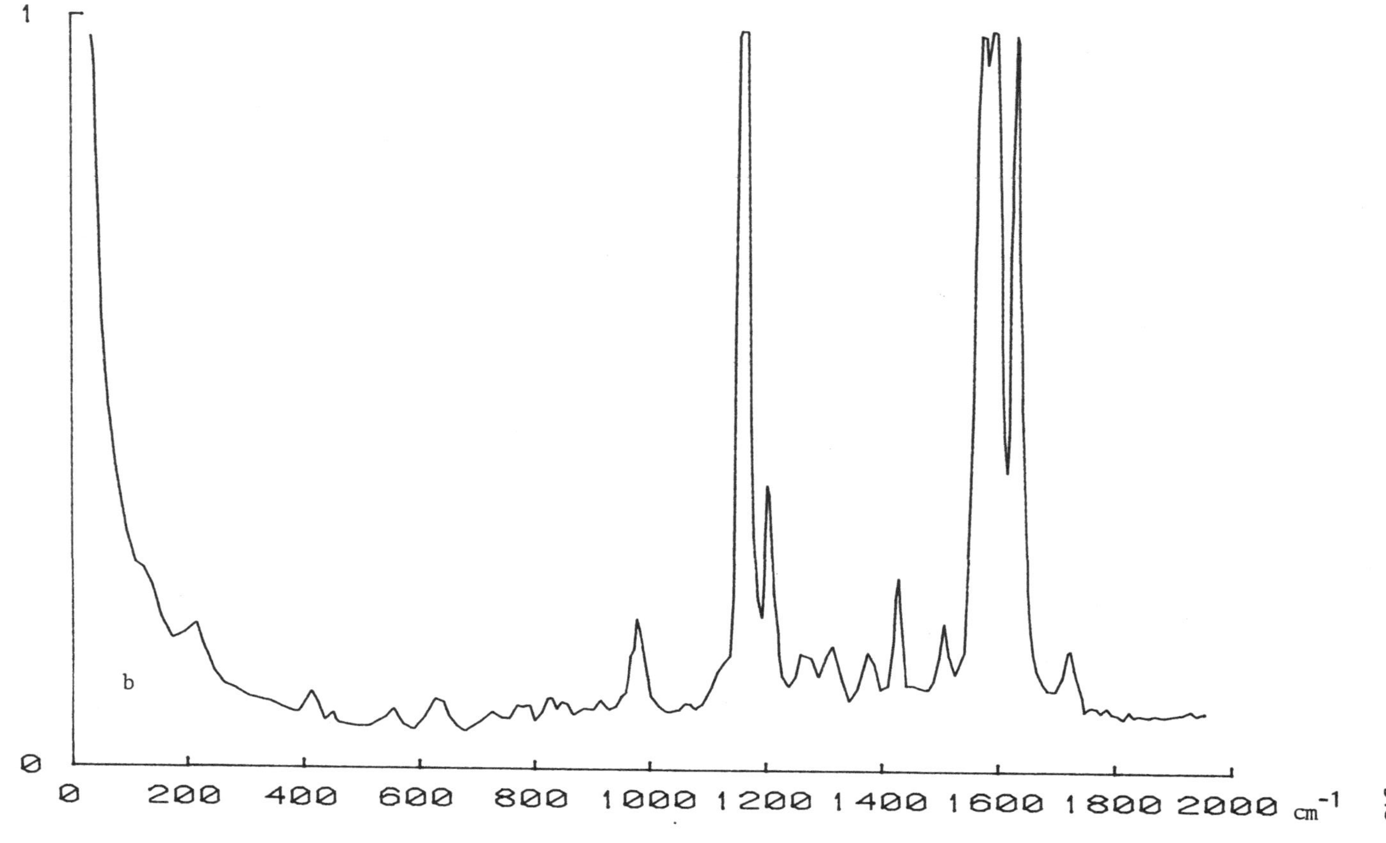

1
0
b
0
200
400
600
800
1000
1200
1400
1600
1800
2000 cm-1

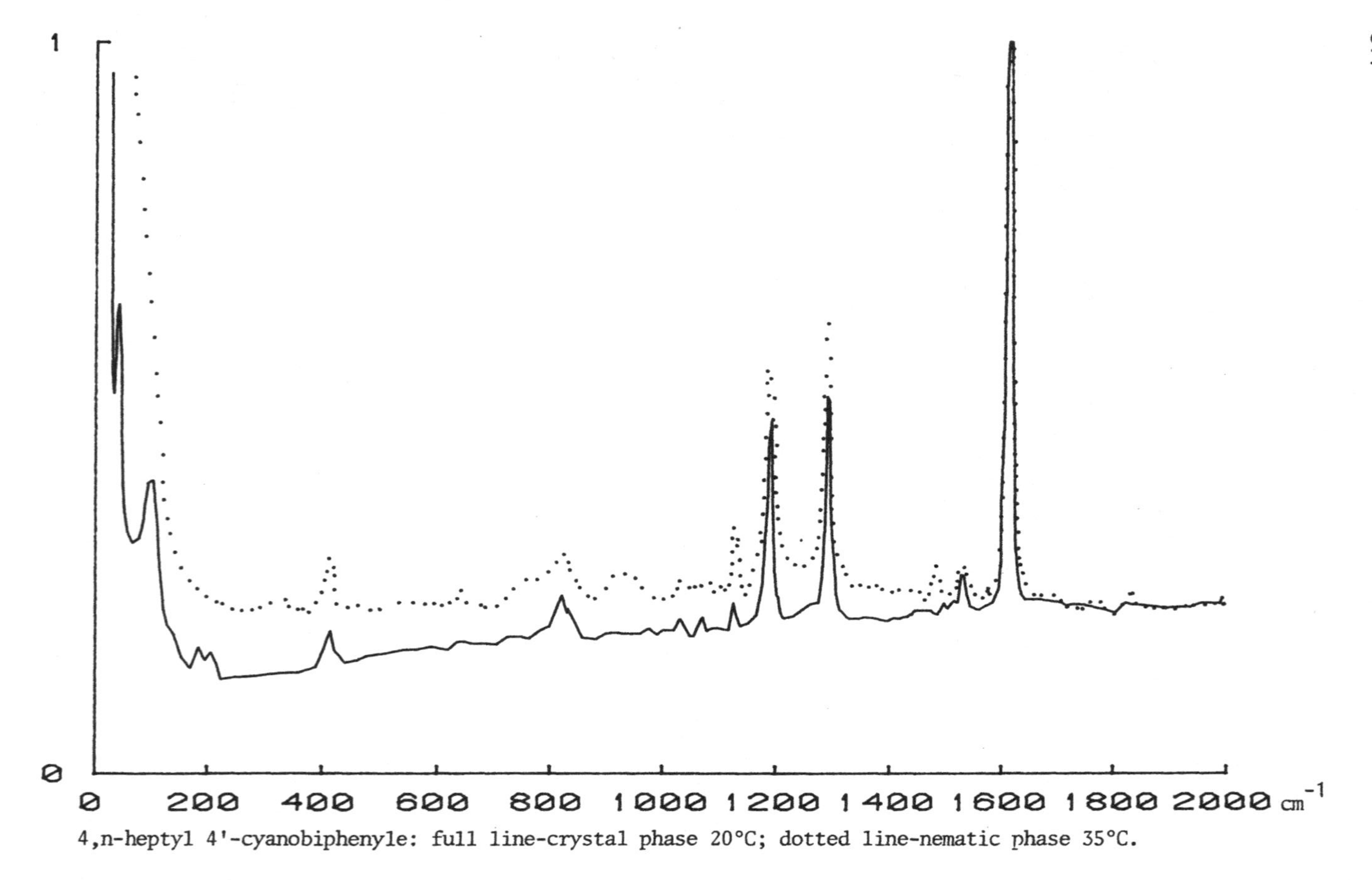

4,n-heptyl 4'-cyanobiphenyle: full line-crystal phase 20°C; dotted line-nematic phase 35°C.

LIST OF COMPOUNDS

List of Compounds

Note: The compounds are listed by number; the numbers on the right refer to the pages in the Atlas in which spectra of a given compound appear: the first numbers refer to Far Infrared spectra, the second to Infrared Spectra, and the third to Raman spectra. For some compounds the spectra were taken in all phases (i.e. crystal, mesophase(s) and isotropic); in most cases the isotropic and most relevant liquid crystalline phases are presented.

Azoxybenzenes

Alkyloxybenzylidenes

30.	N-(4-butyloxybenzylidene)4'-toluidine	69, –, –
31	N-(4-pentyloxybenzylidene)4'-toluidine	–, 153-154, –
32.	N-(4-hexyloxybenzylidene)4'-toluidine	70, 155-156, 297
33.	N-(4-heptyloxybenzylidene)4'-toluidine	–, 157-158, –
34.	N-(4-octyloxybenzylidene)4'-toluidine	–, 159-160, 298-299
35.	N-(4-nolyloxybenzylidene)4'-toluidine	71, 161, –

Benzoates

36.	4,n-pentylacetophenonoxim 4',n-heptylbenzoate	72, –, –
37.	4,n-aminoacetophenonoxim 4',n-heptyloxybenzoate	–, 162-163, –
38.	hydrochinone bis-4,n-heptyloxybenzoate	73, –, –
39.	2-ethylhydrochinone 4,n-hexylbenzoate	74, 164-166, –
40.	4-nitrophenol 4',n-octyloxybenzoate	75, 167-168, 300-302

Alkoxybenzoic acids

41.	4,n-propyloxybenzoic acid	76, 169-170, 303
42.	4,n-butyloxybenzoic acid	–, 171-172, 304
43.	4,n-pentyloxybenzoic acid	77, 173-174, –
44.	4,n-hexyloxybenzoic acid	–, 175-176, –
45.	4,n-heptyloxybenzoic acid	–, 177-180, 305-306
46.	4,n-octyloxybenzoic acid	78, 181-184, 307
47.	4,n-nonyloxybenzoic acid	–, 185-186, –
48.	4,n-decyloxybenzoic acid	79, 187, 308
49.	4,n-dodecyloxybenzoic acid	80, 188-189, –

Esters

50.	4,4'-diethylester of azoxybenzoic acid	81, 190, 309-310
51.	4-methoxyphenylester of 4,n-butyloxybenzoic acid	82, 191-192, –
52.	4,n-hexylphenylester of 4,n-butylbenzoic acid	83, 193-194, –
53.	4,n-hexyloxyphenylester of 4,n-butyloxybenzoic acid	84, –, –
54.	4,n-heptylphenylester of 4,n-butylbenzoic acid	85, 195-196, –
55.	4-methylphenylester of 4,n-hexyloxybenzoic acid	86, 197-198, –
56.	4-methylphenylester of 4,n-hexylbenzoic acid	–, 199, –
57.	4,n-hexyloxyphenylester of 4,n-hexylbenzoic acid	87, –, –
58.	4,n-heptyloxyphenylester of 4,n-hexyloxybenzoic acid	88, 200-201, –
59.	4,n-octyloxyphenylester of 4,n-hexylbenzoic acid	89, –, –
60.	4-cyanophenylester of 4,n-butylbenzoic acid	–, –, 311
61.	4-cyanophenylester of 4,n-hexylbenzoic acid	–, 202, –
62.	4-cyanophenylester of 4,n-heptylbenzoic acid	90, 203-204, –
63.	4,n-hexylester of 4-cyanobenzoic acid	–, 205, –
64.	N-(4-ethoxybenzylidene) 4'-aminophenylester of capric acid	91, 206-207, –
65.	N-(4-ethoxybenzylidene) aminoester of cynnamic acid	92, 208-209, –

66.	4,n-octyloxyphenyle 4'-propyloxyester of cynnamic acid	–, 210-211, 312-313
67.	4-cyanobiphenylester of 4',n-heptylcynnamic acid	93, 212-213, –
68.	N-(4-ethoxybenzylidene)4'-aminophenylester of ennantic acid	94, 214-215, –
69.	n-butyl-4-(ethoxyphenyloxy) carbonylphenylcarbonate	95, –, –

Tolanes

70.	4,n-pentyl 4'-methoxytolane (PMT)	–, 216-218, –
71.	4,n-hexyl 4'-methoxytolane (HMT)	–, 219, –
72.	4,n-heptyl 4'-methyloxytolane	96, 220, –
73.	4,n-butyl 4'-ethoxytolane (BET)	97, 221-222, –
74.	4,n-pentyl 4'-ethoxytolane (PET)	98, –, –
75.	4,n-hexyl 4'-ethoxytolane (HET)	–, 223-224, –
76.	4,n-heptyl 4'-ethoxytolane	–, 225-226, –
77.	4,n-octyl 4'-ethoxytolane (OET)	99, 227-228, –

Cyanobiphenyles

78.	4,n-pentyl 4'-cyanobiphenyle	–, 229, –
79.	4,n-heptyl 4'-cyanobiphenyle	–, 230-231, 314
80.	4,n-heptylbenzoate 4'-cyanobiphenyle	100, –, –
81.	4,n-octyl 4'-cyanobiphenyle (8CB)	101, 232-233, –
82.	4,n-nonyl 4'-cyanobiphenyle (9CB)	–, 234-235, –
83.	4,n-decyl 4'-cyanobiphenyle (10CB)	102, 236-238,–
84.	4,n-pentyloxy 4'-cyanobiphenyle (5OCB)	103, 239-240, –
85.	4,n-hexyloxy 4'-cyanobiphenyle (6OCB)	–, 241, –
86.	4,n-heptyloxy 4'-cyanobiphenyle (7OCB)	104, 242-243, –
87.	4,n-decylbiphenyle	–, 244-245, –

Cholesterics

88.	Cholesteryl acetate	–, 246-247, –
89.	Cholesteryl propionate	107, 248-249, –
90.	Cholesteryl valerate	–, 250-252, –
91.	Cholesteryl butyrate	–, 253-254, –
92.	Cholesteryl decanoate	–, 255-256, –
93.	Cholesteryl undecylate	–, 257-258, –
94.	Cholesteryl miristate	106, 259-260, –
95.	Cholesteryl palmitate	–, 261-262, –
96.	Cholesteryl ennantate	–, 263-264, –
97.	Cholesteryl erucate	–, 265-266, –
98.	Cholesteryl laurate	–, 267, –
99.	Cholesteryl ethylcarbonate	–, 268-269, –
100.	Cholesteryl oleylcarbonate	105, –, –